At The Crossroads

At The Crossroads

THE MINERAL PROBLEMS OF THE UNITED STATES

Eugene N. Cameron

A WILEY-INTERSCIENCE PUBLICATION
JOHN WILEY & SONS
New York • Chichester • Brisbane • Toronto • Singapore

Library of Congress Cataloging in Publication Data:

Cameron, Eugene N. (Eugene Nathan), 1910–
At the crossroads.

Includes indexes.
1. Mines and mineral resources—United States.
I. Title.

TN23.C36 1986 333.8′5′0973 85-29587
ISBN 0-471-83983-3
ISBN 0-471-83982-5 (pbk.)

Printed in the United States of America

10 9 8 7 6 5 4 3 2 1

To Adrienne

Preface

This book is about minerals and mineral resources, about the role they play in the economies of the United States and other countries, and about the problems involved in obtaining adequate supplies of minerals. These problems are important, because civilization as we know it today is dependent on the availability of a wide range of mineral materials and on continuing supply of many of them in very large amounts.

Mankind's use of minerals is very old, dating from some unknown time, perhaps tens of thousands of years ago. When recording of history began, in the fourth millennium B.C., minerals and mineral-derived materials were already being used extensively in many parts of the world, for building materials, for tools and utensils of many kinds, for weapons, and for ornamental and artistic purposes. As the subsequent centuries and millennia passed, knowledge of minerals and of the means of obtaining them and converting them to useful forms increased, and minerals came into ever-wider use. The use of minerals on a modern scale, however, stems from the Industrial Revolution, which began in the eighteenth century. Since that time, consumption of minerals has expanded in both scale and diversity. Expansion accelerated as the nineteenth century progressed. In the twentieth century, expansion has been succeeded by explosion in the rate of use. Since 1900, mankind has produced and consumed more minerals than during all preceding recorded history. There is a direct correlation between the advance of civilization and the growth of mineral use.

During approximately 200 years of life as an independent nation, the United States has changed its economy from predominantly agricultural to highly industrial. The effects on the material well-being of the average American have been dramatic. Through the use of machines made from minerals, his productivity has been prodigiously increased, and through the automobile and airplane so has his mobility. He has available a fantastic variety of manufactured goods. He can communicate almost instantly with his fellow human beings in many parts of the world. To relieve him from hard labor he has a variety of machines—not just the machines of industry but an array of personal machines, from power mowers to snow blowers, dishwashers, washing machines, and dryers, and perhaps an electric pump to change the water in a private swimming pool. By acquiring such items as the electric carving knife and electric toothbrush, he can enlarge his inventory of personal machines to a ridiculous degree.

Thus in the past 200 years, the material basis of life in America, and in the rest of the world, has profoundly improved in quantity, in variety, and in many ways in quality. All this has involved, however, an increasing use of minerals, minerals for the raw materials that are converted into machines, minerals that are converted by the machines into useful forms, minerals that provide the energy to drive the machines that perform the manufacturing processes. Minerals enter into every aspect of American life, quite literally from the cradle to the grave. Our homes, our furniture, our food, our clothing, and the heat and light we demand are either provided directly from minerals or with the aid of machines that are made from minerals. The farmer no longer plods wearily between the handles of a plow. He rides a tractor that can do the work of many horses and many men.

The development of our present industrial civilization is an enormous achievement, but it has its price. The price is a continuing dependence on the availability of large amounts of minerals. The needs of industry must be met. If they are not, industry will die, and civilization as we know it will collapse. This means that we must all be concerned, both as individuals and as a nation, with factors that determine the availability of minerals to the United States and to the world. Those factors are the subject of much of the present work. There are many factors. First and most fundamental are the size, nature, and distribution of the world's mineral resources. Second are the currently available means of finding usable mineral resources and the costs of finding them. Third are the means and costs of extracting and processing mineral raw materials, including energy costs. Fourth are the political and social factors that strongly influence the availability of minerals. In the world today, in effect, each nation sets its own policies with respect to minerals. These policies stipulate the conditions under which the discovery and extraction of mineral resources are permitted and the conditions under which mineral materials may enter the channels of international trade. Since no nation at present produces all the minerals it needs, the policies of each nation are inevitably of concern to all.

This book is especially concerned with the mineral problems of the United States. We live in a period of great change in the nature and basis of the American economy and in the status of the United States as a world power. Part of the change is a deterioration in the mineral position of the United States, defined in terms of our ability to supply our needs for minerals from our own resources. Part of the change is change in our industrial structure, reducing our capacity for converting mineral raw materials into manufactured goods and, by the same token, reducing the diversity of our industrial economy. The enormous strength of America during the first decades after World War II lay not only in the volume of industrial production but in the remarkable diversity and near completeness of American industry, which provided almost all the links of the chain of processes from extraction of mineral raw materials to their conversion into manufactured goods. The period 1940 to 1960 was, for the United States, a golden age, when it drew from its own and the world's resources an unparalleled supply of minerals and built an unparalleled industrial economy.

There are now those who argue that the golden age of industry in the United States is drawing to a close, that the nation's industrial structure in the future will be more limited in scope and restricted to industries ("high technology," for example) in which the United States can maintain or develop special expertise. Those industries will become part of an international, interdependent industrial economy. The trend is already manifest. In the past 15 years, many links in the industrial chain have been lost. More will disappear if the trends of the 1970s and early 1980s continue, encouraged by the trade and other policies of the United States. Whether these trends are desirable or not and whether they could be checked even if we chose to do so are subjects of continuing debate. The stakes in this debate are very high, and the outcome of the debate will be important to every citizen of the United States.

In any event, America is at the crossroads. The nation must decide what path to follow in the future. It can follow the present path, which will lead to further deterioration of American industry and lessened strength and influence in the world, or it can follow another path that offers promise of reinvigoration of American industry and of even greater breadth and strength. However, that other path can only be pursued successfully out of an understanding of the role that minerals play in the economy of the United States, of the strengths and weaknesses of our present mineral position, and of the factors that govern the availability of mineral supplies. If this book contributes to that understanding it will have served its purpose.

The author is indebted to many persons for help in preparing this book. My wife, Adrienne M. Cameron, gave constant encouragement to the project, critically reviewed the manuscript, made many useful comments, and assisted with the editing and indexing. James R. Craig reviewed the Preface and Chapters 3 through 10 and 15 and made numerous suggestions for improvement. Charles C. Hawley and Donald E. Cameron reviewed Chapter 9 and suggested important changes. Robert H. Dott, Jr., made thoughtful comments on Chapter 15. Gerald L. Kulcinski reviewed the section of Chapter 3 on nuclear fusion and supplied two illustrations. Dallas L. Peck, Ronald K. Sorem, Paul Bailly, Stanley R. Riggs, John D. Morgan, and Everett D. Glover kindly furnished illustrations. I am indebted to Michelyn Hass for care in typing and for endless patience with the numerous revisions. Everett W. Smethurst, Robert G. Golden, and other staff members of John Wiley & Sons have expedited publication of this book and aided immensely in editing the manuscript. The help of all these persons is most deeply appreciated.

EUGENE N. CAMERON

Madison Wisconsin
March 1986

Contents

List of Illustrations xiii

List of Tables xix

1 Minerals and Mineral Deposits: Resources and Reserves 1

2 From Discovery to Mineral Production 21

3 The Energy Minerals 38

4 Mineral Raw Materials: The Nonmetallic Minerals 80

5 Mineral Raw Materials: The Metals 108

6 Outlook for World Supplies of the Nonfuel Minerals 144

7 The Mineral Position of the United States 162

8 Mineral Conservation 185

9 Mining Law and Land Policy 204

10 Environmental Regulation 221

11 Taxation of Mineral Industry 238

12 Trade Policy 246

13 Stabilization on the International Scene: Control Schemes in Minerals 262

14 Minerals from the Sea 273

15 Some Thoughts on Our Mineral Future 286

Author Index 301

Subject Index 305

List of Illustrations

1-1 Some examples of mineral deposits 5

1-2 A porphyry deposit of copper 6

1-3 A volcanogenic deposit 6

1-4 A coal seam outcrops around a mesa 9

1-5 Diagram showing the outcrop of a dipping coal seam 11

1-6 Mineral resource classification used by the U.S. Geological Survey and U.S. Bureau of Mines 15

2-1 A *garimpeiro* (with friend), a prospector near Carnauba, Rio Grande do Norte, Brazil 22

2-2 Aeromagnetic map of the Oka complex of igneous rocks, Quebec 23

2-3 Top: A phosphate mine in central Tennessee. Bottom: Aerial view of the Gunnar mine, on the north shore of Lake Athabaska, Canada 25

2-4 Top: A simple quarrying operation near Little Rock, Arkansas. Bottom: Part of the "richest hill on earth," at Butte, Montana 26

2-5 Top: Dragline stripping overburden from a coal seam, near Harrisburg, Illinois. Bottom: The stripped coal seam is being excavated by a power shovel and loaded into a truck 27

2-6 Top: The salt crust at Searles Lake, California. Bottom: A dredge mining the buried beach sand deposits at Trail Ridge, Florida 28

2-7 Vertical sections of two mines, each in a vein deposit, and a plan of one of the mines 30

2-8 Diagrammatic section of coal seam near Middlesboro, Kentucky 32

3-1 Consumption of energy in the United States, 1900–1984, by source 39

3-2 Stages in the formation of a coal deposit 40

3-3 Estimated world recoverable reserves of coal, 1980 43

3-4 Coal fields of the conterminous United States 44

3-5 U.S. production, consumption, and exports of coal 46

3-6 Four examples of the many types of traps in which oil and gas accumulate 50

3-7 (A) U.S. proved reserves of crude oil, 1949–1983. (B) U.S. production of crude oil and natural gas liquids, 1954–1984 52

3-8 World proved crude oil and natural gas reserves, as of December 31, 1984 53

3-9 Value of net U.S. trade in fossil fuels, 1958–1984 54

3-10 U.S. imports of petroleum by country of origin, 1960–1984 55

3-11 World production of crude oil, 1960–1984 55

3-12 Prices of Saudi Arabian light oil, 1970–1984 56

3-13 Rates of change of proved discovery, production, and increase in proved reserves of crude petroleum and natural gas during a complete production cycle 57

3-14 Hubbert prediction, 1956, of future crude oil production in the conterminous United States and adjacent continental shelves 57

3-15 Some areas of active exploration in the United States in 1973 and potential targets of exploration 60

3-16 U.S. proved reserves of natural gas, 1949–1983, and consumption and production of dry natural gas, 1949–1984 63

3-17 World production of natural gas, 1982 63

3-18 Map showing deposits of heavy oil and tar sands in Alberta and Saskatchewan 65

3-19 Principal oil shale deposits of the United States 66

3-20 Distribution of oil shale in the Green River Formation, in Colorado, Wyoming, and Utah 67

3-21 Top view (*a*) and vertical cross-section (*b*) of the NUWMAK thermonuclear fusion reactor, as designed under the Nuclear Fusion Program of the University of Wisconsin 74

4-1 A village in central Madagascar 85

4-2 Part of a clay mine near Mayfield, Kentucky 87

4-3 Part of the Alto Feio pegmatite, Paraíba, Brazil 89

4-4 The "Big Hole," the open pit in the upper part of the famous Kimberley diamond pipe, South Africa 91

4-5 The population explosion 92

4-6 Salt deposit extending across the floor of Death Valley, California 100

4-7 Diatomaceous earth from Lompoc, California 105

5-1 Banded iron formation (taconite) in the wall of the Minntac open pit, Mesabi Range, Minnesota 113

5-2 Part of the open pit of the Hull-Rust-Mahoning mine, Mesabi Range, Minnesota 114

5-3 The Steelpoort chromite seam of the Eastern Bushveld Complex 121

5-4 The Main Magnetite Seam, Eastern Bushveld Complex 124

5-5 Part of the highly productive volcanogenic massive sulfide deposit of Kidd Creek, Ontario 128

5-6 Top: Hydraulic monitor washing gold-bearing gravels, Hunker Creek, Yukon. Bottom: The waste dumps from the deep underground gold mines of the Witwatersrand, South Africa 141

6-1 Relations between prices of metals and amounts consumed in the United States 148

6-2 World per capita consumption of some important mineral commodities in 1964 compared with 1980 150

6-3 World per capita consumption of some important mineral commodities in 1964 compared with 1980 151

6-4 World production of raw steel, 1964–1984 152

6-5 World production of cement and clay, 1964–1984 152

6-6 World production of salt, lime, phosphate, and gypsum, 1964–1984 152

6-7 World production of sulfur, soda ash, and potash, 1964–1984 153

6-8 World production of barite, talc and pyrophyllite, asbestos, fluorspar, sodium sulfate, titanium (for nonmetallic purposes), diatomite, and boron, 1964–1984 153

6-9 World production of aluminum, manganese, copper, zinc, lead, chromium, and silicon, 1964–1984 154

6-10 World production of nickel, magnesium, tin, molybdenum, antimony, tungsten, cobalt, vanadium, cadmium, and mercury, 1964–1984 154

7-1 United States production and consumption of nonfuel minerals, 1900–1979 163

7-2 U.S. production and consumption of sand, gravel, and crushed stone, and 18 other nonmetals 164

7-3 U.S. production and consumption of iron ore and 18 other metals 165

7-4 U.S. net import reliance (imports minus exports) for 32 mineral commodities 166

7-5 Net import reliance of the European Economic Community and Japan in 1983 for supplies of selected nonfuel mineral materials 168

7-6 Net import reliance of the Soviet Union in 1978 for supplies of selected nonfuel minerals 169

7-7 World consumption and U.S. production and consumption of 18 nonfuel minerals, 1930–1984 170

7-8 U.S. imports of bauxite, alumina, and aluminum, 1962–1984 173

7-9 U.S. average annual consumption, 1950–1952, and annual consumption, 1964–1984, of fluorspar, talc and pyrophyllite, asbestos, feldspar, perlite, titanium (for nonmetallic purposes), diatomite, and boron 175

7-10 U.S. average annual consumption 1950–52, and annual consumption, 1964–1984, of cement, clays, salt, phosphate, lime, and gypsum 176

7-11 U.S. average annual consumption, 1950–1952, and annual consumption, 1964–1984, of sulfur, soda ash, potash, and barite 176

7-12 U.S. average annual consumption, 1950–1952, and annual consumption, 1964–1984, of primary nickel, tin, molybdenum, titanium, antimony, tungsten, mercury, cobalt, vanadium, and lithium 177

7-13 U.S. average annual consumption, 1950–1952, and annual consumption, 1964–1984, of primary aluminum, copper, manganese, zinc, lead, silicon, chromium, and magnesium 178

9-1 Percentages of the areas of the western states including Alaska that are Federal public lands 206

9-2 Diagram showing features of a mining claim as required in California 206

9-3 (A) Sketch showing strike, dip, and outcrop of a vein, and a mining claim. (B) Sketch showing cross-sections of veins and other lodes of the kinds sought by prospectors in the 19th century 208

9-4 Three examples of situations in which application of the apex provision causes difficulties 208

9-5 The lead belts of Missouri 210

10-1 Strip mining of coal 224

10-2 Photographs taken in the phosphate-mining area near Columbia, Tennessee 225

10-3 Top: Land mined for coal has been regraded and is ready for restoration of soil stripped off before mining
Bottom: Land dredged for beach sands at Trail Ridge, Florida, has been regraded, recovered with stripped-off topsoil, and planted in grass and pines 226

10-4 A sand-and-gravel operation 227

10-5 Top: A coal seam is exposed above a bench from which another coal seam has already been mined. Bottom: The result (contour mining) of the mining process illustrated above 228

10-6 The total area occupied by mining operations in each western state 230

14-1 Area between the Clarion and Clipperton fracture zones in which manganese nodules contain more than 1.8 percent nickel plus copper 276

14-2 Manganese nodules on the floor of the Eastern Pacific Ocean 277

14-3 Polished section of a manganese nodule showing the concentric layering characteristic of deep-sea nodules 280

14-4 Map showing the Exclusive Economic Zone of the United States 281

14-5 Map showing location of the onshore phosphate deposits of the Aurora district (Lee Creek) of North Carolina and the offshore deposits that occur in the same geologic formation 282

List of Tables

1-1 Composition of the earth's crust 2

1-2 Crustal abundances of certain metals and minimum contents in ores currently being mined 7

1-3 Beryllium: estimated resources of metal 16

1-4 Lithium resources 16

1-5 U.S. reserves and resources of uranium as of January 1, 1976 16

3-1 Composition and heating values of various ranks and subranks of coal 41

3-2 Minable coal reserves of the 10 principal coal-bearing states 45

3-3 World coal production, 1982 46

3-4 U.S. uranium resources, January 1, 1983 70

3-5 Free World production of uranium, 1984 71

3-6 Percentage of various elements and minerals used in domestic energy applications ranked by decreasing percentage 78

4-1 U.S. production of major constructional materials, 1983 and 1984 82

4-2 Consumption of fertilizer components 93

4-3 Production of phosphate rock in 1983 and 1984 and estimates of the reserve base 95

4-4 Average composition of seawater 97

4-5 United States and world production of potash in 1983 and 1984 and estimates of the reserve base 98

4-6 U.S. production of certain saline minerals, 1983 and 1984 99

5-1 World production of iron ore and the reserve base 113
5-2 Major world sources and reserve base of manganese ore 117
5-3 World production of silicon 118
5-4 World production and reserve base of chromite 120
5-5 World mine production and reserve base of nickel 123
5-6 World mine production and reserve base of molybdenum 125
5-7 World mine production and reserve base of copper 129
5-8 World production of aluminum 131
5-9 World production and reserve base of bauxite 132
5-10 World mine production and reserve base of lead 133
5-11 World mine production and reserve base of zinc 134
5-12 World mine production and reserve base of tin 135
5-13 World mine production and reserve base of silver 139
6-1 World production, reserve base, and reserve base/production indices for some important mineral commodities 146
6-2 Annual rates of growth of world production of some important mineral commodities 155
6-3 U.S. nonfuel minerals consumption: comparison of alternative projected annual rates of growth (1970–2000) 156
6-4 1980–2005 consequences of production at 1973–1980 annual growth rates 158
7-1 U.S. chromite supply, 1948 compared with 1981 172
7-2 U.S. manganese supply, 1948 compared with with 1981 172
7-3 U.S. reserve base/consumption indices for nonmetallic minerals 178
7-4 U.S. reserve base/consumption indices for metals 179
7-5 U.S. reserve base/production indices for nonmetals, 1984 and 2005 181
7-6 U.S. reserve base/production indices for metals, 1984 and 2005 182
8-1 Energy consumption per ton of selected mineral commodities in 1973 189
8-2 Old scrap recovered in the United States as a percentage of consumption 192

9-1 Allocation of Alaskan Lands 217
10-1 Land use in the United States in 1980 229
12-1 Mineral commodities in the U.S. stockpile, 1982 257
15-1 Composite data for 10 leading independent mining companies 290

At The Crossroads

1 Minerals and Mineral Deposits

MINERAL RESOURCES AND RESERVES

THE NATURE OF MINERALS

What is a mineral? Actually, two definitions of the word are currently in use. In science, a mineral is defined as a naturally occurring solid substance of inorganic origin, having a particular atomic structure, a fairly definite chemical composition, and a corresponding set of chemical and physical properties. Naturally occurring (native) metals such as copper, gold, silver, and platinum are minerals of simple composition. So is native sulfur. Native copper is easily recognized by its color, bright metallic luster, weight, and other properties, and each of the other native metals has its own peculiar properties. Most minerals, however, are combinations of two or more chemical elements. Halite (the mineral of common salt) is a combination of sodium and chlorine. Pyrite is a combination of iron and sulfur. There is no chance of confusing the two minerals. Halite is white and soft, has a characteristic salty taste, and may form cubic crystals. Pyrite may also form cubic crystals, but it is hard, has a metallic luster, and has a characteristic brassy yellow color. It is sometimes called fool's gold, but it should never be confused with gold. Gold has a pure yellow color. Pyrite is brittle and shatters when struck with a hammer. Gold is rather soft, and it is malleable and ductile, so that it can be hammered into thin sheets or drawn into wire. Pyrite and gold also have quite different chemical properties. So it is with other minerals.

About 3000 different species of minerals have been found in the rocks of the outer part of the earth, and a few dozen others have been found in meteorites or in rocks from the moon. Most of these minerals are rare or uncommon. Fewer than 200 minerals make up the great bulk of the earth's crust, the outer skin of the earth that is accessible to man.

The second definition of minerals is the one that is used in industry. The term mineral is extended to include many materials that are combinations of minerals. In the scientific sense they are rocks, not minerals. Granite, for example, is a mixture of the minerals quartz, feldspar, and mica. In industry, minerals also include the "fossil fuels"—coal, petroleum, and natural gas—which are not of inorganic origin and which have a very wide range of chemical compositions. In this book we must use the industrial definition, because granite, limestone, and other rocks, together with the fossil fuels, are part of our broad mineral problem.

THE SOURCES OF MINERALS—MINERAL DEPOSITS

Our sources of minerals are mostly in the rocks of the earth's crust. The crust ranges from about 3 to 50 miles thick, but only the outermost part is accessible at present. The deepest oil wells go down about 30,000 feet,

TABLE 1-1 Composition of the Earth's Crust (in Parts per Million)

Oxygen	464,000	Cerium	67	Uranium	2.7
Silicon	282,000	Copper	55	Bromine	2.5
Aluminum	82,000	Yttrium	33	Tin	2
Iron	56,000	Neodymium	28	Arsenic	1.8
Calcium	41,000	Lanthanum	25	Germanium	1.5
Sodium	24,000	Cobalt	25	Molybdenum	1.5
Magnesium	23,000	Scandium	22	Tungsten	1.5
Potassium	21,000	Lithium	20	Holmium	1.5
Titanium	5,700	Nitrogen	20	Europium	1.2
Phosphorus	1,050	Niobium	20	Terbium	1.1
Manganese	950	Gallium	15	Lutecium	0.8
Fluorine	625	Lead	12.5	Thulium	0.25
Barium	425	Boron	10	Iodine	0.5
Strontium	375	Thorium	9.6	Thallium	0.45
Sulfur	260	Samarium	7.3	Cadmium	0.2
Carbon	200	Gadolinium	7.3	Antimony	0.2
Zirconium	165	Praseodymium	6.5	Bismuth	0.17
Vanadium	135	Dysprosium	5.2	Indium	0.1
Chlorine	130	Ytterbium	3	Mercury	0.08
Chromium	100	Hafnium	3	Silver	0.07
Rubidium	90	Cesium	3	Selenium	0.05
Nickel	75	Beryllium	2.8		
Zinc	70	Erbium	2.8		

Source: Krauskopf (1967).

but most of the production of oil and gas is from depths less than 20,000 feet. The deepest mines, in the gold fields of South Africa and India, bottom at 12,000 feet or less. Most mines are within a few thousand feet of the surface. When we talk of the mineral resources of the earth, for all practical purposes we must confine ourselves to the outermost part of its outer skin.

What, then, are the materials available to us? As a general guide we can use the average composition of the crust (Table 1-1) as calculated from tens of thousands of chemical analyses of rocks collected from all over the earth. The figures are not perfectly accurate, but the orders of magnitude indicated are undoubtedly correct. The picture is rather startling. Of the 67 chemical elements for which data are given, 9 make up over 99 percent of the crust. The 10 next most abundant elements (phosphorus through chlorine) make up less than 0.5 percent of the crust. All the other elements together make up only about 0.2 percent of the crust.

It is true, of course, that relative abundances do not tell us the whole story. There is also the question of absolute amounts. The element tungsten, for instance, forms only 0.00015 percent (1.5 parts per million) of the crust, but nonetheless an average 13 cubic miles of crust should contain about 58,000 tons of tungsten.* This is more than equal to the annual world consumption of the metal. Beneath the contiguous 48 states alone, the amount of crustal material to a depth of just 1 mile is about 3,000,000 cubic miles; hence an enormous amount of tungsten should be present. However, very little of this tungsten can ever become available. The reason is that extraction of tungsten from average crustal material would be very difficult and far too costly both in dollars and in energy. Sources of tungsten for industry are materials that contain at least 0.3 percent (3000 parts per million) tungsten, about 2000 times the average tungsten content of crustal materials.

The sources of mineral raw materials are mineral deposits. A mineral deposit is a body of rock containing some useful mineral or material. There are many different kinds of mineral deposits, formed by the various geological processes that have shaped the crust of the earth. Some deposits have formed by chemical or biochemical precipitation or by me-

*Data for mineral reserves, resources, production, and consumption, past and present, are expressed in a hodgepodge of English and metric units—short tons, metric tons, and long tons; pounds, kilo grams, grams, troy ounces, carats, barrels of oil (42 gallons) or cement (376 pounds). One unit, the 76-pound flask, is the amount of mercury contained in each of two leather flasks that used to be slung over the back of a mule for transport over the mountains of Spain. In science, the metric system has become standard, but in North American mineral industry and commodity markets, data for the majority of mineral commodities are expressed in English units. In this book, except as noted, English units are used as being in more common use and more familiar to most Americans. Ton means short ton (2000 pounds). It is easily converted to the metric ton (2204.6 pounds), to a close approximation, by multiplying it by 0.907. A long ton (2240 pounds) is 1.12 short tons.

chanical accumulation in surface waters of lakes, rivers, and oceans. Limestones, for example, have formed by chemical and organic precipitation of calcium carbonate on parts of the ocean floors of the present and the past. The deposits are layers, sheetlike deposits that may be thousands of square miles in extent. Sandstones, siltstones, and mudstones consist of mineral particles that were carried by streams and deposited in quiet stretches of river valleys or carried into the oceans and there deposited, mostly as layers along the margins of the continents. Some large salt deposits have formed by evaporation of the waters of saline lakes such as those of Utah, Nevada, and eastern California. These too are sheetlike deposits spread over the basin floors.

An important class of deposits is formed by weathering of preexisting rocks. In central Tennessee there are extensive ancient deposits of a limestone that contains 15 to 25 percent calcium phosphate. Such material cannot be mined economically at present. However, where the limestone has been exposed to weathering, the calcium carbonate has been dissolved out, leaving a residue rich in calcium phosphate (up to 50 or even 60 percent). Such material is an excellent source of phosphate for fertilizer, and the deposits of Tennessee were in the past of great importance to the agriculture of the midwestern states. Weathering has produced workable deposits of many kinds of minerals. It is responsible for the deposits of bauxite that are the world's great source of aluminum metal. As shown in Table 1-1, aluminum is the third most abundant element in the crust of the earth, and a single cubic mile of the crust contains enough aluminum to supply the world for about 65 years at present rates of consumption. Most of the aluminum, however, is tied up in silicate minerals, from which it is far too costly to extract the metal. Our supply of aluminum comes from places where certain rocks have been exposed to intensive weathering, which has removed silica and other elements and left deposits of aluminum-rich material, bauxite, from which the metal can readily be extracted.

At the other extreme are deposits that have formed within the crust in the bodies of molten material (magma) that give rise to the important class of igneous rocks. During the cooling and crystallization of certain magmas, concentrations of valuable minerals may form. Deposits of this class furnish all the world's chromium, important amounts of iron, much of the world's vanadium, and substantial amounts of phosphate, titanium minerals, niobium, and other mineral commodities. Some of the deposits are layers geometrically similar to sedimentary deposits.

Still other deposits consist of minerals precipitated from solutions that have circulated along fractures and other openings in the rocks of the crust. Deposits of this broad group furnish much of the world's supplies of such metals as copper, lead, zinc, molybdenum, and silver. They include important deposits of gold. The deposits have many forms. The simplest is the vein, the filling of a fracture in the crust. The great mining camp at Butte, Montana, is famous for its systems of veins along fractures in a mass of granitic rock that underlies the district.

As a group, mineral deposits have a wide range of both sizes and shapes. They may contain a few tons or billions of tons of material. They range in shape from simple tabular to pipelike to highly irregular in form. Some examples of mineral deposits are depicted in Figs. 1-1, 1-2, and 1-3.

As suggested by the preceding paragraphs, some mineral deposits are bodies of common and more or less widespread types of rocks. Granite, limestone, and sandstone, for example, furnish large amounts of constructional materials. A large share of the mineral deposits currently being mined, however, consists of bodies of rock in which some valuable mineral or minerals are concentrated to an abnormal degree. All the mineral deposits from which the metals are obtained are such concentrations. Copper deposits are an example. The average concentration of copper in the crust is only 0.0055 percent. To be minable today, a deposit must contain at least 0.3 percent copper. In Table 1-2 crustal abundances are compared with contents of the materials that are currently being mined for certain metals.

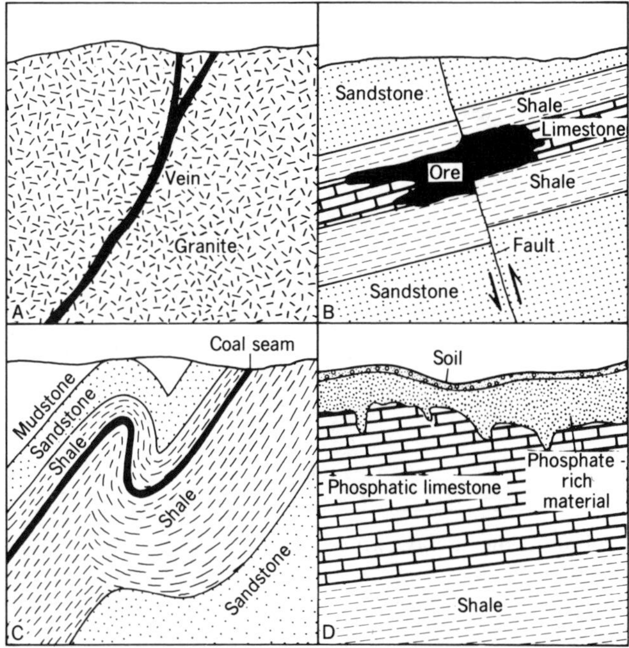

Figure 1-1 Some examples of mineral deposits. (A) Solutions have deposited gold-bearing quartz along a branching fracture. (B) Solutions passing along a fault have caused the replacement of adjacent limestone by a body of lead and zinc sulfides. (C) A folded coal seam enclosed in shale is shown. (D) The upper part of a phosphatic limestone formation has been weathered; calcium carbonate has been leached out, leaving a phosphate-rich material.

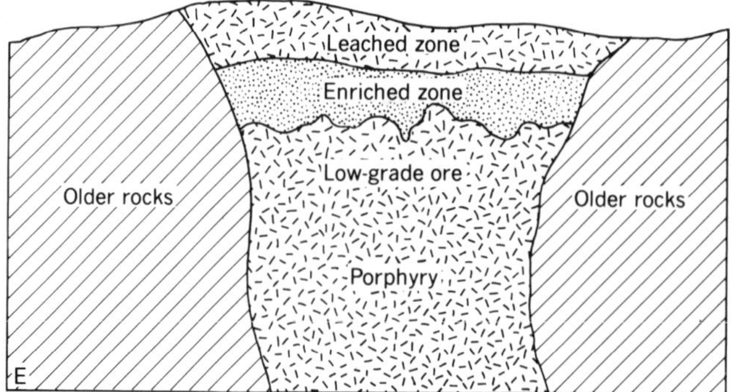

Figure 1-2 A body of porphyry (a kind of granitic rock) has been shattered and impregnated with copper and iron sulfides. Owing to weathering, copper has been leached out of the upper part of the porphyry, carried down in ground water, and redeposited, producing a blanketlike zone enriched in copper. In some deposits of this type, only the enriched material is ore.

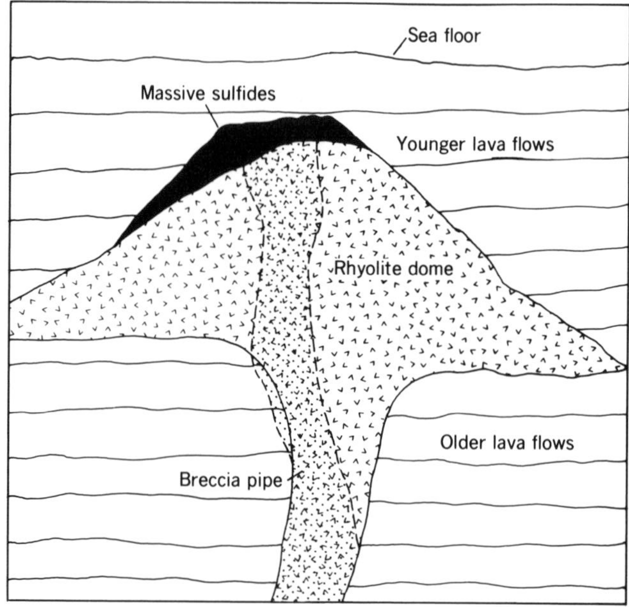

Figure 1-3 A "volcanogenic" deposit, produced during submarine volcanic activity. The older lava flows were formed during the first phase of volcanism. Through these a body of rhyolitic magma (granitic in composition) was erupted, forming a dome on the sea floor. Fracturing of the rhyolite allowed metal-bearing solutions to pass upward, weakly mineralizing the fractured rock (breccia pipe) but forming a massive deposit of copper, iron, and zinc sulfides over the top of the dome. The dome and sulfide deposit were subsequently covered by younger lava flows.

TABLE 1-2 Crustal Abundances of Certain Metals and Minimum Contents in Ores Currently Being Mined (in Percent)

Metal	Crustal Abundance (in percent)	Minimum Content in Ores[a]
Aluminum	8.20	35
Iron	5.60	16
Magnesium	2.30	3.3
Manganese	0.95	30
Chromium	0.0100	25
Zinc	0.0070	6
Copper	0.0055	0.3
Lead	0.0013	3
Tin	0.00021	1
Beryllium	0.0003	0.05

[a]It must be understood that the minimum grade of ore minable varies from mine to mine. Thus ore containing as little as 0.3 percent copper can be mined profitably from some deposits but not from others.

It will be evident that many of the materials present in the earth's crust are simply not useful to us at present as sources of minerals, because the contents of useful elements or minerals are too low or because useful elements are locked up in minerals from which they can be extracted only at very high cost. Most of the metals in the crust are locked up in silicate minerals, which are compounds of metals with silicon and oxygen. Our ability to extract metals from silicates at reasonable cost is very limited. In most cases energy costs are prohibitive. In talking about the availability of minerals from the earth's crust, clearly we must distinguish between bodies of rock from which it is feasible to extract minerals at present and those from which extraction is not feasible. At this point we must introduce several terms that are useful in discussions of mineral resources—*ore*, *grade* or *tenor* of ore, *ore body*, *reserves*, and *resources*.

Ore is material from which one or more useful minerals or metals can be extracted under current economic and technologic conditions in a politically and socially acceptable manner. A mineral deposit or part of a mineral deposit consisting of ore is an *ore body*. Ore bodies range in size from a few tons to more than a billion tons; most of those that contribute significantly to world mineral supply are in the range from tens of thousands of tons to hundreds of millions of tons.

A number of factors determine what is or is not ore. First is the *grade (tenor)* of the mineral material—the content of valuable mineral or minerals. Second is the size and shape of the mineral deposit and its position in the crust. These, together with the physical properties of the material and of the enclosing rocks, determine the methods and costs of mining the material. Third is the composition of the mineral material, which determines the methods and costs of processing the material to recover the valuable substances from it. The location of the deposit is important for many reasons. An unfavorable location may mean high transportation costs, or it may mean that mining has to be done under difficult climatic or topographic conditions. Local costs of labor are an important factor, and availability of water is another. The value of the mineral or minerals that would be produced and the demand for them must likewise be considered.

Even if a mineral deposit satisfies all the economic and technologic requirements for ore, it still may not qualify as ore. A mineral deposit in a national park, however rich it may be in gold or copper or whatever, is not ore, simply because mining in a national park is forbidden by law. A rich mineral deposit in a politically unstable country may not be ore, simply because investment in mining there would be too risky. Mining a particular mineral deposit may have unacceptable environmental consequences; the material of the deposit therefore cannot qualify as ore. The point is that political and social conditions and regulations have much to do with whether mineral deposits can or cannot be mined and therefore must be considered in determining what is and is not ore.

It is essential to recognize that most of the factors that determine whether a mineral deposit contains ore can change with time. Political and social conditions change. The technology of mining and mineral processing changes as research and development of new methods progress. Many mineral materials that are ores today were not ores 50 years ago, simply because economic means of extracting them and converting them to useful forms had not been developed. Technology, however, can have the opposite effect. There are deposits of iron ore in the eastern states that were important at the time of the American Revolution but are not ore today, because the advance of technology and changes in the economics of iron ore production have relegated them to the class of uneconomic deposits.

THE AVAILABILITY OF MINERALS: RESOURCES AND RESERVES

Providing adequate mineral supplies is one of the basic necessities of modern life, and despite all our efforts to conserve, there is little prospect that mineral needs will diminish significantly in the foreseeable future. The question is: What amounts of minerals are available now, and what amounts are likely to be available in the future? The answer to this question is critically important. In any industry, planning for future production must consider whether requisite mineral raw materials and energy will be available in adequate amounts and at reasonable cost. Governments are concerned, because mineral supplies are essential to a healthy economy and become of heightened importance in times of war. Furthermore, federal, state, and local governments set the legal framework within which development of the country's natural resources must be undertaken. Many policy decisions affecting the availability of mineral resources must be made. It is important that decision making be based on (1) the best available information as to the nature, size, and distribution of resources of various minerals and (2) the proportions of those resources that might actually be extracted under present or even future technologic conditions. As a basis for forecasting availability, quantitative data are needed, so that future supplies can be matched against foreseeable demands.

It may seem a rather simple problem. After all, investigation of the mineral resources of the United States has been going on for much more than a hundred years. By this time, surely, we should know precisely what we have. We don't, and there are a number of reasons why we don't. Consider the case of the coal deposit shown in Fig. 1-4. The coal forms a layer that crops out at intervals below the summit of a flat-topped mesa and seems completely to encircle it. We go around the mesa, trenching at regular intervals to expose the coal seam further and then measuring the thickness of the coal in each trench. We find that it ranges from 5 to 8 feet in thickness and averages 7 feet. That is a good thickness for mining. We take many samples; they show that the seam consists of bituminous coal of uniformly excellent quality. So far, so good, but how much coal is present in the deposit? We may *suspect* that the coal seam extends under the whole mesa, but we don't *know* this. Coal seams can vary laterally in quality; the outcrops may not be fully representative. Coal seams can vary in thickness, and in places they may be split by lenses of mudstone or sandstone waste rock that would have to be mined along with the coal and would increase the cost per ton. We must be sure before we invest money and effort in a mining operation. We bring a drill onto the top of the mesa, and holes are put down at regularly spaced intervals over the entire deposit. Cores from the holes give us the thickness of coal present in each hole, and tests of the cores tell us the quality. The results are favorable. There is one small area near the center of the mesa where the coal is too thin to be mined, but elsewhere

Figure 1-4 A coal seam outcrops at intervals around a mesa, occurring along the base of a massive sandstone that forms the mesa cap.

it is of ample thickness. The quality is excellent. We now have data that permit us to calculate the tonnage of minable coal present under the mesa, excluding the small area. It amounts to a little over 40,000,000 tons. Engineering studies and economic evaluation indicate that the deposit can be mined profitably at the present price of coal. Transport costs will be acceptable, and a market for the coal will be available. Further studies show that mining can be done in conformance with environmental and other regulations. The coal now qualifies as ore, and the calculated tonnage of coal constitutes the *reserves* of coal present in the deposit.

In our coal deposit, all the material to be mined is coal, but in most mineral deposits the ore is a mixture of valuable and useless minerals. In a typical copper deposit of Arizona, for example, the copper content of ore may range from 0.4 percent to perhaps 1.2 percent. There will be other material in the deposit that falls below 0.4 percent copper, but that does not qualify as ore and is not included in calculating the reserves for the deposit. Sampling in this case is concerned with determining both the tonnage of ore and the *grade* (i.e., the metal content of the deposit). The reserves of ore established for the deposit can be expressed in either of two ways: (1) in terms of the tonnage and grade of ore or (2) in terms of the tonnage of copper present in the ore.

Now let us suppose that in a given district a series of coal deposits has been investigated. By adding up the reserves established for the individual deposits we can arrive at a figure for the total coal reserves of the district. By adding together the reserves of all the coal districts of the United States, we can calculate the nation's total coal reserves. We still don't know, however, how much coal will be available in the future from the coal deposits of the United States. There are several reasons. One is that our reserve figure must be discounted for losses in mining and processing coal. Depending on the exact nature of each deposit and the manner in which it must be mined, loss may range between 10 and 50 percent. A second reason is illustrated in Fig. 1-5. Here the coal deposit dips into the ground. A prospective mine operator has leased a tract of ground and has found that the coal seam can be followed at the surface all the way across his land, a distance of about 2.5 miles. Based on his economic appraisal, he wants to establish a reserve of coal enough to support production of 250,000 tons per year for 20 years. Investigation of the deposit begins. Surface measurements and sampling are followed by systematic drilling, and the seam is explored from the outcrop to points 3000 feet down the slope (the dip) of the coal seam. The coal seam is found to average 6 feet in thickness, and the coal is of acceptable quality. The drilling and sampling establish a reserve of about 10.5 million tons, and there are no signs that the downward limit of the seam has been reached. Counting on recovery of 50 percent of the coal, this is sufficient. The

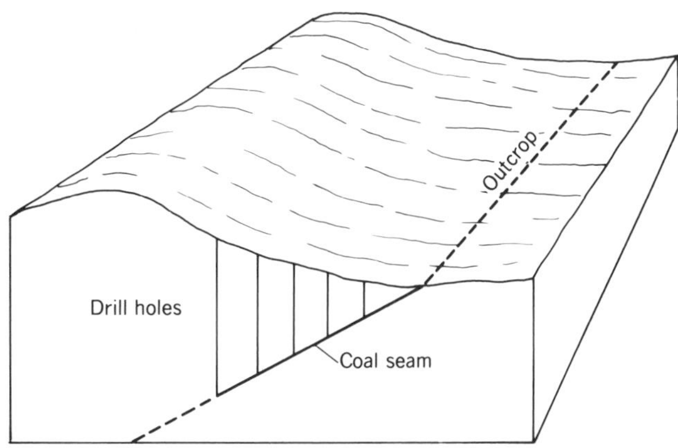

Figure 1-5 Diagram showing the outcrop of a dipping coal seam. Lines of drill holes parallel to the line shown in the front face of the diagram must be used to test the thickness and quality of the seam.

prospective operator is satisfied, and exploration is halted. If engineering and economic evaluation is favorable, the construction of a mine is undertaken.

Why is exploration stopped at this point? Why not explore the whole deposit, or at least the part of it that is covered by the prospective operator's lease? The reason is simply that exploration is expensive, and money invested in additional exploration will not begin to yield a return until the first 20 years have passed. Fifteen years from now, if conditions are still favorable, additional exploration will be affordable. In the meanwhile, however, we cannot know the total reserves that may ultimately be proved to exist down dip in the deposit, and unless and until owners of adjacent lands explore the coal seam on their properties, we cannot know what reserves may ultimately be established for the coal seam as a whole. At present we have only a partial estimate of the amounts of coal available from the deposit. This is the common case in the coal fields of the United States.

Let's consider another situation. A copper deposit has been found, part of it has been thoroughly drilled and sampled, and reserves of copper ore can now be satisfactorily estimated for that part of the deposit. In another part some drilling and sampling have been done, but the information is adequate only for a rough estimate *of* tonnage and grade. Beyond this, however, geologic mapping suggests that the deposit extends beyond the limits of drilling and sampling, so there may be still more ore in the deposit.

In effect, at this stage we have established three categories of reserves for the deposit. The first, in the thoroughly explored part, constitutes the *proved* or *measured reserves*. The second, in the partially explored part, constitutes the *probable* or *indicated reserves*. What about the parts of the deposit that have not been explored? By projection from the other two parts we may choose to calculate an additional tonnage of ore for the deposit, but that tonnage is assigned to a third category of *possible* or *inferred reserves*.

The three categories of reserves—measured or proved, indicated or probable, inferred or possible—reflect a decreasing certainty as to tonnage and grade of material present in a deposit. As we pass from measured to indicated to inferred reserves, our estimates are based less and less on solid fact, more and more on personal experience and judgments. Those judgments may be based on considerable experience with similar deposits, but they are still judgments and are subject to significant error. When we go to the bank for financing our mining operation, our banker will ignore the inferred reserves. He may give some weight to our figures for indicated reserves, but his decision will be based mainly on the quality and size of the measured reserves. They are often referred to as the "bank reserves."

The situations we have just examined are very common, in fact they are typical of mining districts, and this means that at any given time our forecasts of the future availability of minerals are based on reserve data that are incomplete and uneven in quality. This is not the result of any conspiracy to withhold data on reserves. It is simply a consequence of the fact that exploration of mineral deposits is accomplished only over a very long period of time. In the United States, exploration is still far from complete. Our estimates of reserves are inevitably imperfect, the more so because we know very well that we have not yet found all the mineral deposits that exist beneath the surface of the United States. As a matter of fact, we are still learning how to find them, and geologists will hasten to point out that we have much to learn.

There are still further complications. Some of the drill holes put down in the copper deposit considered above indicate that a part of the deposit contains only 0.3 percent copper. This material cannot be mined profitably at present, so it is excluded from the estimate of reserves. If the price of copper rises or techniques of extraction are significantly improved, mining may be feasible in the future, and the material can then be added to the reserves. There is no certainty that this will happen. Demand for copper has been weak for some time, and copper prices have not kept pace with rising costs of production. Some of the material in United States deposits formerly classified as reserves no longer qualifies as ore.

The net result of all the above is that while the concept of "reserves" is a rather simple one, its application to actual mineral deposits is complex. The problems involved must be taken into account in both government and industry planning with respect to minerals.

MINERAL RESOURCES

Thus far we have considered only the question of mineral reserves, but we have mentioned "mineral resources." What is meant by that term? The official definition of a mineral resource, by the U.S. Bureau of Mines and the U.S. Geological Survey, is "a concentration of naturally occurring solid, liquid, or gas material in such form and amount that economic extraction of a commodity from the concentration is currently or potentially feasible." This definition is generally satisfactory for mineral resources if it is understood that in some countries the "economics" of mining is strongly influenced by political considerations. Note that mineral resources, as thus defined, include mineral reserves. They also include huge amounts of materials that do not at present qualify as reserves. Some of them, unfortunately, probably never will.

Our discussion is thus far directed entirely at mineral deposits that have already been discovered. These are called "identified resources." However, our forecasts of future availability of minerals must take into account the fact that exploration of the earth's crust is still incomplete, hence additional mineral deposits will certainly be discovered in the future. In the United States, for example, many mineral deposits have been discovered since World War II. They include a number of large copper deposits, a number of molybdenum deposits, important zinc deposits, new kinds of gold deposits, the world's largest sodium carbonate deposits, a new type of beryllium deposit (the largest in the world), and a new type of vanadium deposit. There are good reasons for supposing that additional deposits will be found in the future and that significant additions to U.S. and world identified resources will be the result. More than half of the world reserves of aluminum ore have been discovered since World War II, more than half of the world iron ore reserves, more than half of the world molybdenum reserves, more than half of the world reserves of potash, and so on for a considerable list of minerals. In talking about the total mineral resources of the earth's crust we must therefore recognize a class of "unidentified mineral resources."

The Bureau of Mines and the Geological Survey recognize two categories of unidentified resources, hypothetical and speculative (Fig. 1-6). Hypothetical resources are those that are similar to known mineral deposits and that may reasonably be expected to exist in the same producing district or region under analogous conditions. Speculative resources are those that may occur either in known types of deposits in favorable geologic settings where mineral discoveries have not yet been made or in types of deposits as yet unrecognized for their economic potential.

The recognition of hypothetical and speculative resources is perfectly logical. The definitions are really statements of three facts: (1) many known mining districts of the United States and the world are still incompletely explored; (2) there are areas that are geologically similar to known mineral districts but are as yet unexplored; and (3) present geological knowledge of the occurrence of mineral deposits in the crust of the earth is incomplete. Fine, but how shall one take undiscovered resources into account in forecasting the future availability of minerals? Should we try to make quantitative estimates of these resources? There have been various responses to these questions. The Geological Survey has published a comprehensive appraisal of the mineral resources of the United States (Brobst and Pratt, 1973). It contains chapters on individual mineral commodities or groups of commodities. The classification of resources shown in Figure 1-6 was used in evaluating resources of the various minerals. Regarding this classification, Dr. V. E. McKelvey, then Director of the Survey, commented:

> *For most minerals, the chief value of the classification at present is to call attention to the information needed for a comprehensive appraisal of their potential [i.e., the potential of undiscovered deposits] for we haven't developed the knowledge and methods necessary for meaningful estimates of undiscovered deposits.*

This is a completely realistic statement. However, in the chapter on beryllium we find a table (Table 1-3) in which numbers are attached to undiscovered resources. Yet the fact is that the basis for estimating undiscovered U.S. and world resources of beryllium is so meager that the numbers attached to them are meaningless. The same is true for other commodities, of which lithium (Table 1-4) is an example.

As another example, suppose we look at one of the estimates of U.S. uranium resources (Table 1-5) made by the Energy Research and Development Administration. The reserves indicated in the table represented the best possible estimates of tonnage and grade; they resulted from study of an enormous amount of data furnished by the uranium mining industry. Total reserves (contained U_3O_8) in the three forward cost classes were 640,000 tons. Down below we have estimates of potential resources, divided into probable, possible, and speculative. Now if we add total reserves to total potential resources, we get a very large figure, more than 3.5 million tons. But when we examine the definition of the various classes of potential resources, we find that about half of the total is for deposits that have not yet been found, and a substantial part of that is in districts not yet known to have significant uranium deposits. This table and the tables for beryllium and lithium illustrate a tendency to assign very large numbers to undiscovered resources, despite our uncertainty over whether they really exist. This is an important matter. Figures for *reserves* may raise questions about the adequacy of future supplies, whereas figures

Cumulative Production	IDENTIFIED RESOURCES			UNDISCOVERED RESOURCES	
	Demonstrated		Inferred	Probability Range (or)	
	Measured	Indicated		Hypothetical	Speculative
ECONOMIC	Reserves		Inferred Reserves		
MARGINALLY ECONOMIC	Marginal Reserves		Inferred Marginal Reserves		
SUB-ECONOMIC	Demonstrated Subeconomic Resources		Inferred Subeconomic Resources		
Other Occurrences	Includes materials that are not considered a resource				

Figure 1-6 Mineral resource classification used by the U.S. Geological Survey and U.S. Bureau of Mines.

TABLE 1-3 Beryllium Estimated Resources of Metal (in Tons)

	In Known Deposits	In Undiscovered Deposits
United States	60,000	250,000
Other countries	32,000	400,000

Source: summarized from Griffitts (1973).

TABLE 1-4 Lithium Resources in Tons of Metal

	Proved and Probable Reserves	Possible Reserves and Conditional Resources[a]	Hypothetical Resources
United States	1,000,000	3,100,000	3,500,000
Canada	129,000	200,000	1,000,000
Africa	94,000	—	1,000,000

[a]Conditional resources designate subeconomic resources in deposits already discovered.
Source: summarized from Norton (1973).

TABLE 1-5 U.S. Reserves and Resources of Uranium as of January 1, 1976, by Forward Cost Classes (in Tons of U_3O_8)[a]

Reserves			
	$10.00	$15.00	$30.00
	270,000	430,000	640,000
Potential Resources			
	$10.00	$15.00	$30.00
Probable	440,000	655,000	1,060,000
Possible	420,000	675,000	1,270,000
Speculative	145,000	290,000	590,000
Totals	1,005,000	1,620,000	2,920,000

[a]Data from the Energy Research and Development Administration, given by Gordon (1977). Note that the figures in any one line are cumulative: $30.00 reserves and resources include $15.00 and $10.00 reserves and resources; $15.00 reserves and resources include $10.00 reserves and resources. Forward costs do not include costs of land acquisition and exploration and are thus less than estimated total costs of production.

for *undiscovered resources* may give an illusion of comfortable abundance. Unfortunately, the distinctions between reserves and the various classes of resources often become blurred when tonnage estimates become the subject of public discussion (Cameron, 1978). The result is confusion in the minds of policy makers. Confusion is compounded by the multiplicity of schemes of classifying resources and reserves. The problem is well stated by Zwartendyk (1981):

In the context of questions about future mineral supply, much can be said for considering the results of mineral resources assessments as working material too undigested to be presented as such to the public. Immense—and immensely misconstruable—figures for mineral resources fail to impart a clear picture of mineral resources adequacy and long-term mineral supply. Such figures are almost routinely mistaken for amounts that will be available at acceptable prices when and where needed, as if the world were economically frictionless.

The author's view (Cameron, 1978) is that numbers should rarely be assigned to unidentified resources. The two categories of undiscovered (unidentified) resources, as McKelvey (1973) stated, have value because they call attention to the fact that disclosure of the earth's resources is incomplete. They also have value as suggesting the likelihood that significant discoveries will be made in the future. However, they should be used only with great caution in the planning of national mineral policy. The uncertainties are far too great. We cannot predict the rate at which such resources might be discovered, the rate at which they might be brought into production, or the costs of production. We cannot predict the consequences of future changes, favorable or adverse, in the technology of exploration for mineral deposits, in mineral extraction, or in mineral use. Furthermore, the record shows that discovery and development of new resources take place only over long periods of time. To place reliance, in public or industrial planning, on undiscovered resources is simply not justified in the light of our knowledge of the mineral deposits present in the crust and in the light of the present limitations on our ability to discover them. The only possible exceptions are undiscovered resources of petroleum and natural gas, discussed in a later chapter, but even here experience has shown that estimates of undiscovered resources are not wholly reliable.

For the nonfuel minerals, our principal sources of mineral resource information are estimates published annually by the U.S. Bureau of Mines in "Mineral Commodity Summaries." They are given as estimates of the *reserve base*, which includes measured and indicated reserves, marginal reserves (on the borderline between economic and uneconomic material), and some subeconomic resources. Only identified resources are included in the reserve base. In effect, the reserve base for a mineral shows the order of magnitude of identified resources that are actually available or could become available with only moderate change in economic or technologic circumstances. In discussing the availability of particular mineral commodities, we shall make extensive use of the estimates of the reserve base. It must be understood, however, that these estimates are very uneven in completeness and quality.

THE VALUE OF MINERAL RESOURCES

We turn now to a very important and much debated question: What is the value of a mineral resource in the ground? There has been much needless misunderstanding about this. The answer is that mineral resources in the ground have no value at all until they have been converted into reserves. They are converted into reserves only (1) when they have been discovered and (2) when means of extracting and processing them economically and in a socially and politically acceptable manner have been devised. The simple fact is that whereas mineral resources are created by nature, mineral reserves are created by man. The history of mining offers countless illustrations of this truth. One of the author's favorite examples is the case of the Potrerillos Mine.

In 1913, William Braden, an American engineer fresh from success in developing the great copper deposit of El Teniente, in the Chilean Andes, visited a group of small copper deposits scattered over a hill in the Andes to the north. Mining had been carried on at various small mines in the hill since 1875, but with little success and small production. The future of the district as a source of copper seemed a gloomy one. Braden, however, developed a new perception of the area. He saw in the hill not just a group of small copper deposits, but a single large copper deposit that embraced them all. He persuaded others of the validity of his perception, and 15 years later, after vast effort and great expense, the Potrerillos Mine of Andes Copper Corporation was brought into production and became an important source of copper. The history of mining is replete with examples of mineral resources such as this that yielded no benefits to society during the long millennia that they lay unrecognized and undeveloped. On the other hand, there are many mineral deposits long known that have no value to society because extraction of minerals from them is not feasible under current or foreseeable conditions. We shall examine some of these in later chapters.

Until resources are converted into reserves, they have no value to society. If society wishes to derive benefits from mineral resources it must establish an economic and social framework that provides the incentives for discovering mineral deposits and converting them from resources to reserves.

In recent years there has developed a concept of "the inherent value of mineral resources" in the ground. For many decades it has been generally accepted that prices of mineral commodities should bear some reasonable relation to the costs of producing them, and the workings of a fairly free international market have ensured that the relationship would be maintained. Under the concept of "inherent value," however, mineral resources, being nonrenewable, have a value that goes beyond the costs of producing them, and prices should reflect this "inherent value." There have been problems in quantifying inherent value. The Organization of Petroleum Exporting Countries, the oil cartel, in effect quantified the inherent value of petroleum by setting prices up to $35 a barrel for oil that cost as little as $2 to produce. In 1974, Jamaica quantified inherent value by raising taxes on its bauxite (the ore of aluminum) by 700 percent. Those have been arbitrary, unilateral actions. However, advocates of the concept of inherent value have felt the need for a formal definition of the term. One definition, by Bosson and Varon (1977), may be cited:

> *Mineral prices have traditionally been determined on the basis of capital, operating, and distribution costs, qualified by bargaining power, with no allowance for the inherent value of the mineral in the ground. A more appropriate yardstick (if adopted by all mineral producers) might be the price just below that of the closest substitute or the cost of secondary (recycled) material. The difference between this price and that based on costs can be taken to represent the intrinsic (inherent) value of the mineral in the ground, which ideally should accrue to the host country.*

This concept is not acceptable in the United States, as illustrated by the debate that led to the imposition of the tax on "windfall" profits from petroleum. The concept of inherent value is very attractive to the developing countries, which would like the highest possible returns from the marketing of their mineral products. The fact is, however, that the definition quoted above is essentially a statement that the price of a mineral should be all that the traffic will bear.

Thus far we have been discussing the value of mineral resources to society, and we can do this in rather general terms. The prospective mine operator, however, must determine the value of a mineral deposit in very specific terms. The creation of a new mine requires first the discovery of a deposit through geological and geophysical investigations, then drilling or some other means of delineating the deposit and determining its tonnage and content of valuable materials. Secure tenure of the land must be acquired, either by purchase or by lease from the owner. Capital needed for constructing the mine, surface facilities (shops, offices, etc.), and a mill for processing the ore must be estimated, and then the costs of operating mine and mill. Local, state, and federal taxes must be estimated, together with transport costs to markets. Demand and market prices for the products must be forecast. If, when the whole process is complete, a reasonable return on investment is indicated, construction of mine and mill can be begun.

Already the prospective mine operator has incurred substantial costs from which as yet he has no income. During the period of mine and mill construction, major capital costs will be incurred, but until mine and mill are complete and put into production, no income will be generated. If the operation is a large one, 10 years or more may elapse between the initial discovery of a deposit and the date at which the miner receives a return on his investment. The long time lag between the beginning of exploration and the beginning of actual mineral production has become one of the most serious problems in the development of the world's mineral resources. There are hazards at every stage of the long process of developing a major mine, because it is difficult, at the stage of valuation, to foresee all the things that will ultimately determine financial success or failure of the enterprise. Demand and prices for minerals fluctuate with time. Costs of labor and equipment may change. Taxes are subject to change, and so are government policies that strongly influence the feasibility of mining operations.

REFERENCES AND ADDITIONAL READING

Bosson, R., and Varon, B., 1977, *The Mining Industry and the Developing Countries.* Oxford University Press, New York, 292 pp.

Brobst, D. A., and Pratt, W. P., eds. 1973, *United States Mineral Resources.* U.S. Geological Survey, Professional Paper 820, 722 pp.

Cameron, E. N., 1978, Resource information credibility. *Mines Magazine,* Vol. 68, No. 2, pp. 7, 8, 28, 29.

Cameron, E. N., 1979, The nature of reserves. *Materials and Society,* Vol. 3, pp. 19–26.

Flawn, P. T., 1966, *Mineral Resources.* Rand McNally, Chicago, pp. 1–38.

Gordon, E., 1977, Uranium—rising prices continue to spur development activity. *Engineering and Mining Journal,* Vol. 178, No. 3, p. 196.

Griffitts, W. R., 1973, Beryllium. In *United States Mineral Resources,* D. A. Brobst and W. P. Pratt, eds., U.S. Geological Survey, Professional Paper 820, pp. 85–93.

Krauskopf, K., 1967, *Introduction to Geochemistry.* McGraw-Hill, New York, pp. 639–640.

McKelvey, V. E., 1973, Mineral resource estimates and public policy. In *United States Mineral Resources,* D. A. Brobst and W. P. Pratt, eds., U.S. Geological Survey, Professional Paper 820, pp. 9–19.

National Academy of Sciences, 1975, *Mineral Resources and the Environment.* National Academy of Sciences, Washington, D.C., 348 pp.

National Academy of Sciences, 1975, *Reserves and Resources of Uranium in the United States.* National Academy of Sciences, Washington, D.C., 236 pp.

Norton, J. J., 1973, Lithium, cesium, and rubidium—the rare alkali metals. In *United States Mineral Resources,* D. A. Brobst and W. P. Pratt, eds., U.S. Geological Survey, Professional Paper 820, pp. 365–375.

Toombs, R. B., and Andrews, P. W., 1976, Minerals and modern industrial economies. In *Economics of the Mineral Industries,* W. A. Vogely, ed., American Institute of Mining, Metallurgical, and Petroleum Engineers, New York, pp. 33–72.

U.S. Bureau of Mines and U.S. Geological Survey, 1981, Principles of a resource/reserve base classification for minerals. U.S. Geological Survey Circular 851, 5 pp.

Zwartendyk, J., 1981, Economic issues in mineral resource adequacy and in the longterm supply of minerals. *Economic Geology,* Vol. 76, pp. 999–1005.

2 From Discovery to Mineral Production

MINERAL EXPLORATION—THE DISCOVERY OF MINERAL DEPOSITS

The traveler through the Rocky Mountain states soon notices that some areas are pockmarked with old mine dumps and that there are many places where mining is in progress. Many of the abandoned mines are inherited from the period following the discovery of gold in California, which stimulated prospecting not only in that state but also in the mountainous regions to the east. Thousands of prospectors roamed the hills. Some were successful, many were not. The methods of the early prospector were very simple and were based on a very limited knowledge of mineral deposits. He depended largely on finding surface indications of deposits. Those he sampled, and if results were encouraging he proceeded to excavate shallow shafts and tunnels to develop and sample the deposit further. He was remarkably persistent and very thorough, and few deposits of the kinds he knew escaped his search. He was successful in discovering many productive mineral deposits, some large, some small, some very rich, but many other deposits that he found were too low in grade to be worth much effort.

Since 1848 the whole region has been scoured by prospectors, parts of it repeatedly, either in the hope of finding deposits that the earlier prospectors overlooked or in the hope of finding deposits of kinds that were unknown or of no interest to the earlier seekers of mineral wealth. However, the day of surface prospecting is largely over. The search now is mostly for deposits that are hidden, being covered by various thicknesses of barren rock. There may be no surface indications. The search is in terms of our knowledge of the geology of various regions of the earth—the kinds and structural arrangements of rocks that are present in each region. Each type of mineral deposit that we know is associated with particular types of rock formations and not with others. In northern Wisconsin, for example, a search for a certain type of metalliferous deposit has been in progress for nearly 20 years. That part of the state is traversed by several belts of ancient rocks that are products of volcanic activity. Prospecting is based on the fact that in eastern Ontario and western Quebec, deposits of zinc, lead, and copper sulfides, sometimes with gold or silver, were formed during similar episodes of volcanism. The volcanic belts of Wisconsin are therefore regarded as favorable terrain for exploration. Several discoveries have actually been made.

The simple prospecting methods of the last century are still used in some places in the world (Fig. 2-1) but in most of the world have given way to a variety of sophisticated techniques that aid in the discovery of hidden deposits. Geological mapping outlines the surface distribution of rock formations in an area and indicates something of their structural arrangement and distribution beneath the surface. Geophysical methods that detect variations in rock density, electrical properties, and magnetism over an area supplement the geological work and may lead directly or indirectly to discovery (Fig. 2-2). Ground surveys may be supplemented by surveys from the air. Geochemical methods may be used to trace patterns of distribution of various chemical elements as indicated by sampling and analysis of soils or ground waters. The patterns may suggest the locations of hidden deposits. The purpose of all these studies of a "favorable" area is to select the part or parts of it in which mineral deposits are most likely to occur. These are the final targets of exploration. The

Figure 2-1 A *garimpeiro* (with friend), a prospector near Carnauba, Rio Grande do Norte, Brazil. His equipment consists of a pick, a shovel, a screen for separating coarse from fine materials of his samples, a miner's pan, a jug of water, a keen eye for traces of the valuable minerals found in the district, and the patience of Job.

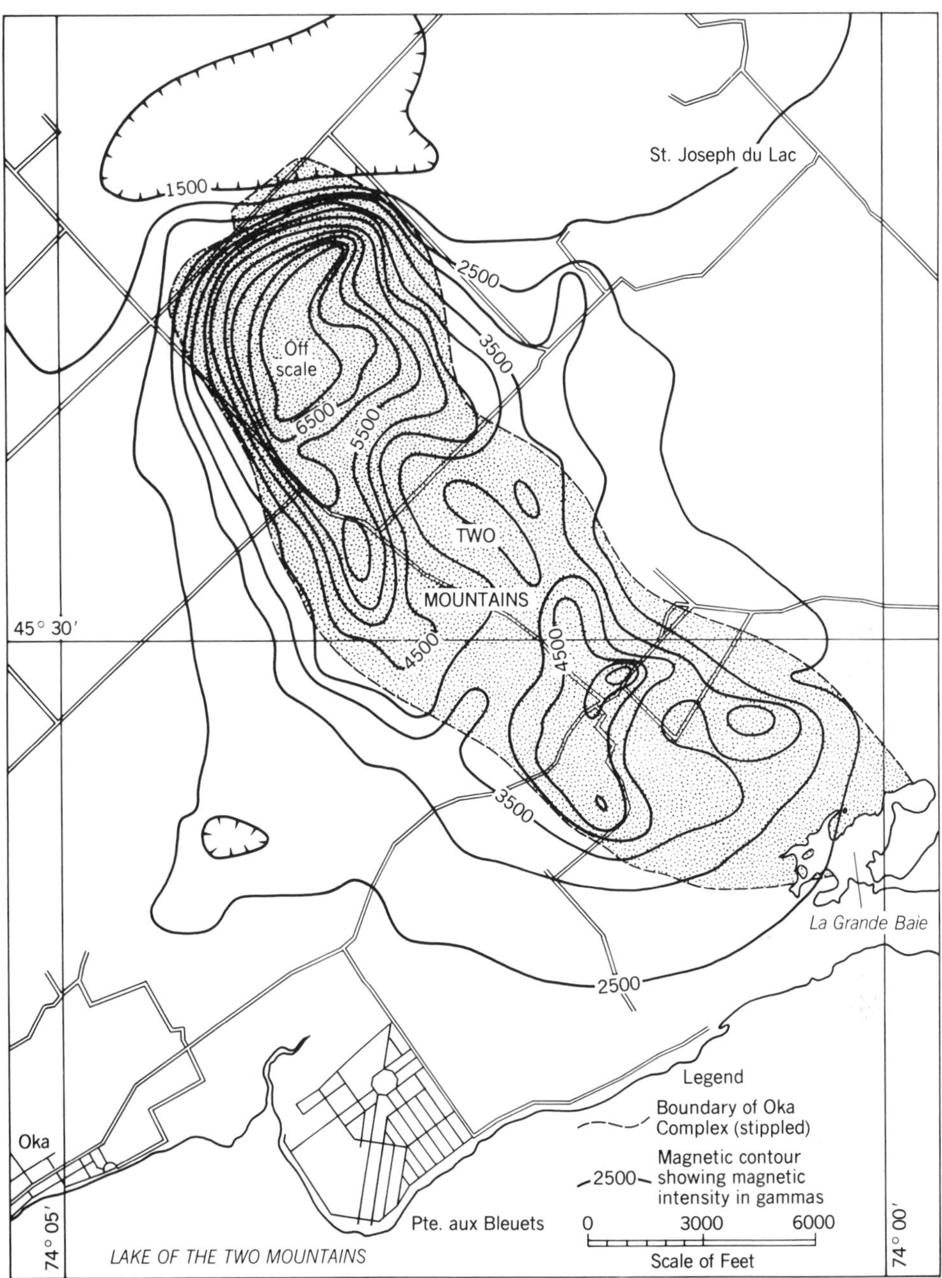

Figure 2-2 Aeromagnetic map of the Oka complex of igneous rocks, Quebec. The magnetic contours are lines passing through points of equal magnetic intensity. The contours portray a strong positive anomaly over the area of the complex, due to the presence of magnetic iron oxide in the component rocks. Based on Figs. 28 and 29 in R. B. Rowe (1958).

next step is usually drilling to find out whether a deposit is actually present. Any deposit discovered must then be drilled out further to obtain the information necessary for satisfactory estimation of tonnage and grade.

Several things are implied in the foregoing discussion. One is that the search for a deposit involves substantial effort over a considerable period of time. There is no quick and easy approach by which the prospector walks up to an outcrop and a few hours or a few weeks later exclaims, "Eureka, I have found it!" The second thing is that there is no advance assurance of success. Geology tells us where to look (within the limits of current knowledge), but in its present state it cannot tell us in advance what will actually be found. And even if a deposit is actually found, there is no guarantee that it will be large enough and rich enough to support a mine. The author recalls asking a colleague, some years ago, about the outcome of an exploration project in which the latter had been involved for some time. "Well," he said, "technically the project was a complete success. Economically the project was a failure." Exploration had been successful in finding a mineral deposit, but drilling and sampling showed that the deposit was too small and too low in grade to support a mining operation. For every success in exploration there are many failures. Mineral exploration is not for the faint in heart.

MINES AND MINING

Mining is the extraction of minerals from the earth. Excavation is usually required. There are three principal types of mines—open-pit or open-cut mines, underground mines, and wells. Figures 2-3 to 2-6 show various types of mining operations. More than 80 percent of the total tonnage of minerals produced annually in the United States comes from open-pit operations. The type of mine developed at a particular deposit depends on the nature of the material, the shape and attitude of the deposit, and the depth to which it extends below the surface. Some mines start as open pits and pass underground as the deposit is followed downward (Fig. 2-3, bottom).

A deposit or part of a deposit that is close to the surface will generally be mined in open pit. If the deposit extends right to the surface, mining of ore begins with first excavation. If the deposit is covered by other material, or if the upper part of the deposit is too lean to be mined, the unwanted material, or overburden, is removed in advance of mining and discarded. Open-pit mining can be done on almost any scale from a few hundred to thousands of tons a day. If the material of the deposit is soft, direct excavation may be possible. If not, machines are used to drill holes in the ore. The holes are loaded with explosives and detonated to shatter

Figure 2-3 Top: A phosphate mine in central Tennessee. The upper, lighter-colored material is overburden being removed by the dragline in the background to expose the darker, phosphate-rich material being excavated by the nearer dragline and loaded into a truck for transport to the processing plant. Bottom: Aerial view of the Gunnar mine, on the north shore of Lake Athabaska, Canada. The mine is in a pipelike ore body produced by solutions passing through fractured rock. The upper part of the deposit was mined in open pit. Deeper parts of the deposit were reached by a vertical shaft (beneath the tall white building near the far right corner of the open pit) from which workings were developed at successively deeper levels.

Figure 2-4 Top: A simple quarrying operation near Little Rock, Arkansas. Behind the rim of the quarry, drill machines are drilling a line of holes, each extending down to the level of the base of the quarry wall. Drill holes are loaded with explosive, and the rock is blasted to the quarry floor, where it is loaded by the power shovel into trucks for transport to a nearby plant for processing into roofing granules. Bottom: Part of the "richest hill on earth," at Butte, Montana. On the skyline is the head frame of one of the shafts leading to the underground workings that honyecomb the hill. Below it is a mill in which concentrates were produced. The white piles are waste dumps.

Figure 2-5 Top: Dragline stripping overburden from a coal seam, near Harrisburg, Illinois. Bottom: The stripped coal seam is being excavated by a power shovel and loaded into a truck. The bucket of the dragline shown in the upper photograph is visible at the lower right.

the rock so that it can be excavated. Power shovels or draglines then excavate the material and load it into trucks or rail cars that carry the ore to the mill. The vehicles range from small trucks to monsters capable of carrying 100 tons or more.

Strip mining is a special type of open-pit mining, applied especially to the mining of flat-lying coal seams. In this mode of operation, mining begins by excavating the overburden and then the coal along a strip ex-

Figure 2-6 Top: The salt crust at Searles Lake, California. In the middle ground are the poles of the power line to one of the wells that tap the brines of the salt body beneath. Brine is pumped through the pipeline to the processing plant beside the "lake." Bottom: A dredge mining the buried beach sand deposits at Trail Ridge, Florida. The beach sands are mined by a cutting head at the far end of the dredge, pumped through the dredge and thence via the pontoon-supported pipeline to a processing plant on the edge of the pond.

tending across the deposit. The overburden and coal are then removed from a second strip alongside the first. Overburden from the second strip is transferred into the excavation of the first strip. Successive strips are mined in the same manner until the deposit is completely mined. The term "quarry" is usually applied to mines from which stone is extracted. Most quarries are open pits, as are the excavations from which sand and gravel are taken.

There are limits beyond which open-pit mining is not feasible either because of rising costs of removing overburden or because there are safety problems with high rock walls. The limits depend on the dimensions and attitude of a deposit, on the nature and soundness of the rocks enclosing it, on the value of the material present, and on the amounts of waste that must be moved in addition to the ore. When the limits are reached, mining must go underground.

An underground mine consists of two kinds of openings or workings, those that are excavated to give access to the deposit and those that are made by removal of the ore. In Fig. 2-7A, for example, a vertical opening, or shaft, has been excavated, and drifts (horizontal, tunnellike openings) have been excavated at several levels to intersect the deposit. These workings serve as means of bringing men, equipment, and materials into the mine and removing ore and waste rock as these are produced in the mining process. Mine safety regulations require that there be a second shaft (not shown) to provide a means of escape should the first become blocked. A system of ventilation must also be provided. Chambers excavated in the ore are called stopes (Fig. 2-7C). Ore stoped (excavated) above a particular level will be dropped to the haulage drift on that level and loaded into cars or trucks that transport it to the shaft for hoisting to the surface. Figure 2-7B shows a similar mine, but here an inclined shaft is used. There are many different patterns of underground mining, each adjusted to the characteristics of the deposit and the rocks that enclose it. The pattern selected for a deposit is hopefully that which will give maximum recovery of ore at acceptable cost.

Open-pit mining has several advantages over underground mining. In general, open-pit mining is cheaper, permits higher recovery of ore, and is considerably safer. Rates of production are also more flexible. The production capacity of an underground mine such as the one in Fig. 2-7A, for example, is limited by the hoisting capacity of the shaft. Beyond this, production can be increased only if another shaft is put down. To increase production from an open-pit mine, it is only necessary to provide additional men and machines for drilling, blasting, excavation, and haulage of ore and waste. In an open-pit mine, 80 percent or more of the ore may be recovered. In underground mines, recovery may be as low as 50 percent, chiefly because part of the ore must generally be left as pillars (as shown in Fig. 2-7C) to support the walls or roof of the ore body and

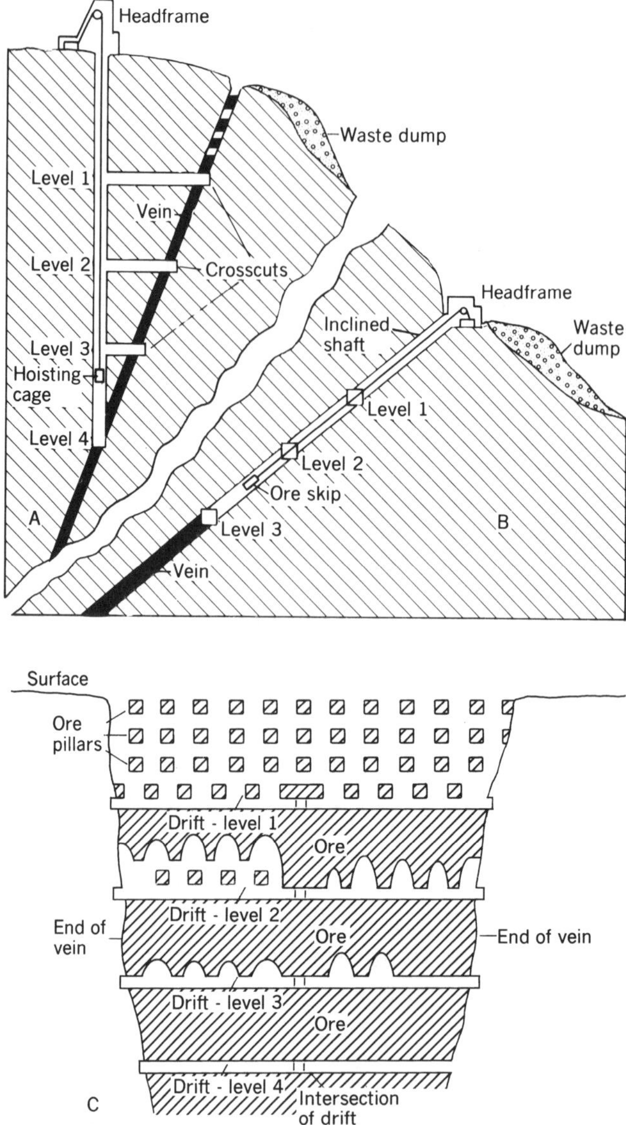

Figure 2-7 Vertical cross sections of two mines, each in a vein deposit. A headframe over each shaft houses hoisting machinery and facilities for transferring ore from the cage or skip to rail cars or trucks. (A) The shaft is vertical, and crosscuts are driven to the vein on the upper three levels. (B) An inclined shaft follows the vein. (C) The plan of workings in the vein of mine A. Stoping of ore has reached various stages on the upper three levels; pillars are left to support the roof. The plan of workings in mine B would be the same, except that the inclined shaft would appear running from top to bottom in the middle of the workings.

to prevent collapse of the drifts and haulageways. There may be parts of the ore body that are structurally too unsound to be mined safely. In places there may be structurally weak wall or roof rocks that also make mining unsafe. Fracture zones that may have heavy flows of water must be avoided or sealed off. Costs of open-pit mining may be only a few dollars per ton, whereas costs of underground mining may be $20 or more per ton. The safety record of open-pit mining is about the same as that of average industry in the United States. Accidents are much more common in underground mines. "Solid" rock is seldom solid. It is fractured to varying degrees. When openings are made in rock underground, the rocks of the walls tend to move into the openings, because support has been partially removed. Depending on the nature of the rocks, their strength and structural characteristics, and the stresses present in the rocks, the movement may be negligible or substantial, gradual or sudden. In some mines, sudden collapses of roof rocks are a danger. In others violent rock bursts periodically take place; pieces of rock bursting from the walls become deadly projectiles. In 1984, a rock burst led to the closing of a section of the famous Creighton Mine at Sudbury, Ontario. Rock bursts are a constant threat in the deep gold mines of South Africa. The behavior of rock masses is not always predictable, even by experienced miners. Many techniques for protecting against such hazards have been developed, but the dangers can never be totally eliminated. Heavy inflows of water along fracture zones encountered in mining are a danger in some mines. In other mines, pockets of toxic or explosive gases are encountered, requiring special means of ventilation and other forms of protection. Underground coal mines are especially hazardous. One reason is that the rocks enclosing coal seams are often structurally weak. A second is that methane (natural gas) poses a constant threat of explosion and fire. A third is that a mixture of coal dust and air is a powerful explosive. Both it and methane are easily ignited. American mining regulations require that elaborate precautions be taken in coal mines to protect the miners, yet since 1900 more than 20,000 miners have been killed in coal mines of the country, and every year adds to the dreadful toll.

In underground mining, openings must be high enough to allow movement of men, materials, and ore. If the thickness of a deposit is such that the minimum height of a stope cannot be provided by excavating the ore, some of the enclosing rock must be removed. This is waste, which must not only be mined but disposed of. The cost per unit of ore mined is thereby increased. If the thickness of waste and ore mined is twice the thickness of the ore, the effect is nearly as though the grade of the ore were halved.

The influence of thickness of ore on the feasibility of mining is illustrated in a number of coal fields of the United States. In a typical coal field several coal seams are present, separated by various thicknesses of mudstone or sandstone. Consider the situation portrayed in Fig. 2-8, which shows diagrammatically an occurrence of coal seams near Middlesboro, Kentucky. The lowest seam, the Jellico seam, consists of high-quality bituminous coal. Over certain areas it averages about 6 feet in thickness and can be mined economically underground. Above it are other seams, but these are only 18 to 24 inches thick, much too thin to be mined economically underground. Portions of the thin seams near the surface, however, can be mined economically in open cut by first stripping the overburden and then mining the coal. When the uppermost seam has been mined, the seam below is stripped of overburden and mined, and so on. Increasing height of overburden as mining progresses into the hill, however, sets an economic limit on the operation.

Open-pit mining has clear-cut advantages over underground mining. It can make reserves out of resources that otherwise could not be mined, either for economic reasons or for reasons of safety. The other side of the coin, however, is that open-pit mining creates much more disturbance of the surface than underground mining. We shall return to this later in considering the environmental consequences of mining.

Not all mineral commodities are recovered from mines such as those described above. Some are recovered from wells, which are the third important class of mines. The most familiar examples are the wells that yield petroleum and natural gas, but there are other mineral-yielding wells. For example, at Searles Lake, California (Fig. 2-6, top), there is a thick salt crust developed by prolonged evaporation of waters draining

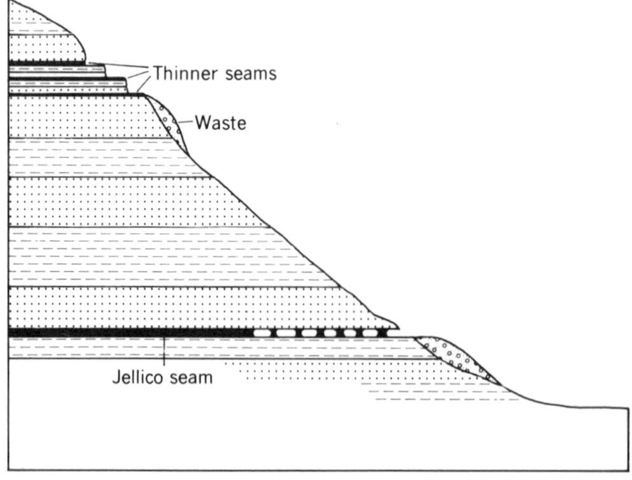

Figure 2-8 Diagrammatic section of coal seam near Middlesboro, Kentucky. The Jellico seam is being mined underground. The three thinner seams higher up are being mined in successive benches from which the rocks overlying the seams have been stripped.

in the past into Searles Lake Basin. The salt crust is very porous, having about 25 percent cavities that are filled with extremely saline brines. Wells are sunk in the crust, and the brines are pumped to a nearby processing plant where potash salts, boron salts, and other valuable salts dissolved in the brines are recovered. The spent brine is then pumped back into the salt crust through other wells. Common salt, magnesium compounds, bromine, lithium, sulfur, and even some uranium are recovered from wells in various districts. In the Paradox Valley, Utah, deep-lying potash deposits were mined for a time from a shaft, by conventional underground methods. An explosion of methane gas that seeped into the workings, however, led to closing the underground mine. The deposit is now worked by wells drilled into the potash-bearing formation. The potash salts are dissolved in water pumped into the wells. The resulting brines are pumped to the surface and treated to recover the potash salts.

A special type of open-pit mining is dredging. Consider the very productive deposit of ancient beach sands mined at Trail Ridge, Florida (Fig. 2-6, bottom), for titanium and other heavy minerals. Here, along the shoreline of a peninsula of the glacial epoch, waves and currents deposited sands containing valuable heavy minerals, the iron-titanium oxide mineral ilmenite being the most important. The area is flat. The water table, the level below which the ground is saturated with water, is only about 10 feet below the surface. For mining, a pit extending to the bottom of the deposit has been excavated and allowed to fill with water. A dredge, equipped with a flexible suction pipe 2 feet in diameter and a cutting head, has been built in the resulting pond. In mining, as the cutter head excavates the beach sands and the overburden, the material is mixed with water and pumped through a pipeline to a processing plant beside the pond. There the heavy minerals, about 4 percent of the material mined, are separated. The remainder is pumped back into the pond behind the dredge. As dredging proceeds, the dredge and the pond literally move through the deposit, which is 19 miles long and up to 2 miles wide. The mined land is regraded and restored to pine forest.

Dredging can be a very efficient, very cheap method of mining unconsolidated materials where conditions favor it. It has been used in mining gold-bearing gravel deposits in many places in the world. Tin-bearing gravels off parts of the coastline of southeast Asia and Indonesia have been mined by this method, and so have sand and gravel, diamonds, and certain other materials. Sometimes suction dredges are used, sometimes dredges with continuous chains of buckets. The most dramatic use of dredging, if it ever becomes politically and economically feasible, will be the dredging of the copper-nickel-cobalt-manganese nodules lying at depths of 12,000 to 15,000 feet on the floors of certain parts of the ocean basins.

PROCESSING OF MINERAL MATERIALS

Some mineral materials that are mined, such as stone and clay, require only simple processing to prepare them for market. At a typical sand-and-gravel operation, material from the pit is simply screened to separate the sand and to recover gravel of various sizes. Many ores, however, are mixtures of a valuable mineral or minerals and waste minerals. In a copper ore, for example, the content of copper minerals may be 2 percent or even less, and the first step in processing is separating the copper minerals from the rest of the material. A mill must be constructed, the design depending on the nature of the material to be treated. Many ingenious methods of separating valuable minerals from waste minerals have been devised. Gold-bearing gravels, for example, may simply be run through a sluice box, essentially a long, sloping, rectangular channel with bottom riffles, which are low crossbars placed at intervals along the length of the sluice. As the gravel cascades over the riffles, the heavy gold and associated heavy minerals settle on the upstream sides of the riffles, whereas the lighter gravel moves on. Periodically the riffles are cleaned out. The heavy-mineral concentrates are removed, and gold is separated from the other minerals. The process simulates the formation of the placer deposits that have formed along channels of rivers draining gold-bearing regions of the world. In the major gold mines of the world, however, the process is quite different. The gold ore is crushed fine and then treated with sodium cyanide solution, in which the gold is dissolved. The solution is then transferred to vats in which the gold is precipitated. Further processing purifies the gold and recovers the silver with which the gold may be alloyed.

One of the most curious, most effective, and most important processes for concentrating ore minerals is known as froth flotation. Consider a typical copper ore in which the valuable minerals are compounds of copper with sulfur or copper with iron and sulfur. The ore is crushed fine to separate the mineral grains from one another and fed into a tank containing a solution full of tiny bubbles. If the bubbles have been produced using the right chemical reagent, they will attach themselves to the particles of copper minerals but not to particles of other minerals. As though hitched to tiny balloons, the copper mineral particles are lifted to the surface of the tank and form a froth, which is swept off by rotating paddles and removed. This can be a remarkably efficient process, recovering 80 percent or more of the copper minerals present in an ore. Its development, early in the present century, made it possible to extract copper from low-grade ores that otherwise would be classified as uneconomic resources. The deposits involved now yield most of the copper produced in the world today. Froth flotation has been applied to many types of ores.

Concentrates produced from many ores must be further treated. Copper concentrates are an example. Concentrates consisting of copper sulfides must be smelted to remove sulfur and other unwanted elements. Mixed with fluxing materials that lower the temperature of melting, they are melted in special furnaces. The sulfur is oxidized to sulfur dioxide gas; silicate impurities rise to form a slag at the top of the melt. Carbon is added to reduce the copper to the metallic state. The molten copper accumulates at the bottom of the furnace, where it is tapped off at intervals. The metal at this stage is still impure and must be subjected to a refining process.

The recovery of a mineral or metal and its conversion to a useful form may thus be very simple or very complex. It may be cheap or very expensive in dollars and in energy consumed. Each new type of ore discovered requires the development of appropriate treatment processes. Years may be required for the necessary research and development. The 20th century record of technological progress in extracting useful minerals from earth materials is a bright chapter in the history of technology. The consumer, who is concerned mostly with the finished goods produced from minerals or with the aid of minerals, should be aware that the provision of those goods on the scale and in the variety to which he has become accustomed is possible only because of the great progress made in processes of extraction.

RATES OF MINING AND MINERAL PRODUCTION

As indicated in Chapter 1, the potential availability of mineral supplies depends fundamentally on the resources of minerals present in known deposits or in deposits that can be discovered in the future, and on success in converting resources into reserves. In terms of these things, we can speak of total amounts of minerals that can be made available to man. But even if the amounts are large, it does not necessarily follow that annual supplies of minerals will be adequate to meet demands. There is also the question of the rates at which they can be recovered from the earth.

Consider a mine of the type shown in Fig. 2-7A, in a vein that has a greater length. Production from the deposit is through a single vertical shaft. As indicated previously, the hoisting capacity of the shaft, in tons per hour, sets a limit on the rate of production of ore. If we desire to double the rate of production, another production shaft must be provided, together with ancillary development workings. This means a large investment that must be repaid from the proceeds of the operation. Suppose the number of tons of ore available in the deposit is not enough to yield a return on the additional capital that must be invested. In that case it is simply not feasible to increase the output of the mine.

Or consider the case of an open pit mine, in which an increase in rate of production may mean only adding to the labor force and the inventory of mining machinery. The more machines in operation, the more the tons of ore that can be produced per hour. Even so, there is a limit on capital investment relative to the total ore that can ultimately be produced, still maintaining an adequate return. Beyond this, we have to consider how the ore will be processed after it is mined. If at a rate of 3000 tons per day the present mill is working to capacity, any increase in production will require additional investment in milling facilities. This may or may not be economically feasible.

The fact is that for every mineral deposit there is a maximum rate of production that cannot be exceeded without reducing the return on investment to an unacceptable level. In talking about availability, then, we must be concerned with possible rates of production. This means that we need data for the production capacities of existing mines, mills, smelters, refineries, and other processing plants and that we need data on the feasibility of increasing, if need be, the capacities of existing plants. Much of the available data is contained in compilations by the U.S. Bureau of Mines, Department of Energy, and Department of Commerce.

THE NATURE OF MINERAL INDUSTRY AND ITS ROLE IN THE ECONOMY OF THE UNITED STATES

The nature of mineral industry in the narrow sense has been indicated in the foregoing discussion. It consists of all the mines, the mills, the smelters, the refineries, and the other processing plants by which the materials mined are converted into forms required by manufacturing industry. The U.S. Bureau of Mines estimates the value of all processed minerals produced in the United States in 1981 at $253 billion, about 7 percent of the gross national product in that year. These figures, however, do not fully measure the role of mineral industry in the economy of the United States, because they do not take into account the total economic activity that is generated by mining industry. Each mine that is opened requires an array of supply and service activities. These range from supply of materials, tools, and machines required for the construction and operation of a mine to the supplies and services that are required for any substantial mining community, from food supplies to educational, medical, and government services. The requirement of supplies and services means that agricultural and manufacturing industries must exist outside the mining community. These services are only partly reflected in the gross value of the products of mineral industry. There is a multiplier effect that must be taken into account. A factor of 2.5 has sometimes been used in calculating the total annual contribution of mineral industry to the national economy. It is only an approximation.

It is important to recognize that the dollar value of its products is not really adequate as a gauge of the importance of mineral industry to the United States and to the world. The wealth of the world is based fundamentally on effective use of *all* its natural resources—mineral, agricultural, forest, fish and wildlife, water, and air. All are essential. They are the base on which all manufacturing and service industries are built. They are the pillars of today's civilization. Without them only a fraction of the world's population could be sustained. In the United States, furthermore, a strong mineral industry is an essential element in the structure that provides for the national security.

ADDITIONAL READING

Lacy, W. C., ed., 1983, *Mineral Exploration*, Hutchinson Ross, Stroudsburg, Pa., 433 pp.

McGraw-Hill, Inc., 1977, *Encyclopedia of Science and Technology*, Vol. 8, pp. 547–592. McGraw-Hill, New York. A series of articles on mining: Mining (E. Just), Open-pit mining (D. P. Bellum), Placer mining (E. H. Beistline), Strip mining (J. Dowd), Underground mining (A. V. Corlett), Undersea mining (M. L. Cruickshank).

Merrill, C. W., 1964, The significance of the mineral industries in the economy. In *Economics of the Mineral Industries*, E. H. Robie, ed., American Institute of Mining, Metallurgical, and Petroleum Engineers, New York, pp. 1–42.

Ohle, E. L., and Bates, R. L., 1981, Geology, geologists, and mineral exploration. In *Economic Geology, 75th Anniversary Volume*, B. J. Skinner, ed., Economic Geology Publishing Company, New Haven, Conn., pp. 766–774.

Rickard, T. A., 1932, *A History of American Mining*, McGraw-Hill, New York, 419 pp.

Rowe, R. B., 1958, Niobium (columbium) deposits of Canada. Geological Survey of Canada, Economic Geology Series No. 18, 108 pp.

Snow, G. G., and MacKenzie, B. W., 1981, The environment of exploration: Economic, organizational, and social constraints. In *Economic Geology, 75th Anniversary Volume*, Economic Geology Publishing Company, New Haven, Conn., pp. 861–896.

3 The Energy Minerals

INTRODUCTION

The Arab oil embargo of 1973 caused great concern about the availability and costs of energy, a concern shared by most of the countries of the world. Modern civilization entails the annual production and consumption of large amounts of energy for provision of heat and light, for transport of people and goods, for conversion of raw materials into finished goods, and for agricultural production. Most of this energy is furnished from mineral fuels, coal, petroleum, natural gas, and uranium. The use of solar and geothermal energy is increasing, but even their more enthusiastic supporters cannot anticipate that they will substantially decrease our dependence on the mineral fuels. The size and availability of reserves and resources of the mineral fuels and the costs of obtaining them are therefore of prime importance in assessing energy prospects for the remainder of the 20th century.

COAL

The Importance of Coal

During the early part of the present century, coal was the principal U.S. source of energy (Fig. 3-1), but the use of petroleum and natural gas increased rapidly in subsequent decades. In 1984 coal furnished only about 23 percent of the total energy consumed, whereas 42 percent was obtained from petroleum and 24 percent from natural gas. Total resources of energy in coal deposits of the United States and the rest of the world, however, are many times the amounts of energy that we can hope to obtain from the other two fossil fuels. In the long run, therefore, coal will be much the most important. Furthermore, there are processes avail-

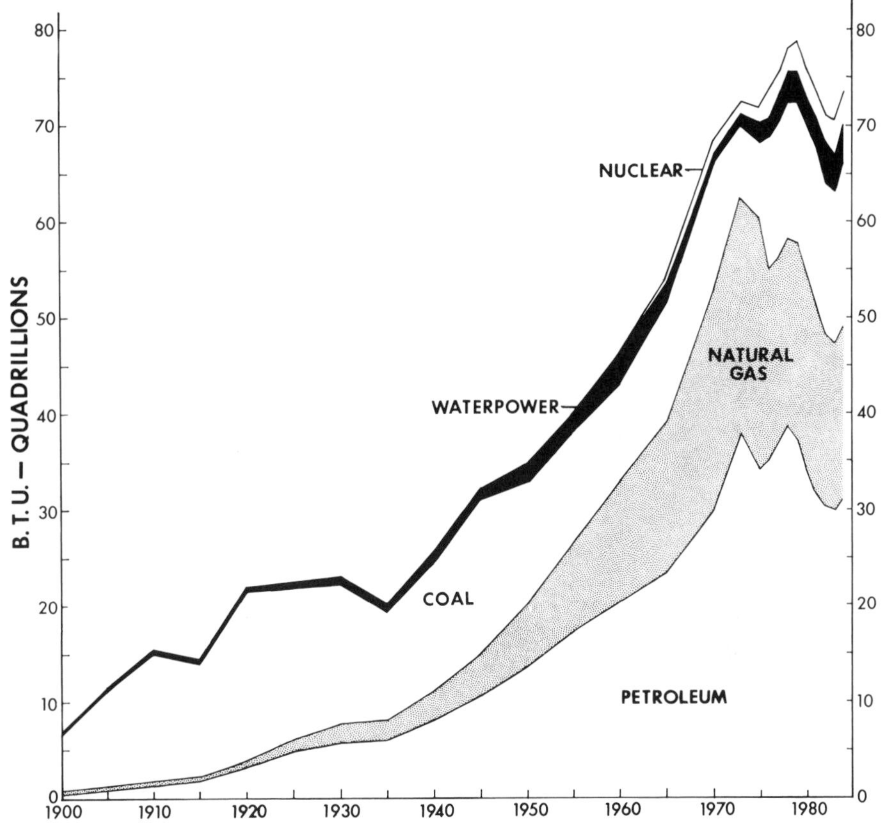

Figure 3-1 Consumption of energy in the United States, 1900 to 1984, by source, expressed in Btu. A Btu is the amount of energy required to raise the temperature of a pound of water by 1° Fahrenheit. Data from Energy Information Administration and U.S. Bureau of Mines.

able for conversion of coal to petroleum and gas when the natural deposits of the two are depleted. Finally, coal is very important as a chemical raw material. It yields a variety of products ranging from plastics to dyes and many chemicals.

The Nature and Formation of Coal

Coal is a complex group of materials with a wide range of properties. Its complexity and its versatility as a fuel and as a chemical raw material are direct consequences of the way coal deposits are formed. Coal formation is a long, complicated process. It begins with formation of a lake, a pond, or a lagoon (Fig. 3-2). Vegetation springs up first in the shallow waters around the margins. If the circulation in the body of water is poor, then as the vegetation grows and dies it is only partially destroyed by decay. The residue accumulates, building up the bottom along the shores. With

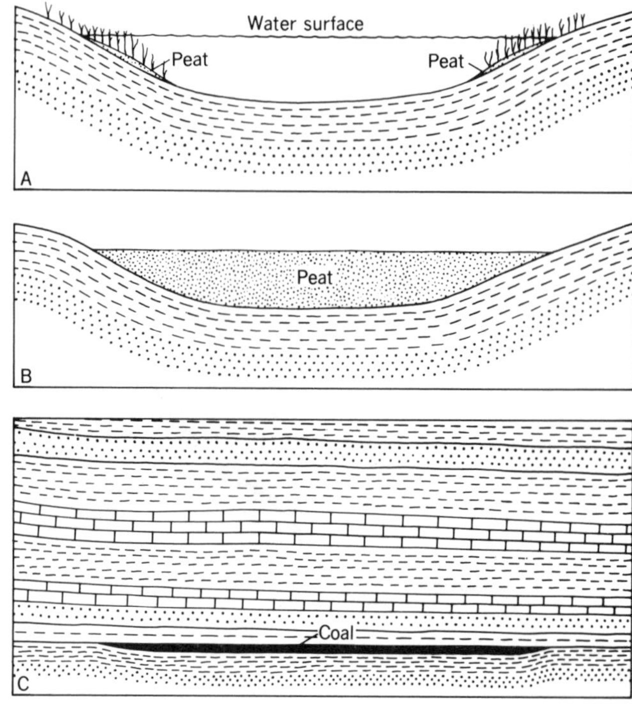

Figure 3-2 Stages in the formation of a coal deposit. (A) Vegetation, changing to peat by partial decay, gradually accumulates along the margins of a lake and extends slowly inward. (B) The lake is filled with peat. (C) The layer of peat is buried beneath a great thickness of sediments, only the lower part of which is shown. The layer is compacted and in time converted to coal by heat and pressure.

time, the belt of vegetation, living and dead, spreads inward, and if the process continues uninterrupted, the lake is ultimately filled with the vegetal remains. The lake becomes a swamp or bog; the filling is peat.

We can observe the various stages of this process at many places in North America and the rest of the world. The glacial period that ended 8000 to 11,000 years ago left many tens of thousands of lakes and ponds in the northern part of the United States and Canada. The majority have already been converted to peat bogs; others are still unfilled. Another great group of peat-bearing swamps is found near the Atlantic coast from New Jersey to Florida. The Dismal Swamp, straddling the Virginia–North Carolina line, and Pogo's home, the Okefenokee Swamp, are well-known examples. Peat bogs range from a few acres to thousands of square miles in extent. Before parts of it were drained and converted to agricultural lands, the Dismal Swamp had an area of 2200 square miles. The layer of peat laid down in it averaged 6 feet in thickness and contained an estimated 672 million tons. There are even more extensive bogs in northern Minnesota and Canada. The peat resources of Minnesota have been estimated at 6.5 billion tons, not quite half the total peat resources of the United States.

Peat is a fuel, though not a very good one, and it has been used as a fuel in Ireland and elsewhere in the world. Peat is not coal, but it is the material from which the coals of the world have been formed during the passage of geologic time. Coal has formed from peat only where the peat deposits have been buried beneath piles of sediments: sandstones, mudstones, and limestones in various proportions. As we go downward in the earth, the temperature rises at an average rate of about 1°C for every hundred feet. As peat is buried deeper and deeper in the crust, the heat and the pressure due to the load of overlying rocks gradually increase, and peat is gradually converted into coal. Plant remains are complex compounds consisting mainly of carbon, hydrogen, and oxygen. Water is present. As heat and pressure rise, water is driven off; likewise much of the hydrogen and oxygen built into the organic compounds. The carbon content rises.

The changes that produce coal are progressive. As they continue, peat is changed to lignite, then to bituminous coal, and then to anthracite. The main changes involved in the progression from lignite to anthracite are shown in Table 3-1. When we examine the world's coal deposits, we find all stages of the general process represented. Much of the coal of the western states is either lignite or on the boundary between lignite and bituminous coal, whereas in the Appalachian and central states the coals range from bituminous to anthracite. The terms *lignite, bituminous,* and *anthracite* designate the *ranks* of coal, from lowest to highest. Each rank except lignite is divided into subranks, as indicated in Table 3-1. The higher the rank of a coal, the higher the heat yield per pound. Of the three ranks of coal, bituminous coals are the most important. They

TABLE 3-1 Composition and Heating Values of Various Ranks and Subranks of Coal

	Fixed Carbon[a] (Percent)	Volatile Matter (Percent)	Moisture (Percent)	Heating Value Btu per lb
Lignite	37.8	18.8	43.4	7,400
Bituminous				
Subbituminous	42.4	34.2	23.4	9,720
Low bituminous	47.0	41.4	11.6	12,880
Medium bituminous	54.2	40.8	5.0	13,880
High bituminous	64.6	32.2	3.2	15,160
Super bituminous	75.0–83.4	22.0–11.6	3.0–5.0	15,420
Anthracite				
Subanthracite	83.8	10.2	6.0	
Anthracite	95.6	1.2	3.2	14,400

[a]Fixed carbon is carbon remaining after coal is heated to red heat. Volatile matter includes carbon, hydrogen, oxygen, nitrogen, and sulfur. Compositions are on an ash-free basis.

have a high heat yield and are excellent fuels. They are also the source of many chemical by-products. Anthracite has a higher heat yield per pound than most bituminous coals, but its by-product yield is very low. It has been used mainly for heating.

The coal seams of the world range from a few inches to more than 200 feet in thickness. Some extend over thousands of square miles, others are of very limited extent. The great Pittsburgh coal seam is of minable thickness and quality over an area of 6000 square miles in Pennsylvania, Maryland, West Virginia, and Ohio. It may originally have extended over an area of 200,000 square miles, but most of it has been lost by erosion.

The classification of coals into lignite, bituminous, and anthracite recognizes major differences in the characteristics of coals, but it by no means indicates their full complexity. The peats that gave rise to coals formed from a wide range of plant materials that are represented in diverse proportions in various coal deposits—reeds, rushes, grasses, woody plants, algae, spores, pollen, and others. Lignite and bituminous coals therefore include a range of chemical materials that have not much in common except a high carbon content. Correspondingly, coals show a range of industrial properties. Besides this, coals differ in content of impurities, which are mainly mud and sand and sulfur. When coal is burned, most of the impurities are left behind as ash, which may range from a few percent to as much as 10 percent by weight of the original coal. A high ash content is obviously undesirable. Sulfur is oxidized to sulfur dioxide. This passes off as a noxious gas, unless means of removing it from the stack gases are provided. Sulfur from the burning of coal has been identified as a major contributor to the problem of acid rain.

The Distribution of Coal Deposits: Resources and Reserves

Although the earth is 4.5 billion years old, all but a small fraction of the world's coal deposits were formed during three great coal-forming epochs distributed over the past 300 million years. The coals range from lignite to anthracite in various fields.

To estimate reserves and resources of coal in the coal fields of the United States and the rest of the world, we obviously need a variety of information. First we need data for the thickness, extent, and quality of individual coal seams. We also need to know their position and arrangement in the crust. Some coal seams lie close to the surface and are potentially available by open-pit mining. Others dip steeply into the earth, and some extend thousands of feet below the surface. Most of the coal in such seams is available only through underground mining. We must have information on the thickness and structural condition of the rocks enclosing the coal seams. All this information is needed for an appraisal of the amounts of coal actually in the ground and the amounts that will be economically recoverable.

There is a vast amount of information. The coal deposits of western Europe have been under investigation for more than 200 years, those of the United States for more than 100 years. Coal deposits in the rest of the world have been studied in varying degrees. Even so, there is much still to be learned about the world's coal resources. It has been evident for many decades that the resources are very large, but as new information becomes available, estimates are periodically revised. The most recent estimate of "technologically and economically recoverable" coal is summarized in Fig. 3-3. The coals range from anthracite to lignite and of course have a range of energy contents. For purposes of the estimate, reserves are calculated in terms of tons of a coal having a standard heat value of 12,600 Btu per pound, corresponding to a typical bituminous coal. In Fig. 3-3, the percentage distribution of recoverable reserves is shown by countries and regions. The concentration of the world's coal reserves in the northern hemisphere is remarkable. It reflects the fact that in much of the southern hemisphere there was only one major epoch of coal formation, whereas there were three in the northern hemisphere. The coal deposits of western Europe are historically very significant, for they provided the fuel for the Industrial Revolution.

Total identified coal resources in the United States are estimated at about 1.74 trillion tons, and there may be an equal amount of resources as yet unidentified in portions of coal deposits that have not yet been explored. Figure 3-4 shows the distribution of the coal fields of the United States. Minable coal is found in 31 states, and coal is actually being mined in 26 states. Reserves of coal (coal potentially minable) are estimated at 483 billion tons, distributed as indicated in Table 3-2. If recovery is 50 percent, 242 billion tons can be produced. This is the energy equivalent

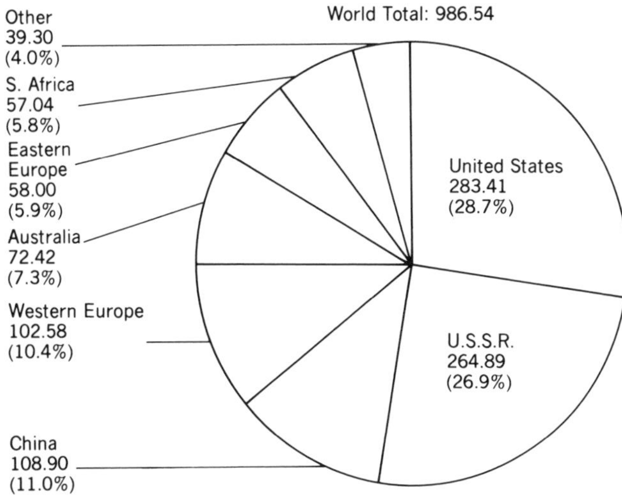

Figure 3-3 Estimated world recoverable reserves of coal, 1980, in billions of tons. Source: Energy Information Administration (1985).

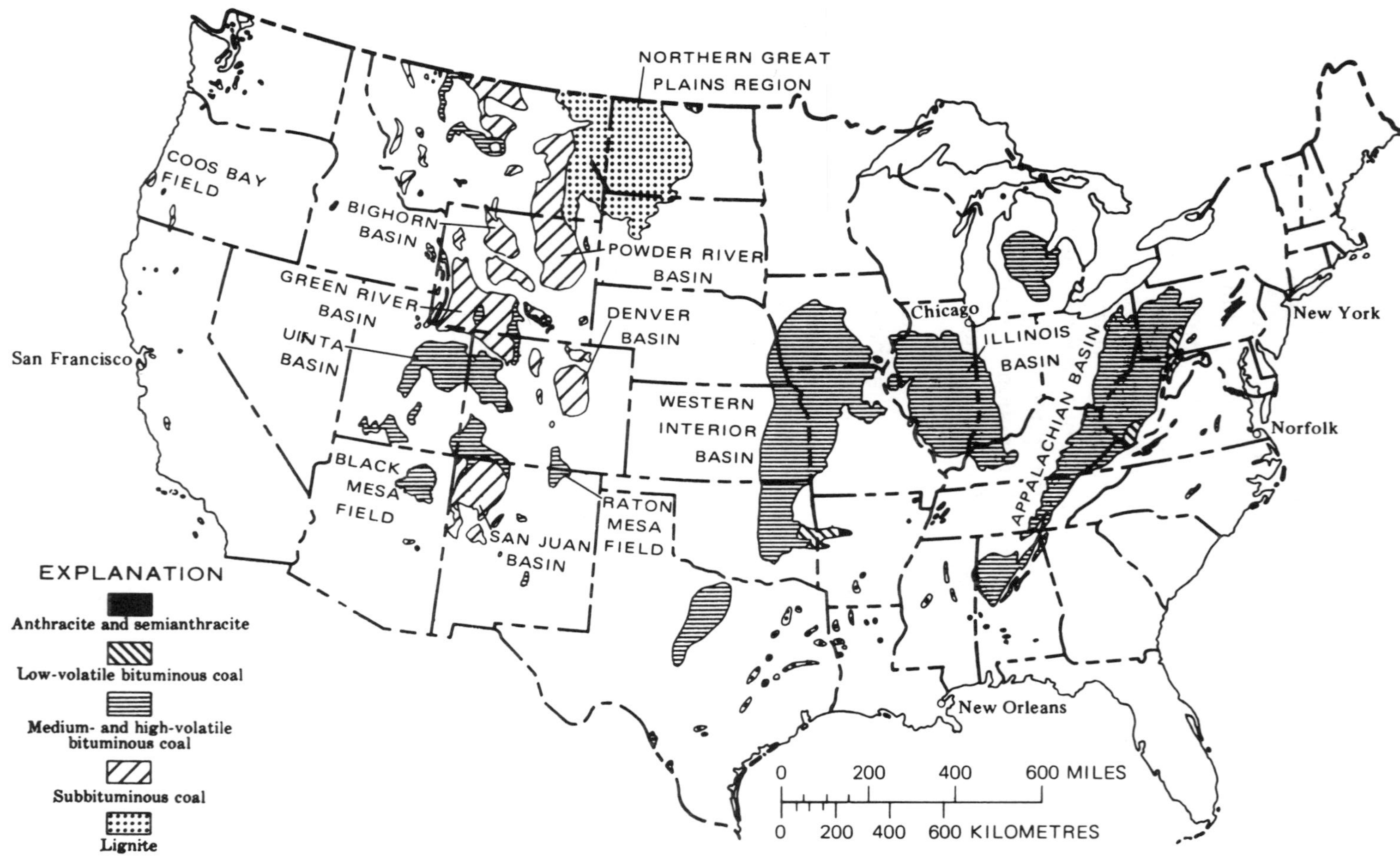

Figure 3-4 Coal fields of the conterminous United States. Furnished by U.S. Geological Survey, courtesy of Dallas L. Peck.

TABLE 3-2 Minable Coal Reserves of the 10 Principal Coal-Bearing States (Billions of Tons), January 1, 1982

State	Reserves	Percent of Total
Montana	120.3	24.9
Wyoming	69.7	14.4
Illinois	78.8	16.4
West Virginia	39.3	8.1
Kentucky	33.5	6.9
Pennsylvania	30.0	6.2
Ohio	18.9	3.9
Colorado	17.2	3.6
Texas	13.5	2.8
Indiana	10.5	2.2
Other states	51.3	10.6
Total	483.0	100.0

Source: Energy Information Administration, *Annual Energy Review 1983*, April 1984.

of about 980 billion barrels of oil, roughly 35 times the total proved reserves of recoverable crude oil in the United States. The 50 percent figure for recovery may be conservative, since it is based on experience with underground mining. At present about 60 percent of U.S. coal production is from open-pit mines, at which recovery ranges up to 80 or 85 percent. However, there is much uncertainty about future recovery percentages (see Schmidt, 1979, pp. 62–70).

As indicated in Fig. 3-4, most of the coal in the eastern and midwestern states is bituminous coal, whereas the bulk of the coal in the western states is subbituminous coal and lignite. There are marked variations in sulfur content. Averitt (1975) reports that sulfur in U.S. coals ranges from 0.2 percent to 7 percent. Coals of the eastern and midwestern states are mostly medium-sulfur to high-sulfur coals, with 1.1 to more than 3 percent sulfur. Those of the western states are mostly low-sulfur coals, with less than 1.1 percent sulfur. The difference is important at a time when emissions of sulfur from the burning of coal are a matter of great concern. Restrictions on such emissions by the Environmental Protection Administration in recent years have encouraged greater use of western low-sulfur coals.

Production of Coal

Table 3-3 shows world coal production by countries in 1982. The United States was the leading producer of coal; Russia was second. The only

TABLE 3-3 World Coal Production, 1982 (Millions of Short Tons)

United States	838
West Germany	247
United Kingdom	134
Yugoslavia	60
Czechoslovakia	138
East Germany	294
Poland	250
Soviet Union	791
South Africa	154
Australia	161
China	734
India	118
Others	463
Total	4382

Source: Energy Information Administration, *Annual Energy Review, 1984.*

important producers of coal in the southern hemisphere are South Africa and Australia, although production of coal in Zimbabwe is important to that country and its immediate neighbors. Figure 3-5 shows U.S. production, consumption, and net exports of coal from 1960 to 1983. Increases since the Arab oil embargo of 1973 have been stimulated by shortages and increased prices of petroleum.

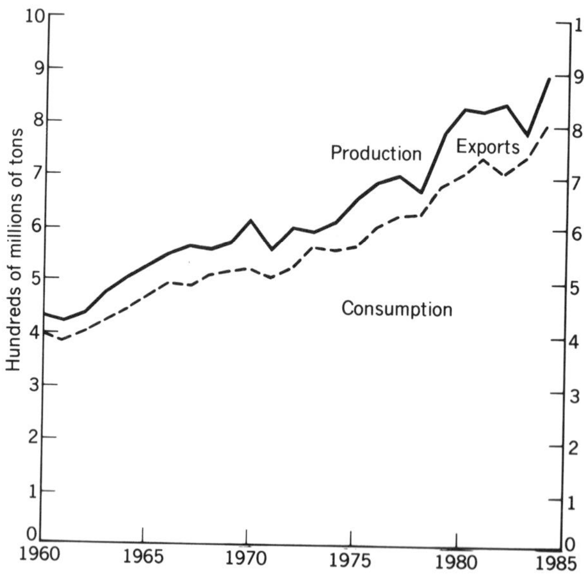

Figure 3-5 U.S. production, consumption and exports of coal. Source: Energy Information Administration (1985).

Coal resources of the United States and the world are capable of sustaining present production, or even greater production, for many decades, perhaps hundreds of years. However, the cost of coal will gradually rise as the better and more cheaply minable coal deposits are depleted.* About half of the estimated minable coal in the United States is available by surface mining, whereas surface mining at present yields about 60 percent of the coal produced annually. As time goes on, the proportion of coal produced from underground mines will rise, and average costs of coal will increase.

Coal as a Source of Other Fuels

During World War II, Germany was hard pressed to maintain adequate supplies of petroleum, particularly after the oil fields and refineries in Romania were bombed. Plants for the manufacture of gasoline out of coal were therefore built and operated. The Fischer-Tropsch process that was used, like other processes for producing petroleum from coal, involves adding hydrogen to the carbon of coal to produce the compounds of the two elements, the hydrocarbons, that make up natural petroleum. A number of products ranging from gasoline to oils and other lubricants can be made. The cost is high, and synthetic gasoline is no longer made in Germany, but South Africa has accepted the cost as the price of reducing dependence on imports of petroleum. No economic deposits of petroleum have yet been found in South Africa, despite an intensive search. A substantial part of the country's supply of petroleum products is now produced from coal.

Burnable gas (water gas) was produced in various cities from coal for many years, until it was replaced by natural gas. To make water gas, coal is treated with steam to produce a mixture of carbon monoxide, carbon dioxide, methane, and hydrogen. The gas is low in heating value, 100 to 200 Btu per cubic foot compared to 900 to 1200 Btu per cubic foot for natural gas. Water gas cannot bear the costs of pipeline transport. Processes for the production of high-Btu gas (950 to 1050 Btu per cubic foot) are under development. The Great Plains project for gasification of coal, in North Dakota, is designed to produce 125 million cubic feet of gas per day; total cost has been estimated at $2 billion. In May 1985, the plant was in operation, but costs of operation were high, and the future of the project was uncertain. The plant is subsidized by the Synthetic

*There is an ambiguity in the use of the term "depletion" in the literature of mineral resources. It is used to mean progressive reduction in the tonnage of ore in a deposit through mining. It is also used as a synonym for "exhaustion." The first usage is the more appropriate one, because few mineral deposits are exhausted by mining. What happens is that as the higher-grade material (ore) and more easily accessible material (ore) is mined, costs rise, and an economic limit of mining is reached. The parts of deposits that cannot be mined profitably are left in the ground. In this book the term "depletion" is used in the first sense, not in the sense of exhaustion.

Fuels Corporation created under the administration of President Carter. The production of liquid fuels by liquefaction of coal is in the research and development stage.

Developing processes for producing other fuels from coal is not just important as a means of supplementing and replacing petroleum and natural gas as fuels. Some processes under consideration also offer means of removing and disposing of the pollutants that are one of the chief drawbacks to the use of coal as a source of energy.

A substantial fraction of the total coal resources of the United States is in beds that lie too deep or are too thick or too thin to be recovered by conventional mining methods. The Department of Energy, in collaboration with private industry, is therefore sponsoring research that is aimed at developing new recovery techniques. One of these is underground coal gasification, the technical feasibility of which has been proved in Britain and in the Soviet Union where underground coal gasification plants have been under development for more than 20 years. Air is injected into holes drilled into a coal bed, and the coal is set afire. Air and steam or steam alone is then injected. Methane, carbon dioxide, and hydrogen are produced and removed through production wells. The Department of Energy estimates that as much as 1.8 trillion tons of otherwise unminable coal might be produced by underground gasification, roughly tripling the amount of energy that could be recovered from the coal resources of the United States. The practice of underground coal gasification would also eliminate the safety problems that have plagued underground coal mining since its inception. However, underground coal gasification is still in the research and development stage. Its future as a source of energy is uncertain.

PETROLEUM AND NATURAL GAS

Importance of Petroleum and Natural Gas

Petroleum and natural gas are the principal sources of energy in the United States. They provide the fuels for our transport systems and for most of our home heating, and they provide a substantial share of the fuel used to generate electricity in our power plants. They have many advantages. Petroleum and natural gas are produced from wells that cause minimum damage to the environment. Natural gas is a relatively clean fuel, yielding a minimum of pollutants. Fuels derived from crude oil can be used to power the great engines that drive the ships of the world and the small engines that drive our automobiles, power mowers, and boats for recreation. Petroleum furnishes the raw materials for a vast array of plastics, pharmaceuticals, and other chemicals that are in daily use in industry, agriculture, and our homes. Natural gas is the major source of hydrogen for the manufacture of ammonia, which is in turn used in manufacturing the essential nitrogen-bearing ingredients of fertilizers. As chemical raw materials, no other naturally occurring substances, even coals, have the versatility of petroleum and natural gas. They are the basis of the giant petrochemical industries of the world. Petroleum products are also the principal lubricants of the world's machinery. The availability and cost of adequate supplies of petroleum and natural gas therefore concern every country of the world.

The Formation and Accumulation of Petroleum and Natural Gas

We are concerned in this book primarily with the availability of petroleum and natural gas, but their availability is best understood in terms of the manner of their formation and accumulation. The crust of the earth has a basement consisting of hard crystalline rocks. Over large areas of the earth the basement is overlain by a skin of sedimentary rocks, laid down as successive layers of mud, limy mud, sand, and gravel. The skin ranges from a few hundred feet to tens of thousands of feet in thickness. Accumulations of petroleum and natural gas are in the thicker parts of the skin. Formation begins with the accumulation of organic muds laid down on portions of the sea bottoms of the past. With time, the muds become buried beneath other layers of sediments, muds, silts, sands, and limestones. The deeper the burial, the greater the heat and pressure on the organic muds. The organic material in them is gradually transformed into hydrocarbons, combinations of hydrogen and carbon that range from viscous asphalt to fluid oil to natural gas. Some of the hydrocarbons are squeezed out and migrate into porous formations, usually sandstones or limestones, and move through those rocks. Some hydrocarbons escape to the surface and are forever lost, but some become trapped in various favorable arrangements of rocks. Thus reservoirs of oil and gas are formed. There are many types of traps, a few of which are illustrated in Fig. 3-6. The petroleum reservoirs of the world, then, are bodies of porous sedimentary rocks saturated with oil and gas that have been prevented from escaping by impervious confining rocks. Exploration for oil and natural gas consists of finding the traps in the sedimentary accumulations on the continents and on the continental shelves and tapping the oil and gas in them by means of wells. The fluids are piped or otherwise transported to refineries, where impurities are removed and oil and gas are separated. Gasoline, kerosene, diesel oil, and many other types of oils and greases are produced from the crude oils, which are always mixtures of various hydrocarbons. Sulfur is an important impurity in many oils. It is removed and marketed.

An area that is underlain by one or more accumulations of oil and gas

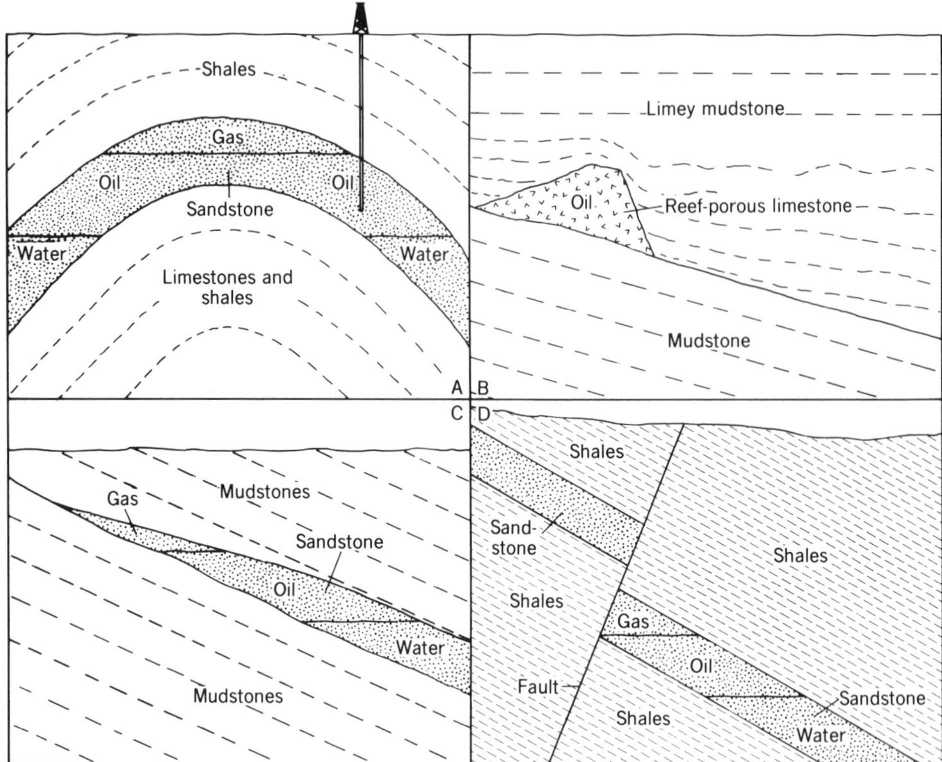

Figure 3-6 Four examples of the many types of traps in which oil and gas accumulate. In each case oil and gas have accumulated in a porous rock (reservoir rock) beneath an impervious rock, usually a shale or mudstone, that prevents upward escape of the fluids.

is termed an oil or gas field. Fields range from small to very large. There are thousands of oil fields in the United States alone. The largest fields, the giants, are those in which reserves in excess of 100 million bbl of oil (a barrel is 42 gallons) have been proved by exploration. Supergiants are fields with reserves in excess of 1 billion bbl. Only four U.S. oil fields, Elk Point, California (1.1 billion bbl); East Texas (1.9 billion bbl); Santa Maria Basin, California (1.2 billion bbl); and Prudhoe Bay, Alaska (10 billion bbl.), qualify as supergiants. The world's greatest oil field is the Ghawar field of Saudi Arabia, with 75 billion bbl—nearly three times the proved reserves in all the oil fields of the United States. The next largest, the Burgan field of Kuwait, has proved reserves of 66 billion bbl.

Reserves of Oil and Gas

There has been a great deal of confusion over the meaning of figures for reserves of petroleum and natural gas. The American Petroleum Institute, which compiled reserve data annually for many years, defines proved reserves as the amount of crude oil shown by geological and engineering data to be reasonably certain of recovery from known reservoirs under operating conditions at the time of estimate. The American Gas Association uses a similar definition for proved reserves of natural gas.

Why use these definitions? Why are estimates of proved reserves revised every year? Suppose we consider a hypothetical oil field. It was discovered in 1975 after geological and geophysical investigations indicated a possible reservoir approximately 7200 feet below the surface. The first well drilled disclosed a thickness of 50 feet of sand with payable amounts of recoverable oil. A single well, however, does not make an oil field, and it is not a basis for calculating proved reserves of oil. Wells were therefore drilled into adjacent parts of the reservoir, and the extent of the reservoir began to be defined. Core samples were taken from the drill holes, and tests of porosity, permeability, and other properties of the reservoir rock were made. These indicated what proportions of the oil and gas present would be recoverable. Exploration continued, and by late 1976 a first estimate of oil and gas reserves in the field was possible. The estimate covered only the part of the field explored to date. The amounts were stated as the proved reserves for the field. Obviously, the figure announced at that time did not indicate the total amounts of oil and gas that might be recovered from the field. Production began early in 1977, drilling was extended westward from the part of the field originally explored, and additional reserves were proved. At the end of 1977, a new estimate of proved reserves was given out. The amount was calculated by adding original and newly proved reserves and subtracting the amounts of oil and gas produced during the year. This procedure has been continued in intervening years, and the limits of the field have gradually been defined. What about the future? At some point the limits of the field will be reached by exploration. Beyond this there will be no additions to reserves. As production continues, proved reserves remaining in the field will drop and will ultimately decline to zero.

This is the history of a typical oil or gas field, and it is the key to understanding the nature of estimates of proved reserves. Over any period of years, estimates of proved reserves of oil and gas in the oil fields of the United States must be revised to take into account both the depletion of reserves previously proved and the additions to proved reserves through further exploration of known oil fields and through the discovery of new oil reservoirs. At no time will it be possible to give a precise figure for the amount of oil and gas that will ultimately be produced from all the fields.

There is another complication that is expressed in the definition of proved reserves of oil as recoverable oil. Proved reserves are never equivalent to total oil in the ground. In every oil reservoir there is a fraction of the oil that cannot be recovered economically using current technology. The rocks of some oil reservoirs are highly permeable and allow oil to move freely through them. Recovery of oil from such reservoirs may exceed 90 percent. In other reservoirs, the rocks are so tight that the flow of fluids through them is negligible. Possible recovery may be less than 5 percent; well production is uneconomic. Average recovery of oil from all the oil fields of the United States is estimated to be somewhere between 30 and 40 percent at the present time. Massive efforts are being made to improve the technology of petroleum production; increase in the recovery percentage has been slow but nonetheless significant. As recovery percentages change, estimates of proved reserves must be revised.

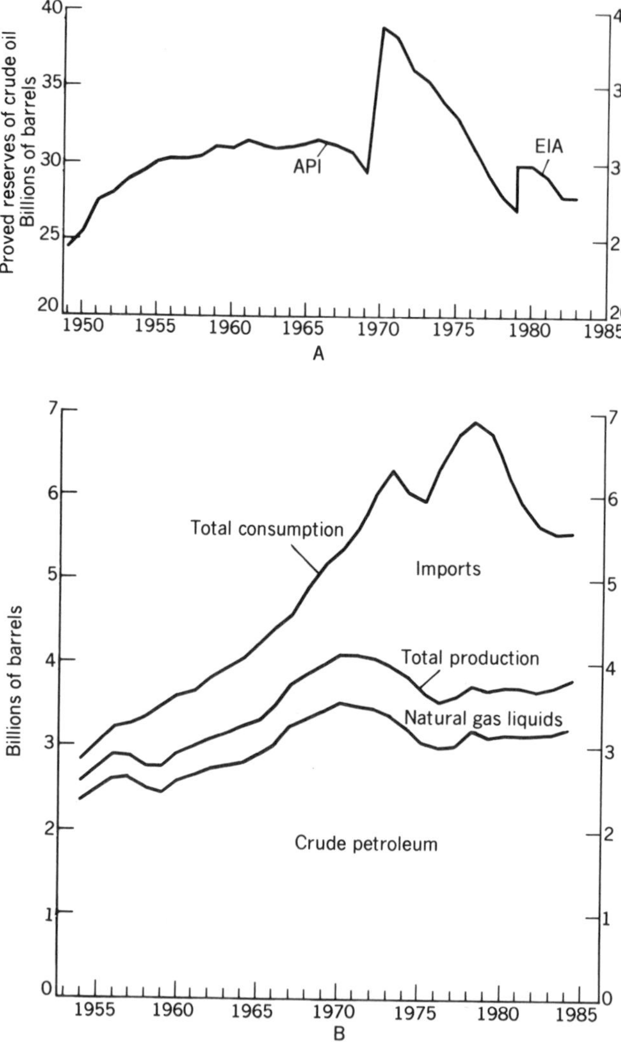

Figure 3-7 (A) U.S. proved reserves of crude oil, 1949 to 1983. Estimates for 1949 to 1979 are by the American Petroleum Institute; those for 1979 to 1983 by the Energy Information Administration. The marked increase in reserves in 1970 is due to discovery of the Prudhoe Bay field. (B) U.S. production, total consumption, and imports of crude oil and natural gas liquids, 1954 to 1984. Source: Energy Information Administration (1985).

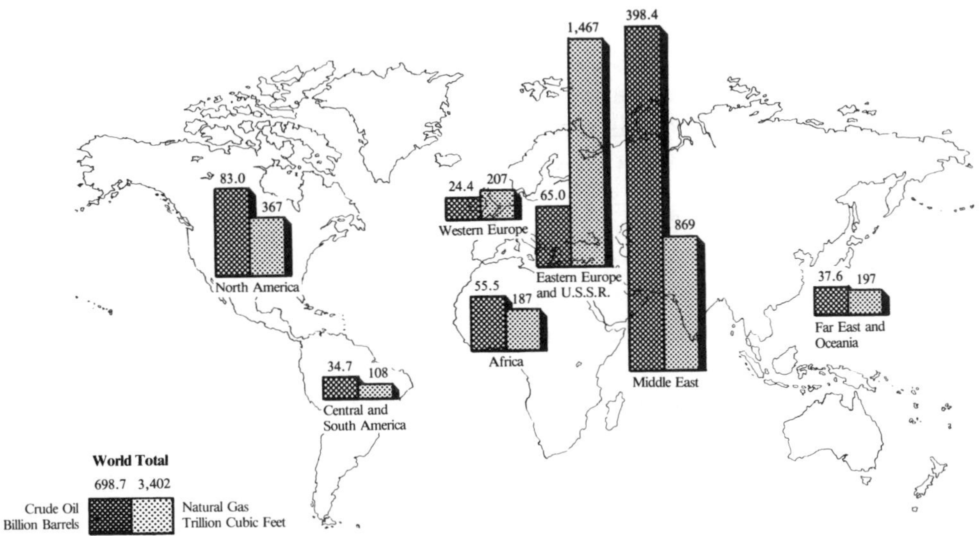

Figure 3-8 World proved crude oil and natural gas reserves, as of December 31, 1984. Energy Information Administration (1985). Bars are scaled in proportion to the Btu content of the reserves. One billion barrels of crude oil equals approximately 5.3 trillion cubic feet of wet natural gas.

In Fig. 3-7A, changes in U.S. proved reserves of crude oil from 1949 to 1983 are shown. Reserves rose from 1949 through 1960, leveled out during 1960–66, then declined from 1967 to 1969. In 1970, proved reserves for Prudhoe Bay were added, bringing total reserves to an all-time high of 39 billion bbl. Although intensive exploration has continued since 1970, proved reserves have almost steadily declined. This means that discovery of new oil has not kept pace with production.

In Fig. 3-8, the worldwide distribution of proved reserves of crude oil is indicated. There is a high concentration of reserves in the Middle East. Total reserves indicated in the figure for Middle Eastern countries are 398.4 billion bbl, or 57 percent of total world proved reserves. Of that amount, 172 billion bbl is in Saudi Arabia; that country therefore has a key role in supply of petroleum from the Middle East. Note that the Soviet Union has large reserves. Owing to successful exploration in recent years, Mexico now holds 41 percent of the total proved reserves of the western hemisphere. In South America, Venezuela holds 25.8 billion bbl out of the total of 34.7 billion bbl for the whole continent. Total proved reserves for Africa are 55.5 billion bbl; two-thirds are in Libya and Nigeria. Total reserves for the Far East and Oceania are 37.6 billion bbl, of which half is held by China and another quarter by Indonesia. The world's proved reserves of oil are thus very unevenly distributed.

Production, Consumption, and Price of Petroleum

United States production, consumption, and imports of petroleum during the period 1954 to 1984 are shown in Fig. 3-7B. From 1954 to 1970, production (including liquids recovered from natural gas) increased rapidly, but it has declined irregularly since. Production from Prudhoe Bay, beginning late in 1977, has checked the decline temporarily. Meanwhile consumption has reached record levels and remains high despite a decrease during 1979–83. As a consequence, imports of petroleum have escalated. The bill for those imports (Fig. 3-9) has been staggering; imports of petroleum accounted for more than half the mounting trade deficits of the past 10 years. Despite efforts toward conservation, imports of petroleum accounted for more than one-third of total supply of petroleum in 1984.

Included in total imports is crude petroleum purchased for the Strategic Petroleum Reserve (SPR). Target for the SPR is 750 million bbl; as of December 1984, the SPR contained 450.51 million bbl. Sources of imports from 1960 to 1984 are indicated in Fig. 3-10. A striking feature of the chart is the decline in the importance of the OPEC countries as a source of U.S. supply. The OPEC countries are the members of the Organization of Petroleum Exporting Countries, the oil cartel discussed in a later chapter. They are Algeria, Ecuador, Gabon, Indonesia, Iran, Iraq, Kuwait, Libya, Nigeria, Qatar, Saudi Arabia, United Arab Emirates, and Venezuela.

World production of petroleum from 1960 to 1984 is shown in Fig. 3-11. The importance of the OPEC countries as sources of world petroleum supply has also declined. In 1973 they produced more than half of the world total; in 1983 their share had fallen to one-third.

Changes in the price of oil since 1970 are shown in Fig. 3-12. The impact

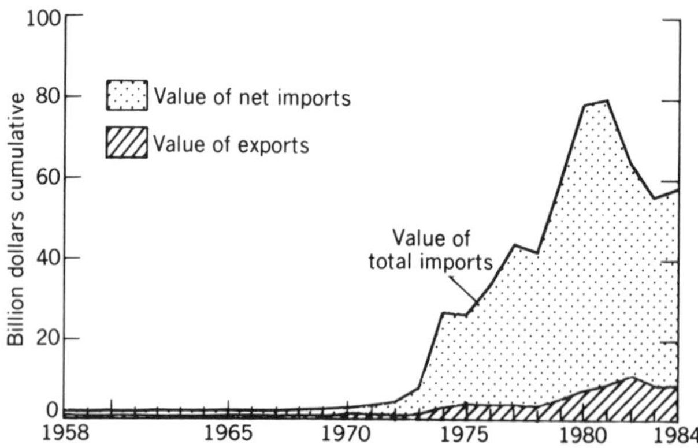

Figure 3-9 Value of net U.S. trade in fossil fuels 1958 to 1984. Energy Information Administration (1985).

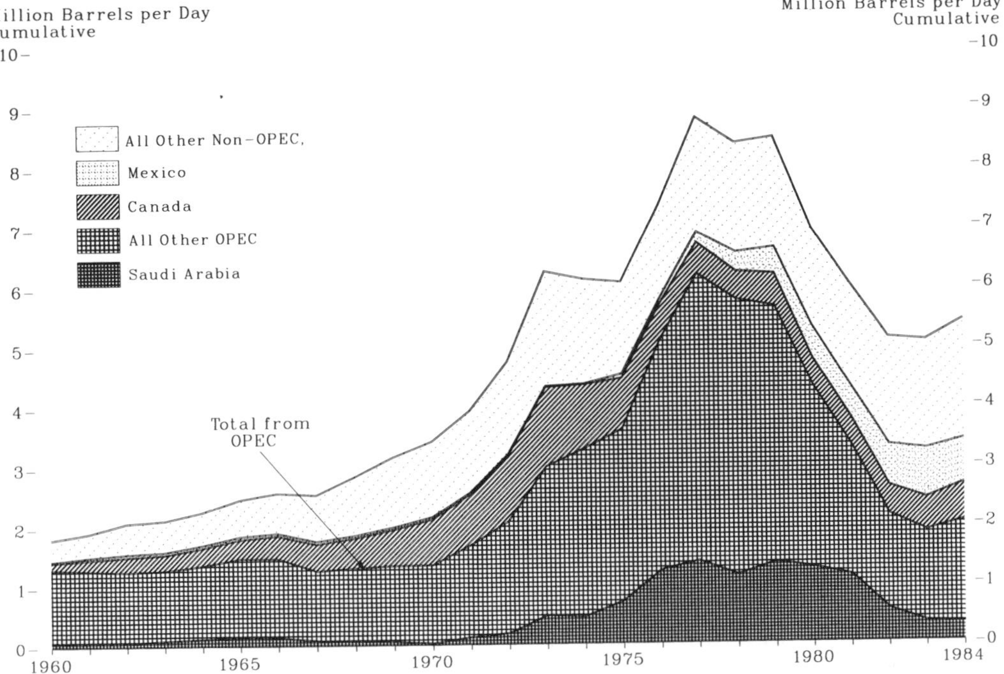

Figure 3-10 U.S. imports of petroleum by country of origin, 1960–1984. Energy Information Administration (1985).

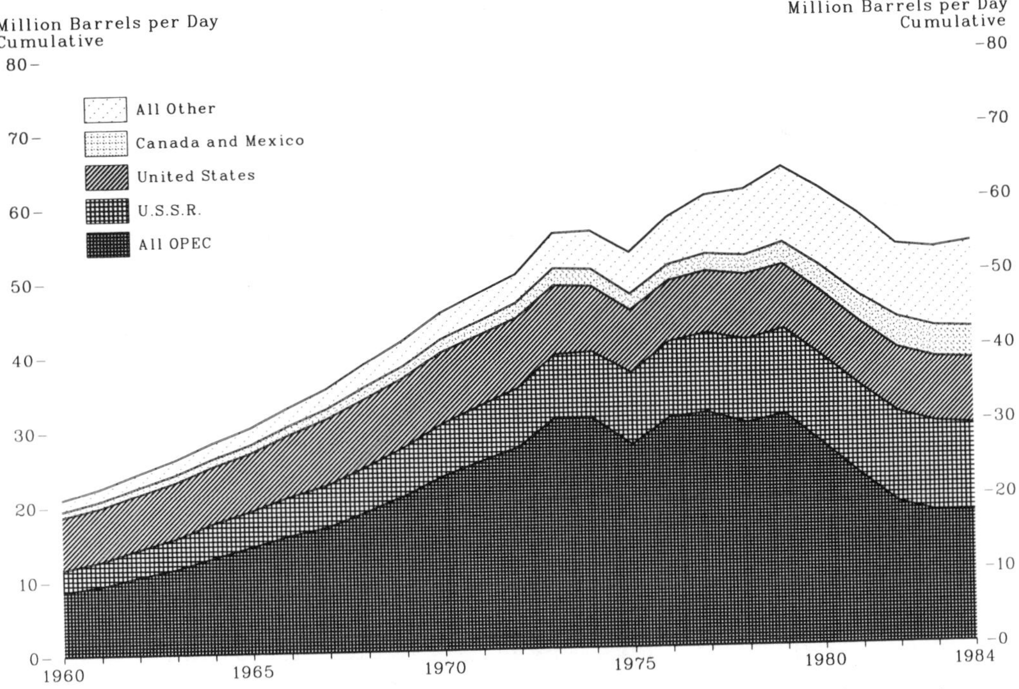

Figure 3-11 World production of crude oil, 1960–1984. Energy Information Administration (1985).

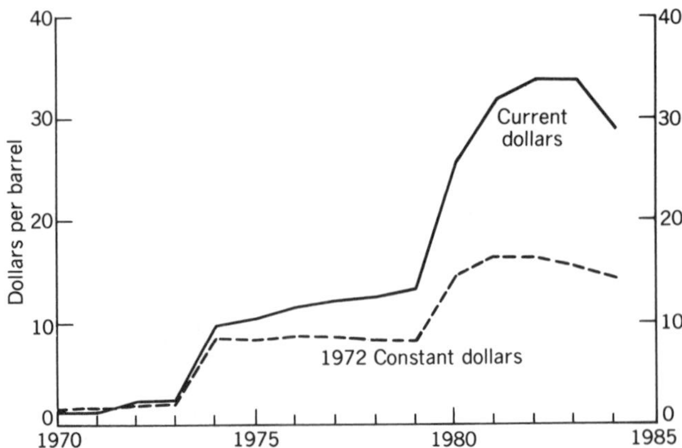

Figure 3-12 Prices of Saudi Arabian light oil, 1970 to 1984. Based on data from the Energy Information Administration (1985).

of the increase in price has been felt throughout the world. All countries are affected, but the impact on the developing countries has been especially severe, since most of them depend heavily on imports for their oil supplies.

Future Availability of Petroleum

In formulating national energy policy, it is desirable to forecast the future availability of petroleum from domestic resources. In doing this we start with proved reserves in known oil fields, but we must add an estimate of the amounts of recoverable oil in oil fields that have not yet been discovered. This is indeed a formidable task, but there have been a number of efforts to arrive at a reasonable estimate.

The most interesting forecasts have been the work of M. K. Hubbert. Hubbert points out that rates of discovery and production of any finite resource tend to follow bell-shaped curves (Fig. 3-13). In the early years after discovery and first use of such a resource, the rate of addition of new reserves rises, reaching a peak as new deposits become more difficult to find, and thereafter declines. Production begins after discovery, but follows a similar curve. As reserves increase, the production rate can also be increased. However, after the rate of discovery begins to decline, a point is reached at which no further increase in production rate is possible. Beyond this point production in turn declines. The relation of discovery to production is shown by the upper two curves of Fig. 3-13. The bottom curve shows the rate of change in proved reserves. At first discovery outstrips production, and reserves increase. At the time when the upper two curves intersect, however, discovery rate and production rate are equal, and the change in size of reserves is zero. Beyond this point, as

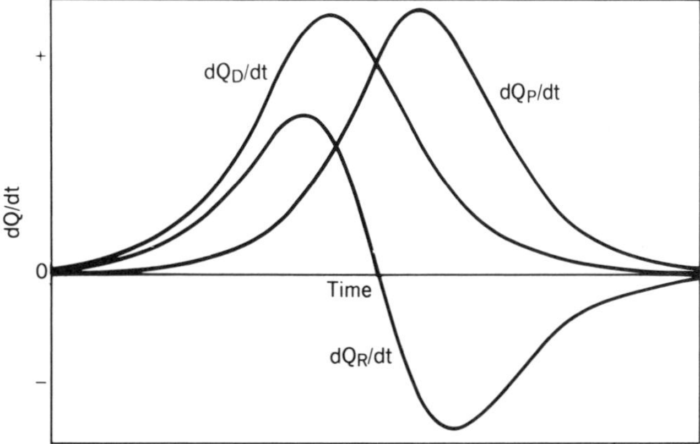

Figure 3-13 Rates of change of proved discovery (dQ_D/dt), production (dQ_P/dt), and increase in proved reserves (dQ_R/dt) of crude petroleum and natural gas during a complete production cycle (Hubbert, 1962, Fig. 24).

production outstrips discovery, reserves decline and ultimately approach zero.

Beginning in the 1950s, Hubbert plotted annual data for proved reserves and production and showed that both were following the predicted pattern (Fig. 3-14). From the curves he forecast that oil production in the United States would peak in the late 1960s; as noted earlier, the peak actually came in 1970. Hubbert's prediction was remarkably close, considering all the variables that would inevitably influence the exact timing of the peak. A major discovery such as that of Prudhoe Bay can shift the curve to the right, as illustrated in Fig. 3-7B. The same kind of shift can be produced by an increase in price, which accelerates the rate of explo-

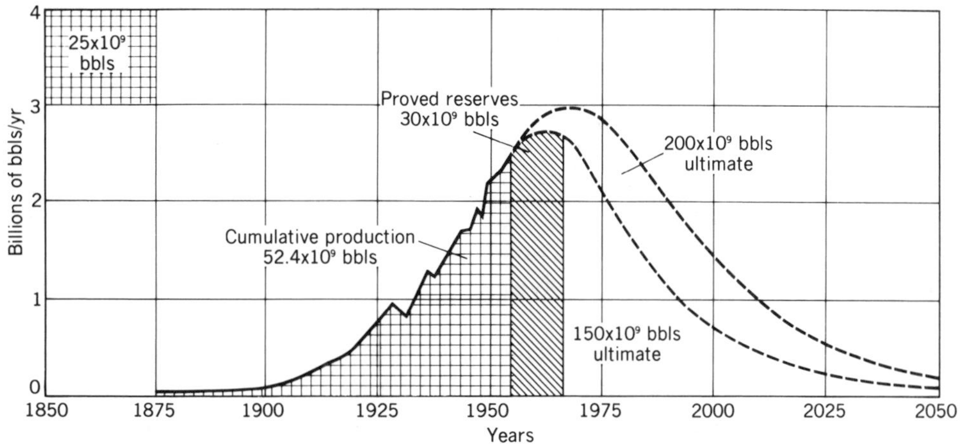

Figure 3-14 Hubbert prediction, 1956, of future crude oil production in the conterminous United States and adjacent continental shelves (Hubbert, 1956, by permission of the American Petroleum Institute).

ration. That actually happened in 1979 and 1980, when domestic oil prices were partly deregulated and the price of oil on the world market rose. Increase in price also makes it possible to produce more expensive oil and to undertake exploration for deep-lying reservoirs. Improved techniques of production can also increase the amounts of oil recovered. All these factors can cause shifts of the discovery and production curves to the right, but they cannot change the essential form of the curves. The oil and gas resources of the United States are finite and nonrenewable; discovery and production must inevitably follow the curves of Fig. 3-13. From an analysis of discovery rates, most recently in 1978, Hubbert concluded that approximately 72 billion bbl of recoverable oil remained to be discovered in the United States. Adding proved reserves to this gave a total of about 106 billion bbl of recoverable oil.

In forecasting the availability of a natural resource, there is a tendency among economists to regard supply as simply a function of price. The higher the price, the more of the commodity will become available. There is obviously a great deal of truth in this. For oil, we can safely predict that as the price of oil rises (in real terms), it will be possible to produce oil that cannot be produced economically now. Some of the roughly 60 percent of total oil left in the ground will be recovered. The problem is that price is a two-edged sword. As price rises, more of a commodity can be economically produced, but at the same time its use must be restricted. We already see the effects of this on the use of oil in the United States. Since the Arab oil embargo of 1973, and especially since the doubling of the price of oil in 1979, U.S. consumption of oil has dropped from about 19 million bbl per day to about 16 million bbl per day. The decline is good from the standpoint of conservation of natural resources, but it means that we can no longer afford to use petroleum to the extent we did before. The consequences are being felt throughout the economy. Our homes are no longer as comfortable as they were in the cold of winter and the heat of summer. Future homes will have to be smaller and less consumptive both of energy and of building materials, which become more costly as the price of energy rises. The rate of growth of our electric utility industry has markedly decreased. We can no longer afford to use as much steel as we did before 1973. This is one factor in the decline of the American steel industry. Travel, whether by air or by automobile, becomes steadily more expensive. The cost of transporting goods has risen.

Hubbert's estimate of undiscovered recoverable oil in the United States rests on mathematical analysis of the vast amount of data on discovery and production accumulated since petroleum production began in 1859. There are other estimates that employ a geological approach. A great deal has been learned in the past hundred-odd years about the nature of the earth's skin of sedimentary rocks. We also know a great deal about the occurrence of petroleum and natural gas in the sedimentary skin. All this knowledge gives a basis for estimating how much recoverable oil remains to be discovered in sedimentary basins that are as yet unexplored or only partly explored. Several different geological estimates have been made, and attention was focused on them in late 1973 and 1974 as a consequence of the Arab oil embargo. The estimates differed widely, from a U.S. Geological Survey estimate of 400 billion bbl to an oil company estimate of 89 billion bbl. The various estimates, together with the Hubbert estimate, were reviewed by a committee appointed by the National Academy of Sciences (NAS, 1975). It was concluded that the best estimate of undiscovered recoverable oil possible at that time was 103 billion bbl (crude oil plus natural gas liquids).

Subsequent developments suggest that even those estimates may have been too optimistic. Figure 3-15 shows areas of active exploration in 1973 and areas then considered worthwhile future targets of exploration. Of the latter, the eastern Gulf of Mexico, the Baltimore Canyon area, offshore California, the Gulf of Alaska, and the Beaufort Sea offshore from the Alaskan North Slope were considered particularly promising. In 1974, a group of oil companies paid the federal government $632 million for a lease covering a prime target in the eastern Gulf of Mexico, a large, apparently favorable structure (the Destin dome) indicated by geological and geophysical work. Seven holes, all dry, were drilled, and the project had to be abandoned. In the late 1970s, exploration in the Baltimore Canyon area was abandoned after expenditures of $1.7 billion. There has been no success in the Gulf of Alaska. In the summer of 1983, it was announced that the latest hole drilled there, to a depth of 18,000 feet at a cost of $44 million, was dry. The biggest disappointment of all, however, was the December 1983 announcement that the first hole in the Mukluk prospect, offshore from the North Slope, was dry. The prospect, in the words of one oil company official, was "the largest unexplored low-risk prospect in the U.S." The consortium of oil companies involved spent $1.6 billion in acquiring leases and $140 million in drilling the first hole. What further drilling will be undertaken is uncertain. The only one of the five major potential target areas that has fulfilled its promise is offshore southern California, where a major oil discovery (Point Arguello) was announced in 1983. In the meantime, a 1980 estimate by the U.S. Geological Survey gave undiscovered recoverable crude oil as 64.3 to 105.1 billion bbl, with a 93 percent probability for the lower amount but only a 5 percent probability for the higher amount. In May 1985, however, government estimates of undiscovered offshore oil resources were drastically reduced.

It is evident that exploration for petroleum and natural gas involves

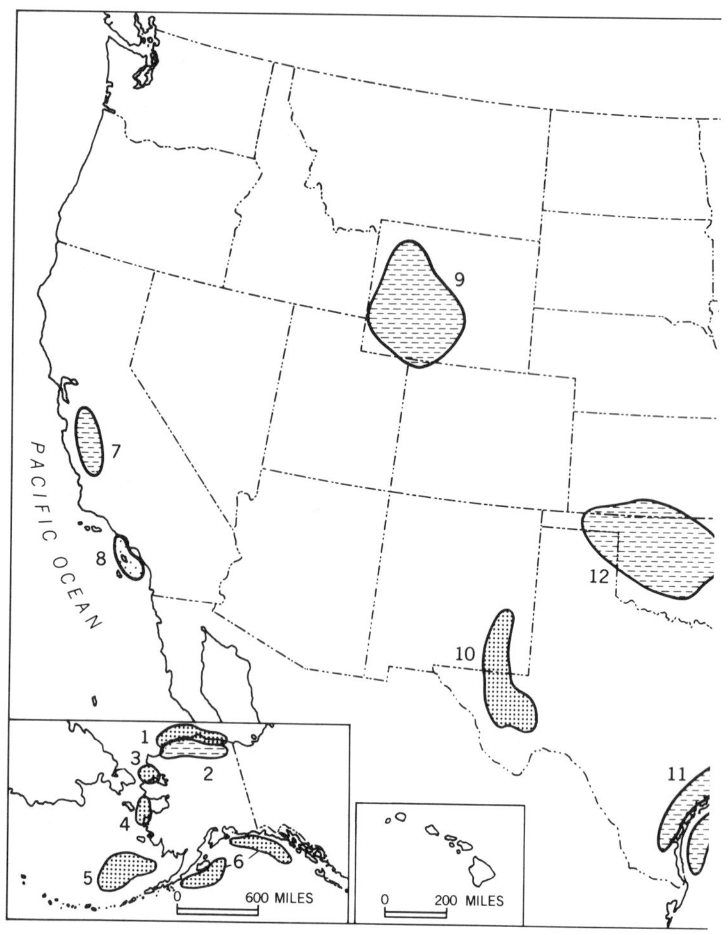

Figure 3-15 Some areas of active exploration in the United States in 1973 (striped areas) and potential targets of exploration (stippled areas). 1, Offshore North Slope; 2, onshore North Slope; 3, Chukchi; 4, Norton Sound; 5, Bristol Bay; 6, Gulf of Alaska; 7, California basins; 8, offshore California; 9, Washakie–Green River basins; 10, North Basin and Del-

substantial financial risks as well as physical risks. Geology and geophysics can tell the petroleum explorer where to look, but they cannot tell him what he will find.

Although there have been major disappointments in oil exploration in the United States in recent years, there have been discoveries in onshore fields. Also, discoveries continue in the middle and western parts of the Gulf of Mexico in U.S. territorial waters, even though no giant new fields have been found. Exploration for natural gas has been greatly influenced by partial deregulation of gas prices, and in 1980 the decline in proved reserves of gas was arrested temporarily. We can expect more oil and gas fields to be discovered in the United States with passage of time, but

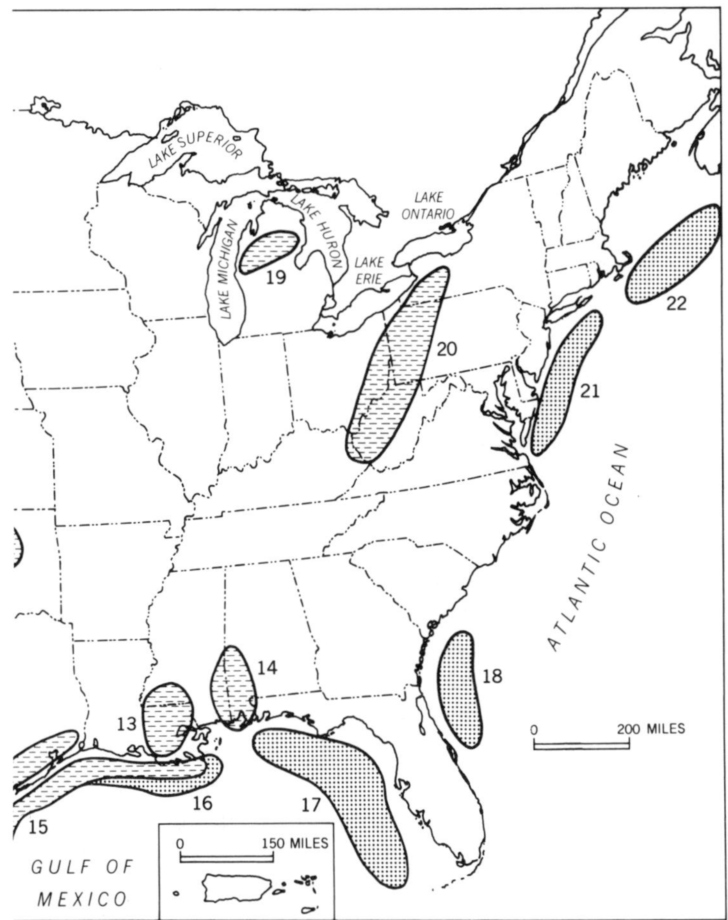

Figure 3-15 *(Continued)*
aware; 11, onshore Texas Gulf Coast; 12, Anadarko Basin; 13, southern Louisiana; 14, Alabama-Florida-Mississippi; 15 and 16, offshore central and western Gulf Coast; 17, offshore Florida Gulf Coast, including the Destin Dome; 18, Georgia embayment; 19, Michigan basin; 20, Appalachian Basin; 21, Baltimore Canyon; 22, Georges Bank.

the cost of discovery is rising. On land the shallow portions of the sedimentary basins have mostly been explored. The search now is mostly for deeper fields. The average depth of hole drilled in 1949 was 3842 feet; the average depth in 1983 was 5199 feet. In 1953 the deepest producing well bottomed at 17,122 feet. In 1984 the deepest producing well bottomed at 26,536 feet. Holes 15,000 to 30,000 feet in depth are becoming common. The latter depth is near the bottom of the zone in which oil is stable in the crust; below the zone only gas can be expected. Drilling to such depths is very expensive. Stress on drilling equipment becomes very severe; equipment must be heavier, stronger, and more resistant to heat and corrosive liquids. Drilling at sea is more expensive than drilling on land, even though most holes offshore are drilled in relatively shallow waters. Both drilling and production platforms must be strong enough to withstand the buffeting of strong seas and strong gales, even hurricanes. The capital costs of offshore production may be five to ten times those on land. All this is reflected in the rising costs of drilling and the rising costs of discovering oil. The average cost per foot of drilling (1972 constant dollars) rose from $18.94 in 1949 to $52.56 in 1982. A study by the accounting firm of Arthur Andersen & Co. (Wall Street Journal, June 10, 1984) showed that the average cost of discovering a barrel of oil or natural gas energy equivalent rose from $10.99 in 1980 to $16.52 in 1982. To sum up, the prospect for the future in the United States is one of declining rates of discovery and rising costs of petroleum and natural gas.

Natural Gas

Natural gas is recovered as a by-product of petroleum production, but there is increasing production from wells that yield only natural gas. Until the 1930s, when gas pipelines were constructed in the United States, natural gas was largely wasted, burned off at the wells for lack of means of transporting it to market. Much natural gas is still wasted in the oil fields of the world, for example in the Middle East, but pipelines are now being constructed in some countries. In 1980, Saudi Arabia completed the world's longest pipeline (725 miles) to transport natural gas from the Shedzim gas plant of the Eastern Province to the port of Yanbu on the Red Sea. Cooled to very low temperatures and subjected to sufficient pressure, natural gas is liquefied and can be transported in special tankers.

Changes in U.S. production and consumption of natural gas from 1949 to 1984 and changes in reserves from 1949 to 1983 are shown in Fig. 3-16. Consumption of natural gas has been largely tailored to the availability of gas from domestic fields, although about 1 trillion cubic feet of natural gas has been imported annually during recent years, mostly from Canada and Mexico. Production of natural gas in the United States peaked in 1972 and 1973 and has been declining since. Changes in reserves are parallel to changes in reserves of crude oil. Reserves of natural gas peaked in the late 1960s and have since declined sharply, although the decline was checked temporarily by discoveries during the intensive exploration of 1979–82. Undiscovered recoverable natural gas resources, as of 1980, have been estimated by the U.S. Geological Survey as between 594 and 729 trillion cubic feet.

World production of natural gas is indicated in Fig. 3-17. Outside the United States, the Soviet Union is by far the largest producer and possesses the largest share of estimated proved reserves (Fig. 3-8).

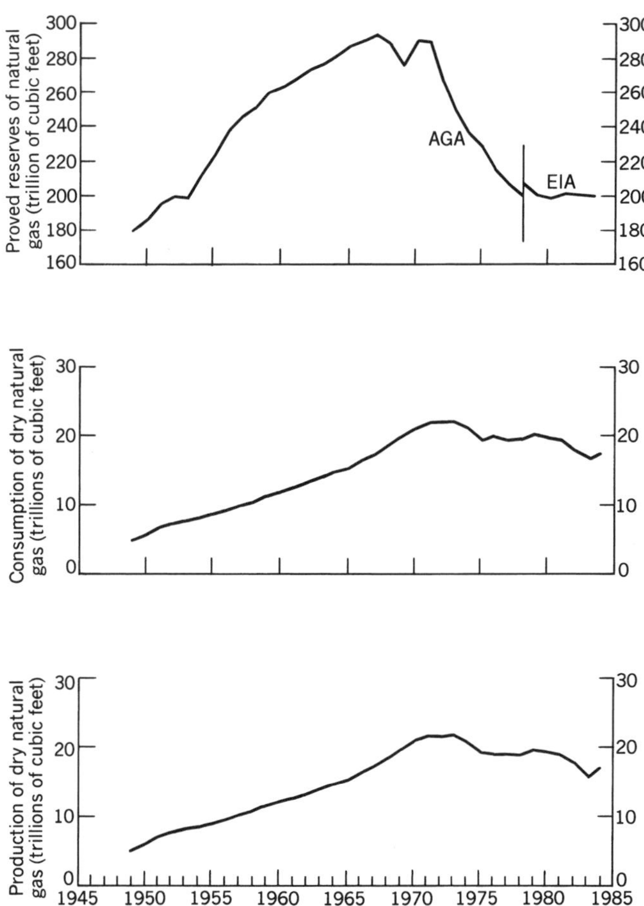

Figure 3-16 U.S. proved reserves of natural gas, 1949–1983, and consumption and production of dry natural gas, 1949–1984. Data from Energy Information Administration (1985). AGA - American Gas Association.

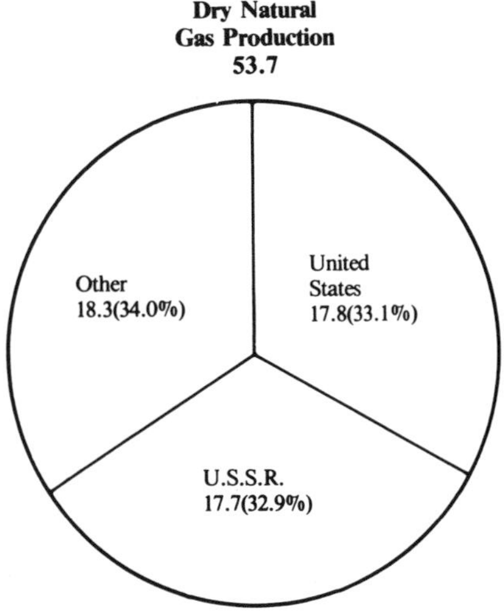

Figure 3-17 World production of natural gas, 1982, in trillions of cubic feet. Energy Information Administration (1985).

Alternate Sources of Petroleum

It has been mentioned that average recovery of oil in the United States is somewhere between 30 and 40 percent. This means that with present technology about 600 billion bbl will remain in the ground, roughly twice the amount of oil that will actually have been produced. Some of this is in shallow fields and could be recovered by mining the oil-bearing rocks and treating them to separate the oil. Still other resources could be mined underground. Preliminary estimates indicate costs of $10 to $60 per barrel, depending on the nature and situation of the deposits. As the cost of recovering oil by conventional methods rises, interest in oil mining, particularly in shallow fields, is stimulated. Getty Oil Company has announced that studies of the feasibility of mining large deposits of oil-bearing rocks in the McKittrick field in California are under way. It seems probable that this and similar projects will add to future petroleum reserves of the United States, but there are many obstacles to be overcome.

Conventional petroleum deposits consist of lighter hydrocarbons that are fluid and that, under suitable conditions of rock porosity and permeability, will flow from or can be pumped from wells. Related to the conventional deposits is a spectrum of deposits of heavier hydrocarbons ranging from those that flow only very slowly (heavy oils) to those that are so viscous that they do not flow at all at ground temperatures. Deposits at the one end of the spectrum are called heavy oil deposits, those at the other end are called bitumen deposits or tar sand deposits. Oil can be pumped from such deposits only if it is thinned with lighter oil or heated in the ground by injecting steam or by setting parts of the deposits on fire. Alternatively, the oil-bearing or bitumen-bearing sands may be mined, the rock crushed, and oil separated from sand by treating the crushed material with steam. Oil is then separated from the resulting mixture of oil and water and treated in a suitable refinery to produce usable oils, gasoline, and a variety of other petroleum products.

The largest known heavy oil and tar sand deposits are those of northern Alberta and adjacent Saskatchewan (Fig. 3-18) and those of Venezuela. The tar sand deposits of the Athabaska, Peace River, and Wabasca areas and the heavy oil deposits of the Cold Lake area are estimated to contain about 1 trillion bbl of bitumen and heavy oil (McRory, 1982), of which about 245 billion bbl might ultimately be recovered. That amount would yield about 185 billion bbl of synthetic crude oil. Tar sands are currently being mined on a substantial scale in the Athabaska area, and pilot projects involving in situ extraction from tar sand or heavy oil deposits are under way in the Peace River, Wabasca, Cold Lake, and Lloydminster areas. Deposits of heavy oil and tar sands in the Oficina-Temblador region of Venezuela, north of the Orinoco River, have been variously estimated to contain 700 to more than 1,300 billion bbl of hydrocarbons. In the

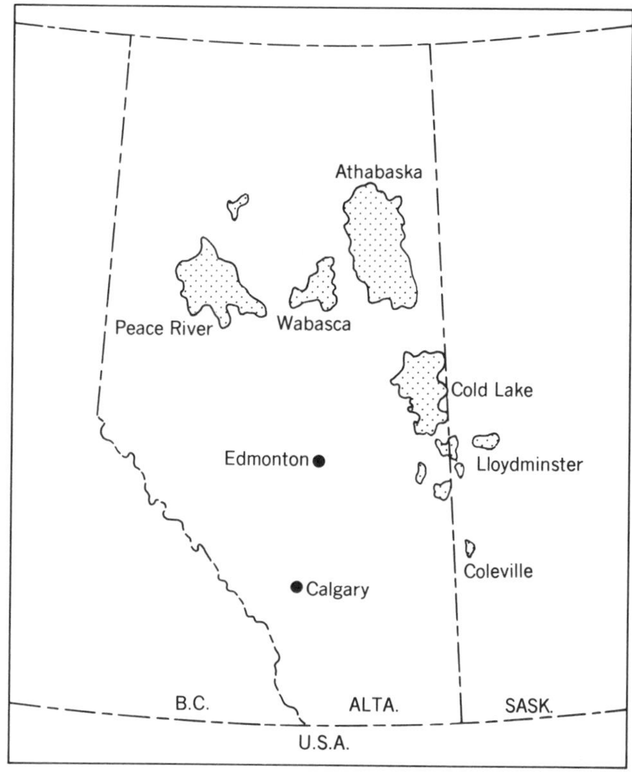

Figure 3-18 Map showing deposits of heavy oil and tar sands in Alberta and Saskatchewan.

United States, tar sand deposits containing about 29 billion bbl of bitumen are known.

At the present time, the only production of heavy hydrocarbons is from Canada, from the Athabaska sands and the deposits of the Cold Lake and Lloydminster areas. The largest operations, in the Athabaska sands, are in effect subsidized by the Canadian government. Yet interest in heavy oil sands is increasing, and research and development of the sands are being pursued in the United States and Venezuela. At stake are amounts of oil several times the size of proved world reserves of conventional petroleum.

A second major alternative to conventional petroleum lies in the "oil shales" of the world. The name is a misnomer but is so firmly established that it is not likely to be changed. Oil shales are sedimentary rocks of diverse compositions that contain solid organic matter that is not oil but will yield oil when heated and processed. Deposits of oil shales are known in many countries of the world and underlie certain extensive areas of the United States (Fig. 3-19). The most important deposits are in the Green River formation, in sedimentary rocks deposited in lakes that occupied

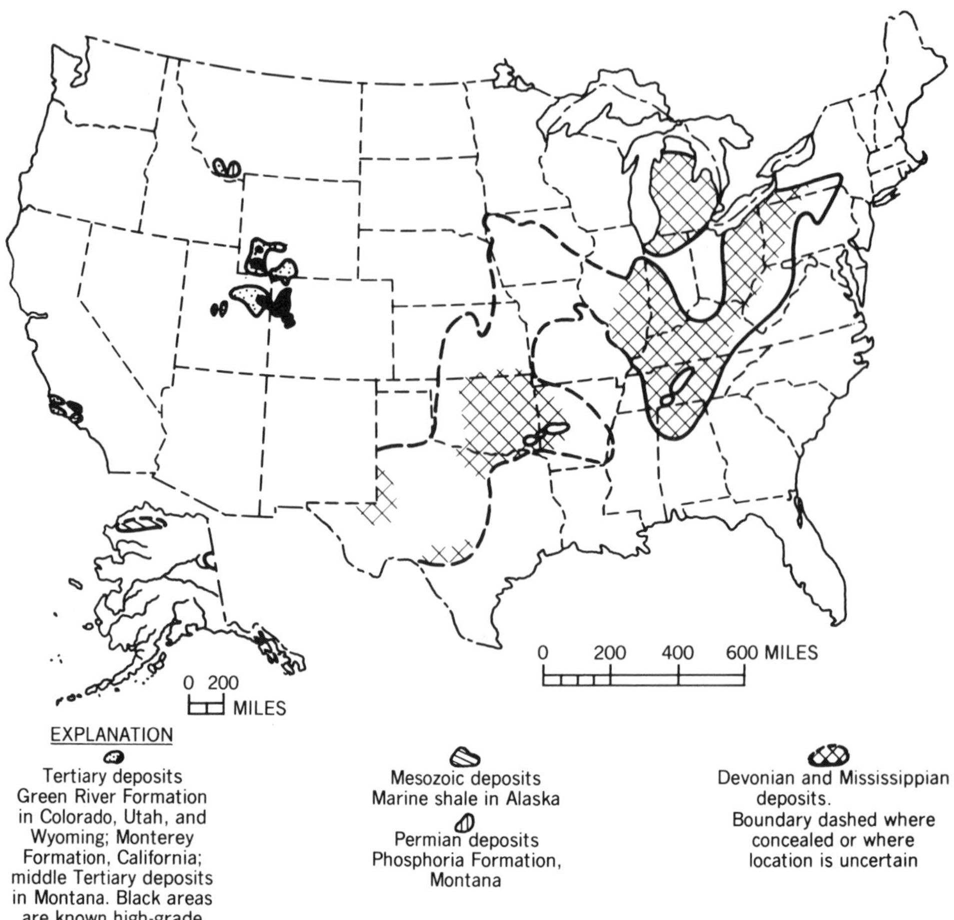

Figure 3-19 Principal oil shale deposits of the United States. Boundaries are dashed where covered or where presence of oil shales is uncertain. From map by Duncan and Swanson (1965).

parts of Wyoming, Colorado, and Utah (Fig. 3-20) about 40 million years ago. The oil-bearing rocks will yield a few gallons to as much as 100 gallons of oil per ton. The amount of oil potentially available is enormous, 418 billion bbl in shales yielding 25 to 100 gallons per ton, and an additional 1.4 trillion bbl in rocks yielding 10 to 25 gallons per ton. Research and development of methods of recovering oil from the rocks have been under way since the 1930s. Several technologies have been proved, yet thus far no viable commercial operation has been established. There are several obstacles. One is the huge capital cost, more than $1 billion for a plant that would produce 55,000 gallons of oil per day. Another is the environmental damage that could result from the large-scale mining that would be necessary. A third obstacle to some of the oil shale projects is

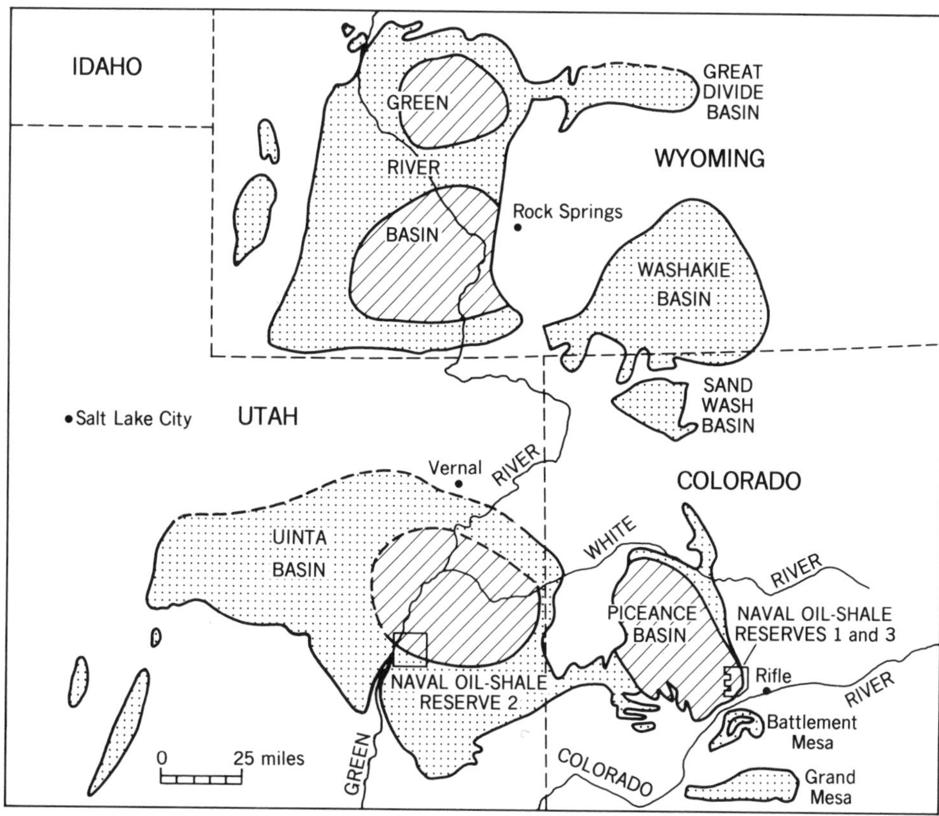

Figure 3-20 Distribution of oil shale in the Green River Formation, in Colorado, Wyoming, and Utah. Striped areas are underlain by oil shale more than 10 feet thick, yielding 25 gallons or more per ton of shale. Stippled areas are underlain by Green River Formation, but oil shale in them is unappraised or contains less than 25 gallons per ton. From Duncan and Swanson (1965).

water supply in a region in which population is already pressing on available supplies. A fourth is the instability of the world petroleum market. The enormous capital investments that would be necessary to develop significant production are not likely to be made unless there is assurance of stable prices high enough to permit reasonable returns on investment. Only government can give that assurance, and thus far the federal government has been unwilling to do so.

Production of oil from oil shales is not new. In Latvia, as much as 100 million bbl of oil per year has been produced from oil shale. Kerosene was produced from Scottish oil shales for many decades. In Brazil, development of an oil shale industry is part of a drive to relieve that country's heavy dependence on imported oil. The Iratí oil shale of southern Brazil yields only 15 to 25 gallons of oil per ton, but reserves in shales averaging 18.5 gallons per ton are estimated at 600 million bbl, resources in the hundreds of billions of barrels. Large oil shale deposits are also known in other countries of the world. In the oil shales, as in the heavy oil sands, the world has enormous resources of energy. Their utilization, however, will mean higher costs of energy, and some damage to the environment will have to be accepted.

ALTERNATIVES TO THE FOSSIL FUELS

We turn now from the fossil fuels to examine briefly three of the alternative sources of energy—nuclear, geothermal, and solar. We begin with nuclear energy, which in 1984 furnished 4.6 percent of the energy supply of the United States.

There are two ways of obtaining energy from the nuclei of atoms—nuclear fission and nuclear fusion. Nuclear fusion is the source of the energy of the hydrogen bomb, but only nuclear fission is used today as a source of industrial energy. For energy from nuclear fission, we are dependent on elements that spontaneously undergo fission (splitting) and are present in the earth's crust in sufficient concentration to be recoverable as sources of energy. There are two such elements, uranium and thorium. Although resources of thorium may be large, we are not currently producing energy from the element. All nuclear energy is produced from uranium, and here we will concern ourselves only with the use of uranium and the size and nature of uranium resources.

Many uranium-bearing minerals are known, but a few are particularly important as sources of the element:

Uraninite and pitchblende	U_3O_8
Carnotite	$K_2O \cdot 2UO_3 \cdot V_2O_5 \cdot 3H_2O$
Tyuyamunite	$CaO \cdot 2UO_3 \cdot V_2O_5 \cdot 8H_2O$
Brannerite	$(U,Th,Ca)(Ti,Fe)_2O_6$
Coffinite	$[U(SiO_4)_{1-x}(OH)_{4x}]$

We are dependent today on materials in which one or more of the above minerals are present in sufficient amounts to give uranium contents of 0.05 percent or more.

Now the potential energy available in uranium is very large. Complete fission of a pound of uranium would yield the energy equivalent of 1300 tons of good bituminous coal. Unfortunately, only a small part of that energy is readily available. The reason is that natural uranium is composed of three isotopes of the element (i.e., atoms of different atomic weights), in the following proportions:

^{234}U	0.006%
^{235}U	0.711%
^{238}U	99.283%

Only ^{235}U undergoes significant spontaneous fission, so only 1/140 of the potential energy in a pound of uranium is readily available. The commercial reactors in operation in the United States today use ^{235}U as the fuel. To serve as a fuel, U_3O_8 (uranium oxide) as obtained from ores must be treated (enriched) to produce a concentrate containing 5 to 6 percent ^{235}U. The Canadian Candu reactor, however, does not require enriched uranium.

To tap more of the energy in a pound of uranium, ^{238}U must be converted to ^{239}U, plutonium, which undergoes spontaneous fission and therefore can be used as a fuel. This can be done by using ^{235}U as a source of neutrons, as follows:

(1) $^{235}U \rightarrow$ energy + neutrons + fission products

(2) ^{238}U + neutron $\rightarrow$ $^{239}U \rightarrow$
^{239}Np (neptunium) + electron $\rightarrow$ ^{239}Pu (plutonium)

The second reaction is the "breeder reaction," and the vessel in which the reaction is brought about is called the breeder reactor. The breeder reactor is technologically feasible, but, although much research and development of breeder reactors has been done in the United States, Great Britain, France, and the Soviet Union, only the French Super Phenix reactor has reached the commercial stage. In the meanwhile most of the world is dependent on ^{235}U as a source of industrial atomic energy.

Although only ^{235}U is the natural fuel, it is customary to discuss the availability of uranium for nuclear energy in terms of the size of reserves and resources of U_3O_8 (containing all three isotopes in the proportions indicated above) and the rates at which U_3O_8 can be produced. For the United States, the standard estimates are those prepared annually by the Department of Energy (DOE) and those prepared earlier by the Atomic Energy Commission (AEC) and the Energy Research and Development Administration (ERDA). Uranium resources are divided into "forward-cost classes." Forward costs are those costs incurred by the producer after land acquisition, exploration, and economic evaluation, hence they do not represent total costs of production. As a rule of thumb, a factor of 1.7 has sometimes been used to convert forward costs to total costs per ton. The AEC-ERDA-DOE system is unique in the manner of calculation of resources and reserves.

Reserves as shown in Table 3-4 are calculated from information furnished annually by all uranium mining companies in the United States, including records of drilling, sampling, and estimation of tonnage and grade. As indicated earlier, the estimates of reserves are considered highly reliable, but the estimates of potential resources are another matter. The bulk of the uranium estimated under the three classes of potential resources is in deposits that have not yet been discovered, and some of it

TABLE 3-4 U.S. Uranium Resources, Jan. 1, 1983 (Thousand Short Tons, U_3O_8)

	Forward Cost (Dollars per Pound)[a]		
Class	$30 or Less	$50 or Less	$100 or Less
Reserves[b]	180	576	889
Potential resources	1127	2066	3381
Probable	654	1167	1887
Possible	257	508	842
Speculative	216	391	652
Totals	1307	2642	4270

[a]Forward costs are those costs yet to be incurred and, therefore, do not represent prices at which U_3O_8 will be sold.

[b]Does not include 140,000 tons of U_3O_8 estimated to be available as a by-product of phosphate and copper production during the 1980–2010 time period.

Source: U.S. Department of Energy, Grand Junction Area Office, Colorado; Statistical Data of the Uranium Industry, Report No. GJO-100(83), Jan. 1, 1983.

is estimated for districts that are not yet known to contain uranium deposits. The reliability of the estimates of potential resources is highly questionable.

The adequacy of U.S. reserves and resources is a function not only of their size but of the rates at which new reserves can be added and the rates of production that can be achieved. In the late 1960s and first half of the 1970s, there was a rapid increase in the number of nuclear plants under construction or planned, and concern arose over the future adequacy of supplies of uranium from domestic sources. It was predicted (NAS, 1975) that in the 1980s a serious shortfall of supply would appear.

Forecasts of shortages have been invalidated by subsequent events that have dimmed the future of nuclear power in the United States. Opposition to nuclear power, strengthened by the accident at Three-Mile Island, Pennsylvania, in 1979, and by rising costs of building nuclear reactors, have effectively reversed the trend of increase in use of nuclear energy in the United States. The adequacy of reserves has become virtually an academic question. U.S. production of U_3O_8 fell from a peak of 21,850 tons in 1980 to 10,600 tons in 1983. The merits and demerits of nuclear power will not be debated in this volume. Arguments based on scientific and engineering grounds are meaningless in the face of public fear of a nuclear accident.

In the remainder of the world the development of nuclear power continues, though at a somewhat slower pace than anticipated in the 1970s. Free World production of uranium in 1984 is shown in Table 3-5. Free World reserves of uranium are large. Discoveries of uranium deposits in Australia and Canada in recent years have resulted in major additions

TABLE 3-5 Free World Production of Uranium, 1984 (Millions of Pounds of U_3O_8

Australia	10.3
Canada	27.9
Central Africa	10.8
France	8.2
Namibia	9.0
South Africa	15.0
Other nations[a]	2.2
United States	15.6
Total	99.0

[a]Includes Argentina, Belgium, Brazil, Portugal, and Spain.

Source: Engineering and Mining Journal, May 1985.

to reserves. Even without the development of breeder reactors, reserves are considered adequate for a period extending well into the 21st century.

Nuclear Energy from Fusion

The fusion of atoms (more properly thermonuclear fusion) is the sun's source of energy, but most of the fusion reactions that take place in the sun require temperatures and pressures far beyond what present technology can achieve. Current research and development are concerned with reactions that could be produced at about 100 million degrees Celsius, particularly a reaction involving two isotopes of hydrogen, deuterium and tritium. The nucleus of ordinary hydrogen consists of one proton, that of deuterium consists of a neutron and a proton, and that of tritium, two neutrons and a proton. The fusion reaction is

(1) ^{2}H (deuterium) + ^{3}H (tritium) $\rightarrow$ neutron + ^{4}He (helium) + energy [17.6 million electron volts (MeV)]

Deuterium is obtainable from seawater. Tritium must be produced from the two isotopes of lithium, ^{6}Li and ^{7}Li, by the reactions

(2) a. Neutron from (1) + $^{6}Li \rightarrow$ ^{3}H (tritium) + ^{4}He (helium) + energy (4.5 MeV)

b. Neutron from (1) + $^{7}Li \rightarrow$ ^{3}H (tritium) + ^{4}He (helium) + energy (−2.87 MeV)

The available supply of deuterium in seawater is essentially unlimited. Lithium is obtainable from several types of deposits. Discoveries in recent years indicate that U.S. and world lithium reserves and resources are large relative to foreseeable demand.

Nuclear fusion is potentially capable of providing enormous amounts of energy for many centuries to come. It has several advantages over nuclear fission. One is that there is no possibility of a meltdown in a fusion reactor. In nuclear fission the problem is to control the reaction well short of a meltdown. In nuclear fusion the problem is to keep the reaction going long enough to produce energy. If anything goes wrong in the reactor, the reaction simply stops. The products of the fuel reaction are not radioactive and therefore present no long-term disposal problems. The reactions given above, however, do produce neutrons, and these will cause damage to the walls of the reactor. Absorption of neutrons by the reactor components will produce radioactive isotopes which must be disposed of. The isotopes, however, have much shorter half-lives than those produced by fission and present much less serious disposal problems. Current fusion concepts call for the waste structural material to be stored at surface for a few hundred years, when radioactivity will have fallen below any conceivable hazard level.

The design of a fusion reactor is an enormously difficult undertaking, far more difficult than getting a man to the moon. The fuel elements must be heated to about 100 million degrees Celsius to convert them to a plasma, a high-temperature fluid composed of charged particles. The density of the plasma must be above a certain critical level, and both temperature and density must be maintained for a sufficient time to allow the fusion reaction to take place. The plasma must be confined in such a way that it will not come in contact with the walls of the reactor. A blanket in which tritium is produced from lithium encircles the chamber. In it neutrons and photons (light energy) are absorbed. Systems for cooling and heat exchange must be provided. Energy produced by the fusion reaction must be captured in the form of heat, which is in turn used to generate power. All the above structures must be surrounded by a shield to absorb remaining neutrons and radiation. Each step in design presents formidable problems. A major one is to reduce the mass of materials, metallic and nonmetallic, required for the reactor structures. Otherwise the energy produced in the reactor may be less than that required for mining and transport of materials and for construction, operation, and maintenance of the reactor. Several alternative designs for reactors have been subjects of intensive study. One of these, NUWMAK, a version of a Tokomak reactor, is illustrated in Fig. 3-21.

The potential importance of thermonuclear fusion as a source of energy is the basis for the intensive research and development that is being carried on in the United States, England, France, Germany, Italy, the Soviet Union, and Japan—an enormous international effort. Up to 1985, $14 billion had been spent on fusion research worldwide, and some 6400 professional scientists were involved. Many more years or perhaps decades of work will be required to determine whether fusion on a commercial basis is possible.

Solar Energy

Much of the energy currently produced by man is from the fossil fuels, in which solar energy has been stored by life processes. The fossil fuels are thus fossil solar energy. At low latitudes in clear weather, solar energy received is 6 to 8 kilowatt hours per square meter per day. Under these ideal conditions, a collector with a total area of about 12 square meters (about 14.5 square yards) will receive the amount of energy (about 30,000 kilowatt hours) used annually by the average American household. In practice, however, there are problems that thus far have restricted the use of solar energy largely to heating and air-conditioning. Solar energy is intermittent, since at best it is received only during daytime hours. Part of the energy received must therefore be stored against periods in which no energy is received or, owing to clouds, the energy received is reduced. For production of large amounts of energy, very large collection areas must be provided. Offsetting these disadvantages is the fact that once the capital costs of installation of collection and storage equipment have been paid, solar energy is nearly free.

The use of solar energy is growing, especially in countries where there is a high percentage of sunny days. In Israel all new housing units must be equipped with solar water heaters. Solar units in use in the United States have been gradually increasing in number. Solar energy is widely used in Australia. There are no figures for the percentage of total energy consumption represented by solar energy. It is certainly much less than 1 percent, but it is too early to predict the future importance of solar energy. Introducing any major new technology requires a long time, during which technology must be improved, institutional barriers overcome, necessary capital generated, and equipment gradually installed. Even where it is economically feasible now to use solar energy, existing energy-producing systems cannot be scrapped and replaced overnight. The cost would be prohibitive, and such drastic change will simply not take place. Over coming decades, as present energy-producing systems become worn out or obsolete, solar energy will find its place in the spectrum of energy sources available to man.

Geothermal Energy

Geothermal energy is energy that is released by processes operating in the crust of the earth. The deeper we go in the earth, the higher the temperature. On the average, the boiling point of water at atmospheric

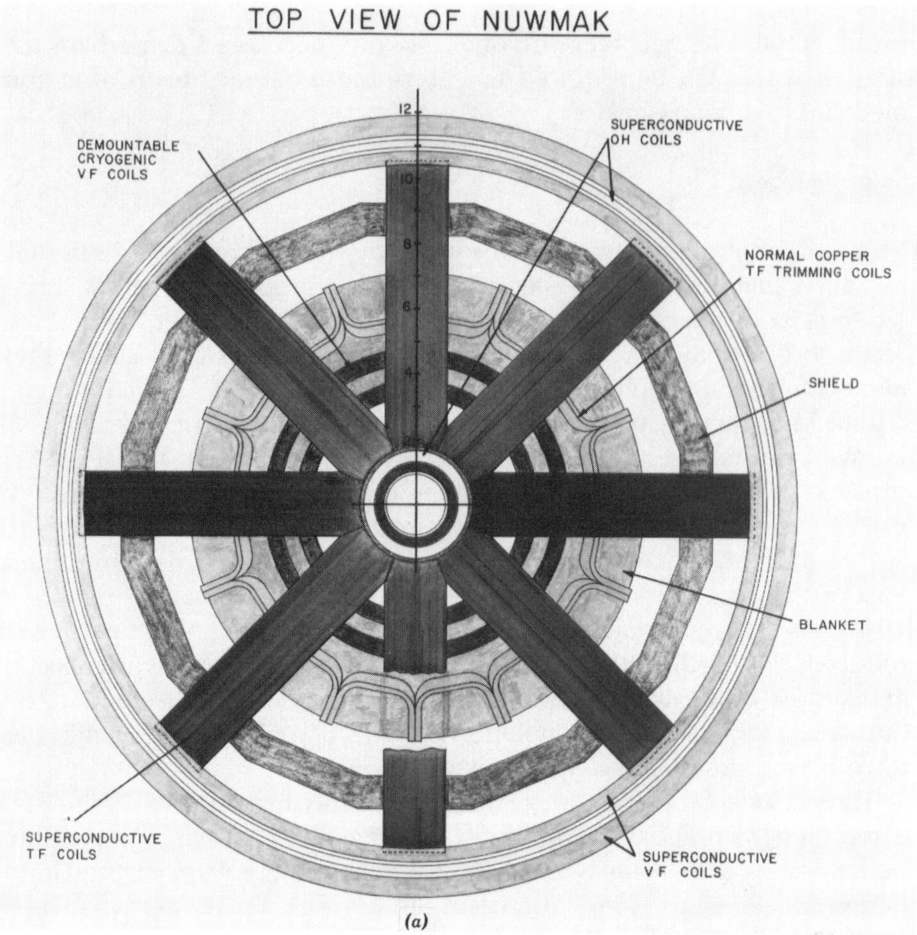

(a)

Figure 3-21 Top view *(a)* and vertical cross section *(b)* of the NUWMAK thermonuclear fusion reactor, as designed under the Nuclear Fusion Program of the University of Wisconsin. The deuterium-tritium plasma would circulate through a doughnut-shaped chamber. Energy to heat the plasma to fusion temperature would be supplied by the RF power supply. Toroidal and vertical magnetic fields (TF and VF) would be generated by several sets of coils and would confine the plasma to a doughnut shape and compress the plasma to the required density. Kinetic and light energy produced by the fusion reaction would be absorbed by the stainless-steel blanket and converted to heat. Heat

pressure is reached at about 9000 feet below the surface. Most of this internal heat, for the present at least, is completely unavailable for generating power. There are certain areas of active or recent volcanism, however, in which abnormal amounts of heat are being generated. Waters circulating through the heated rocks in these areas may become heated to such temperatures that they become steam when brought to the surface by means of natural vents or boreholes. The steam can be used to generate electricity. The most important area in the United States is the Geysers area 100 miles north of San Francisco. Wells of the principal operator, Union Oil Company, produce enough steam to support a generating capacity of nearly 1 million kilowatts of electricity, about the requirements

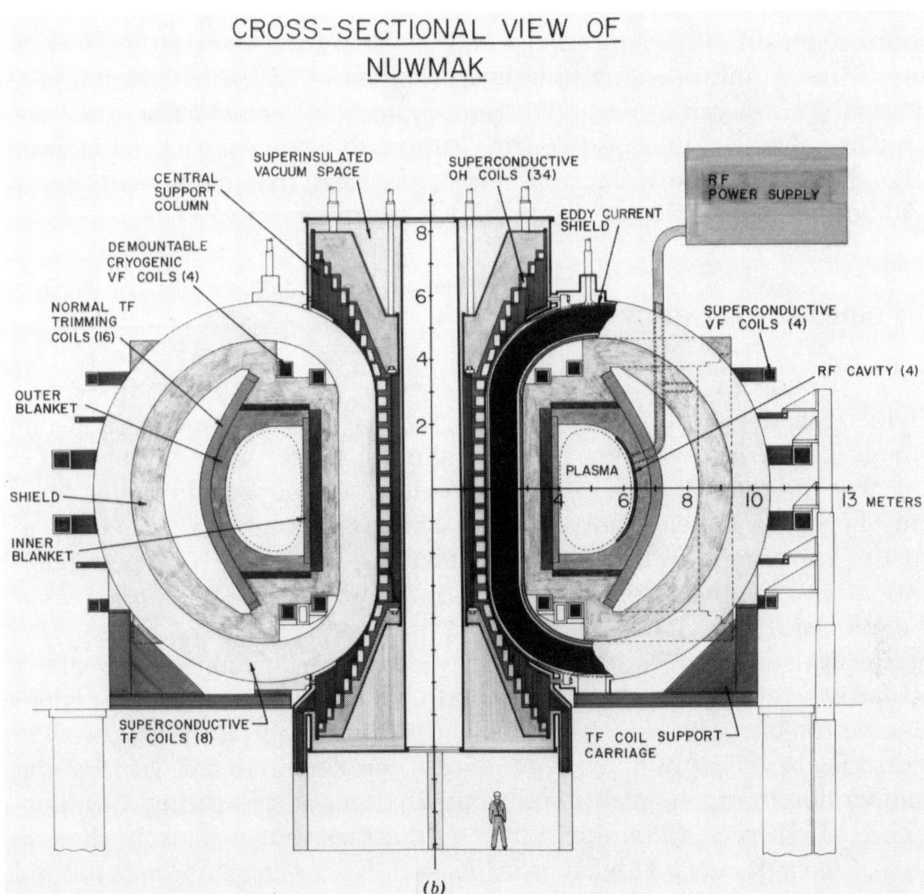

Figure 3-21 *(Continued)*
would be transferred to a lead-lithium alloy in the blanket, then to water circulating through the blanket. Steam would be generated, sent to steam drums where it would be separated from water, and used to drive the turbines that would generate electric power. Steam would be condensed to water and recirculated through the blanket. Tritium generated in the blanket would be fed into the plasma for reaction with deuterium. Radiation would be absorbed in the shield, constructed of lead, boron carbide, and stainless steel. Diagrams from University of Wisconsin Report No. UWFDM-330, March 1979, courtesy of G. L. Kulcinski.

of a city of 1.25 million people. When fully developed, the field will supply twice the present generating capacity. A more recent development, at Roosevelt Hot Springs, Utah, is supplying power to the grid of the Utah Power and Light Company. The geothermal system of the Salton Sea basin in southern California is being actively developed as a source of electric power. Other geothermal areas of the United States are under investigation. Abroad, geothermal energy is produced in 14 other countries, notably Italy, New Zealand, Iceland, Japan, and the Philippines. Geothermal development in Hawaii is under consideration.

Geothermal power can have substantial advantages over other energy sources. Steam is a clean energy source. Some thermal waters, low in salts, present minimal environmental problems. To the extent that geothermal sources can be discovered and brought into production, they can provide valuable additions to energy supply. It is too soon to predict how large the contribution will be, but it is not foreseen that geothermal energy will become a major component of total energy supply.

THE OUTLOOK FOR ENERGY

To most people in the United States, and to many of our political and financial leaders, the present world glut of petroleum means that our energy problems are over. This is a short-sighted view. A longer perspective indicates that the energy problems of the United States have only begun to develop and that we will rue a failure to take steps to ensure our energy future. Prospects for long-term supply of petroleum and natural gas from domestic fields are bleak. Despite the intensive exploration of 1979–83, discoveries of natural gas have barely made up for production, and proved reserves of petroleum are declining. After falling off for several years, consumption and imports of petroleum are rising. U.S. petroleum-refining capacity is being reduced, and the petrochemical industries based on production of refined petroleum are facing increasing competition from new plants being built in other oil-producing countries. U.S. coal reserves are ample for many decades, but increasing the use of coal to make up for deficiencies in supply of petroleum and natural gas poses environmental problems that have yet to be resolved. One major utility estimates that imposition of new regulations for control of emissions from coal-fired generating plants may increase the cost of electricity to its customers by 20 to 25 percent. The oil shales contain our second largest resources of fossil energy, but tapping them on a substantial scale seems farther off now than it did 10 years ago. Present perception of the dangers of nuclear fission precludes its use as a major source of energy in the United States. U.S. sources of hydroelectric power are near full development. The future of solar energy and geothermal energy is uncertain, but it seems clear that they cannot supply a major fraction of U.S. energy supply.

Under the circumstances, there are no grounds for complacency with regard to future availability of energy at reasonable cost. We saw in the early 1970s that hasty response to a crisis is a poor substitute for addressing the energy problem on a long-term basis. Yet once again there is a lapse into the complacency that has repeatedly frustrated efforts to formulate and implement long-term U.S. policies with respect to energy and mineral raw materials.

The problem of future energy supply is certainly far from simple. It is not a question of resources. The energy resources available on and in the earth are adequate for centuries to come, and success in nuclear fusion could extend the life of energy resources into the indefinite future. The problems are those of utilization—how to produce energy from the various sources at reasonable cost and how to produce and consume energy minerals in acceptable ways. Too many countries are deficient in energy resources and are therefore drastically affected by developments that influence the cost or distribution of world energy supplies. Energy-deficient countries will be dependent on maintenance of stability in the world and on control of energy prices through competition among energy producers in world markets. The United States and the world must ultimately face the problems of dwindling resources and rising costs of petroleum and natural gas. The cost of coal will also rise.

The energy problem can only be solved through intensive research and development of new or improved technologies of energy production, particularly from coal, oil shales, and heavy oil sands, which jointly contain the really large amounts of energy in the fossil fuels. Those efforts will require massive support over an extended period of time. Time is truly of the essence. No quick results can be expected, and only long-term commitment of money and technical and scientific resources will be effective. For this reason, the failure of the United States to develop a long-range energy policy is dismaying to those concerned with future energy supplies.

ENERGY SYSTEMS

The size and availability of resources of the energy minerals are basic to consideration of the energy problem. Beyond this, however, the *provision of energy* requires creating a series of facilities. There must be the mines and wells by means of which the energy minerals are extracted. There must be the facilities for converting fuel minerals into useful forms of energy. There must be the means of transporting the fuel minerals—by barge, rail, truck, tanker, or pipeline—to the conversion plants. Finally, for large-scale conversion plants there must be the means of distributing energy to the consumer. Together these facilities constitute the *energy systems* of the United States and of the world. An electric power system based on coal includes all the equipment used in finding and mining coal deposits, and the trucks, railways, or barges that carry the coal to the generating plant. It includes the concrete, steel, and other minerals used in constructing the power plants, together with the turbines, generators, switchboards, and transformers that generate and control the electric power. It also includes the transmission lines that carry electricity to factories, business and government establishments, and individual households. All these elements of the system are constructed largely from mineral-derived materials. Every power system, from those powered by coal, oil, or natural gas to solar and geothermal systems, has its bill of mineral raw materials. The larger the energy system, the larger the bill.

In succeeding chapters, important groups of nonfuel mineral raw materials and their many applications are examined. We use them to produce a wide range of consumer goods, from pots and pans to jet aircraft, but sizable percentages of many nonfuel minerals or elements derived from them are used in activities related to energy production (Table 3-6). These materials are therefore integral parts of today's energy systems. In considering the total energy problem, we are therefore much concerned with the availability of the "nonfuel" materials.

TABLE 3-6 Percentage of Various Elements and Minerals Used in Domestic Energy Applications Ranked by Decreasing Percentage

Element or Mineral	Percent	Element or Mineral	Percent
Quartz crystal	100.00	Bentonite	31.0
Uranium	≤100.00	Indium	31.7
Sheet mica	95.6	Antimony	25.0
Gallium	90.6	Gold	24.8
Barium	88.1	Molybdenum	23.4
Hafnium	83.3	Nickel	22.0
Rhenium	82.2	Cadmium	21.8
Beryllium	73.1	Diamond	20.2
Thorium	69.3	Niobium	16.2
Thallium	66.7	Tin	15.7
Tantalum	66.2	Cesium	15.1
Platinum	65.7	Rubidium	14.2
Lead	65.1	Tungsten	11.0
Strontium	62.1	Aluminum	10.5
Bromine	60.2	Zinc	10.3
Germanium	59.1	Silicon	9.4
Copper	58.0	Sulfur	8.0
Rhodium	56.3	Nitrogen	7.4
Palladium	51.0	Iron	6.7
Cobalt	48.9	Manganese	5.3
Mercury	47.7	Garnet	5.3
Rare earths	44.9	Lithium	4.1
Hydrogen	42.9	Graphite	3.5
Scandium	40.0	Zirconium	2.6
Selenium	34.9	Sodium	0.8
Silver	32.8		

Source: Goeller, H.E., (1980).

REFERENCES AND ADDITIONAL READING

Averitt, P., 1975, Coal resources of the United States, January 1, 1974. U.S. Geological Survey Bull. 1412, U.S. Government Printing Office, Washington, D.C., 131 pp.

Cameron, E. N., Conn, R. W., Kulcinski, G. L., and Sviatoslavsky, I., 1979, *Minerals Resource Implications of a Tokamak Fusion Reactor Economy*, University of Wisconsin, Madison, Fusion Research Program, UWFDM-313, 68 pp.

Cuff, D. J., and Young, W. J., 1980, *The United States Energy Atlas*. Macmillan, New York, 415 pp.

Duncan, D. C., and Swanson, V. E., 1965, Organic-rich shale of the United States and world land area. U.S. Geological Survey, Circular 523, 30 pp.

Energy Information Administration, 1984, *Annual Energy Review 1983*. U.S. Government Printing Office, Washington, D.C., 259 pp.

Energy Information Administration, 1985, *Annual Energy Review 1984*. U.S. Government Printing Office, Washington, D.C., 275 pp.

Goeller, H. E., 1980, Future energy supply: Constraints by nonfuel mineral resources. Oak Ridge National Laboratory, Oak Ridge, Tenn., ORNL-5656, 108 pp.

Hubbert, M. K., 1956, Nuclear energy and the fossil fuels. In *Drilling and Production Practice*, American Petroleum Institute, New York, pp. 7–25.

Hubbert, M. K., 1962, Energy resources, a report to the Committee on Natural Resources. National Academy of Sciences, National Research Council, Publication 1000-D, Washington, D.C., 141 pp.

Hubbert, M. K., 1973, Survey of world energy resources. *Canadian Mining and Metallurgical Bulletin*, Vol. 66, No. 735, pp. 37–54.

Hubbert, M. K., 1974, U.S. Energy resources, a review as of 1972, Part 1. Committee on Interior and Insular Affairs, U.S. Senate, Ser. No. 93-40 (92-75), U.S. Government Printing Office, Washington, D.C., 201 pp.

Kellogg, H. H., 1977, Sizing up the energy requirements for producing primary materials. *Engineering and Mining Journal*, Vol. 178, No. 4, pp. 61–65.

Kulcinski, G. L., 1974, Fusion power, an assessment of its potential impact in the United States. *Energy Policy Journal*, Vol. 2, No. 2, pp. 104–125.

Kulcinski, G. L., 1984, Fusion reactor design: The road to commercialization. *Atomkernenenergie-Kerntecknik*, Vol. 44, pp. 1–5.

Kulcinski, G. L., Kessler, G., Holdren, J., and Häfele, W., 1979, Energy for the long run: Fission or fusion? *American Scientist*, Vol. 67, No. 1, pp. 78–89.

McRory, R. E., 1982, *Oil Sands and Heavy Oils of Alberta*. Alberta Energy and Natural Resources, Edmonton, 94 pp.

National Academy of Sciences, 1975, *Mineral Resources and the Environment*. National Academy of Sciences, Washington, D.C., pp. 79–126.

National Geographic Society, 1981, *Energy: Special Report*. National Geographic Society, Washington, D.C., 115 p.

Phizackerley, P. H., and Scott, O., 1978, Major tar sand deposits of the world. In *Bitumens, Tar Sands, and Asphalts*, G. V. Chilingarian and T. V. Yen, eds., Elsevier, Amsterdam, pp. 57–93.

Ruedesili, L. C., and Firebaugh, M. W., 1975, *Perspectives on Energy*. Oxford University Press, New York, 527 pp.

Schmidt, R. A., 1979, *Coal in America*. McGraw-Hill, New York, 458 pp.

Squires, A. M., 1976, Chemicals from coal. *Science*, Vol. 191, No. 4227, pp. 689–700.

Steinhart, E. C., and Steinhart, J. S., 1974, *Energy*. Duxbury Press, Scituate, Mass., 362 pp.

4 Mineral Raw Materials

THE NONMETALLIC MINERALS

INTRODUCTION

The mineral raw materials used in industry today are very diverse; it is not possible to discuss them all or all the many thousands of uses to which they are put. We can only examine the nature and distribution of the principal materials, indicate their importance, and point out some of the problems of securing adequate supplies. They are often referred to as the nonfuel minerals. In industry and in discussions of mineral resources, it is customary to recognize two major groups of mineral raw materials, metallic and nonmetallic. The latter are often described as the industrial minerals and rocks. The convention will be followed here, although the line between the two groups cannot be sharply drawn. Of the two groups, the nonmetallic minerals are the more important in terms of tonnage annually produced and consumed. The annual tonnage of nonmetallics is actually greater than that for metallic minerals and fuels combined, but the dollar value of mineral fuels produced annually exceeds that of both metallic and nonmetallic minerals.

One of the most fortunate features of the environment in which man arose some millions of years ago and has developed since is the abundance of many kinds of nonmetallic materials. Man is still learning how to use them. The earliest materials were probably stones used for building crude shelters and making equally crude weapons, but with time man's use of nonmetallic materials progressed along three paths. One path was selection among materials. Some rocks and stones were found to be much better than others for particular purposes. Flint and obsidian proved to be far superior to most other stones for spearpoints, arrowheads, and certain household tools, because they could be chipped so as to create hard, sharp points and edges. A bowl was difficult to make out of hard granite, but it was easy to carve a bowl from soft soapstone. Clays could molded into brick, but some clays made better brick than others. The second path of progress was toward modifying nonmetallic materials to improve their usefulness. Bricks and pots made from clay and dried in the sun were usable, but they were not very durable, and they were porous. Firing them in kilns made them both stronger and impervious. The third path was that of extracting useful materials from materials that were not very useful, or not at all useful, in their natural state. Salt scraped from salt pans could be dirty and impure. Dissolving the salt and then evaporating the solution gave a clear, pure product. Ground limestone was quite useless, but burning it gave quicklime that could be used as plaster or a primitive cement.

The history of man's efforts to tap the nonmetallic mineral resources of the earth is one of progress along all three paths. Progress continues. We are still learning to select, to modify, and to extract. The end of learning is certainly not in sight, but the accumulated knowledge and experience that is our heritage from the past has already given us access to a remarkable array of nonmetallic minerals and mineral-derived materials. Roughly a hundred nonmetallic minerals are in use in industry and in our daily lives today. So heterogeneous are they, and so varied in their uses, that they are a most unwieldy subject for discussion. Each of the minerals has its own problems of availability. Some are widely distributed in large deposits; others are very restricted in their distribution. Each has its own history of discovery and application to the needs of man. Some have been used since the beginnings of civilization, but some have only recently come into use. In this volume we cannot possibly do justice to them all, but we will consider the following important groups of nonmetallic raw materials.

Sand, gravel, and stone
Cement materials
Clays
Other ceramic materials
Mineral abrasives
Fertilizer minerals
Saline minerals
Sulfur
Fluorspar
Talc and pyrophyllite
Asbestos and some other nonmetallic minerals

Examining these groups will indicate something of the range of nonmetallic minerals, the functions that they serve today, and the size and distribution of important nonmetallic mineral resources of the United States and the rest of the world.

SAND, GRAVEL, AND STONE

Sand, gravel, and stone, mundane materials that they are, are of paramount importance as constructional materials and have a long history of usefulness to man. The tonnage of these materials produced and consumed in the United States is greater than that of all other mineral raw materials. They find massive use in the construction of highways, large buildings, dams, and other works and structures, and widespread use for driveways and foundations. The enormous scale of use is due to their cheapness, their ready availability in many places in the country, and their versatility as constructional materials. Production of these and other major constructional materials in the United States in 1983 and in 1984 is shown in Table 4-1.

Sand and gravel are in part used directly in road building, either for surfacing or as the substrate for asphalt or concrete roadways. Where sand and gravel are unavailable or relatively expensive to produce, crushed stone is used for the same purpose. However, much of these materials is used in concrete, the most important single constructional material of modern times, or in the mortar that bonds stone or brick or concrete block into strong, solid structures.

In ancient times, dimension stone (blocks and slabs) and brick were

TABLE 4-1 U.S. Production of Major Constructional Materials, 1983 and 1984 (in Thousands of Tons)

Material	1983	1984[a]
Sand and gravel	655,100[a]	680,000
Crushed stone	863,000	953,000
Dimension stone	1,186	1,224
Cement	71,350	79,000
Common clay	27,066	28,500
Lime[b]	1,490	1,610
Gypsum[c]	11,000	12,300

[a]Estimated.

[b]Constructional lime is about 10 percent of total lime production.

[c]Approximate amount produced for constructional purposes.

Source: U.S. Bureau of Mines, *Mineral Commodity Summaries*, 1985.

the major materials for heavy construction. In the United States and in general in industrial countries, dimension stone has been largely supplanted by concrete, but nearly 1.5 million tons of dimension stone is annually used in the United States in the construction of buildings and monuments and for other purposes. Marble, granite, and certain other stones are still prized highly for their beauty and as symbols of what is lasting among the works of man. Stone has an aesthetic and traditional appeal that concrete can never have, and high-quality stone continues to command a market despite its higher cost.

In colonial times and in the earlier days of the American republic, building stone was quarried from local sources, many of which yielded stone of inferior quality. With time, production has moved to sources of stone of special beauty and quality—the marble quarries of Vermont, Georgia, and Colorado; the granite quarries of Vermont; the limestone quarries of the Bedford district of Indiana; and the sandstone quarries of Tennessee are famous examples. But there are a number of other localities at which dimension stone is produced.

Most of the larger countries of the world have ample supplies of sand and gravel or crushed stone. In the United States in 1982, sand and gravel were produced from about 6000 pits in the 50 states, and crushed stone was produced from about 1400 quarries. There is no prospect of a shortage of these materials. There is, however, a problem of rising costs as cheaply minable deposits close to some places of use become depleted. Transport is an important element in the cost of such bulk materials to the consumer. As distance to the point of use increases, costs of transport rapidly equal or exceed the costs of mining and processing. Urban and suburban sprawl is steadily restricting the sites available for extraction. Real costs of the materials will therefore rise with time.

Almost any mineral material has a special history of use. One crushed-stone product is a conspicuous example. When asphalt came into use for roofing material around the turn of the century, it was found to deteriorate under the actinic rays of the sun. The problem was finally solved by coating sheet roofing and asphalt shingles with small granules of stone. At first the colors of shingles were simply the colors of the rocks that were crushed to produce the granules, but soon processes were developed for giving granules colored ceramic coatings. Thus were born the varicolored roofs that cover most houses in America today.

CEMENT MATERIALS

Since very early times, man has been concerned with binding bricks or stone blocks together so as to make his buildings and other structures strong and impervious to wind and rain. There is a long history of development of various cements; simple lime cement was already in use when the pyramids were built in ancient Egypt. Plaster of Paris was also made in early times by calcining gypsum. However, plaster of Paris is very soft, and lime cement will not set under water. One that would, a "hydraulic cement," was made by the Romans by adding slaked lime to volcanic ash. It was much stronger than either lime or gypsum cement and ultimately became widely used in Europe. However, in 1756 John Smeaton, in England, made a cement by burning an argillaceous limestone at temperatures above about 1470°F to produce a clinker, which was then ground fine. When water was added, chemical compounds crystallized, and the final product was a solid material of great strength. Mixed with sand, it made a mortar that would firmly bind brick or stone. Concrete could be made by mixing the cement with sand and gravel or crushed stone. This type of cement became known as natural hydraulic cement. Limestone beds having suitable compositions were identified at certain localities, and these became centers of cement production. Manufacture of natural cements spread to America. Rosendale, New York; Louisville, Kentucky; Milwaukee, Wisconsin; and an area near Lehigh, Pennsylvania, became important sources of cement.

Natural cements were not entirely satisfactory, largely because the limestone beds that were quarried differed somewhat in composition and individual beds varied in composition. The resulting cements were not uniform or constant in properties. There was much experimentation, especially in England, and it culminated in 1824 with the award of a patent to Joseph Aspdin covering the manufacture of a cement that he called portland cement. It resembled Portland stone, a building stone in common use in England at the time. Aspdin produced the cement from a controlled mixture of limestone and clay. A uniform product was thus assured. During the 19th century it steadily replaced the natural hydraulic cements.

The advent of portland cement was a major advance in the development of constructional materials. In the intervening years, many varieties of portland cement have been developed, each with properties that make it especially suitable for certain purposes. Close control of the mix of raw materials is one reason for this success. It has made possible the rise of concrete to its foremost position as a material for heavy construction of many kinds. There are other types of cements in use, and we still use the older lime and gypsum cements in the interiors of buildings. The dry-wall construction in common use today employs wallboard panels made from gypsum.

Materials for the manufacture of portland cement are widely distributed over the world. Pure limestone is desirable as a base material, since other constituents can be added to it in controlled amounts to make whatever type of cement is desired. Deposits of pure limestone, however, are not found everywhere, and access to some very good deposits is now restricted or prevented by the growth of cities. The price of cement has been stable for some years but is likely to rise in some areas as lower-cost sources are depleted. Gypsum deposits are somewhat more restricted in distribution than deposits of materials used in portland cement, but there are a number of large deposits in the United States. In 1982 gypsum was mined at 70 mines in 22 states, whereas cement was produced at 152 plants in 40 states. Fifty percent of the cement production, however, was from Texas, California, Pennsylvania, Michigan, Missouri, and Florida. Some countries are deficient in cement raw materials and must import their requirements, but portland cement is manufactured in many countries of the world, and in the less-developed countries concrete is gradually supplanting more primitive building materials.

CLAYS

Like stone, clay has been used as a constructional material since long before recorded history. Widely distributed, easy to mine, and readily used in construction, in many areas it may well have been the first-used constructional material. It lends itself to both simple and sophisticated use. Mixed with water it can be plastered over frameworks of sticks or brush, or it can be molded into bricks that need only to be dried in the sun. It is still used extensively in this simple manner in tropical and subtropical areas to build walls, huts, and houses that may not be very durable but that are very easily repaired (Fig. 4-1). Mud brick was in use

Figure 4-1 A village in central Madagascar. The houses are built of sun-dried brick plastered with a mixture of white clay and cow dung.

in Anatolia by 6500 B.C. When bricks are fired in kilns, very durable building materials are produced, and their quality can be enhanced by proper selection of materials and careful firing. The common clays used in making brick are also used in making sewer pipe, drain tile, and roofing tile, and they were much used in making pottery in ancient times. Brick was an essential building material in ancient Mediterranean civilizations, from Egyptian and Mesopotamian to the later civilizations of Greece and Rome.

The clays of the world are a very complex group of materials, and as their nature and properties have become better understood, many special uses of clays have developed. As *materials,* clays are generally defined as very fine grained (particle size less than 0.2 mm) and consisting predominantly of one or more clay *minerals.* There are two main groups of clay minerals. Minerals of the kaolin group are simple hydrated aluminum silicates. Those of the smectite group have, besides alumina, silica, and water, various combinations and proportions of calcium, magnesium, sodium, potassium, iron, and lithium. The industrial properties of a clay material are functions of the kinds and proportions of clay minerals present, but they are also functions of the amounts and kinds of impurities present and the clay mineral particle sizes. Such fine-grained materials are very difficult to investigate, and progress was slow until the 1920s, when x-ray techniques began to be applied to mineralogical analysis of clays. It has taken more than 50 years of intensive research to arrive at a reasonable understanding of the nature of clays and the causes of their physical and chemical behavior, and there is still much work to be done.

Kaolin clays of great purity have been used for centuries in making chinaware and other ceramic products, but their major uses today are as a filler and extender in paper, plastics, rubber, and paints. So useful have they become that U.S. annual consumption of kaolin has grown from about 500,000 tons in 1930 to nearly 8 million tons in 1981, with a value of nearly $200 million. Nearly half the total production is used in paper products, of which kaolin constitutes up to 30 percent. It fills the interstices of the mat of paper fibers and coats the paper surface, giving gloss, brightness, opacity, smoothness, and printability. Much kaolin is used as a filler in rubber, improving the strength, abrasion resistance, and rigidity of rubber and lowering the cost. It is used in large amounts as an extender in paint, being especially suited for the latex paints that have come into such widespread use. It is used as a filler in plastics, imparting smoothness, stability, and resistance to chemical attack. There are other uses too numerous to mention here.

The United States is fortunate in having some of the world's largest deposits of high-grade kaolin in a belt extending from central Georgia northeast into South Carolina. Reserves in the deposits are in excess of 5 billion tons. There are also substantial reserves and resources of kaolin in England, France, Germany, Czechoslovakia, the Soviet Union, Spain, and other European countries; in a number of countries in Asia, Africa, and South America; and in Australia. There are no estimates of total world resources, but they certainly are adequate for centuries to come.

Closely related to the kaolins are the ball clays, found in the United States mainly in western Kentucky and western Tennessee and also in Mississippi. Kaolinite is the main clay mineral in ball clays, but other clay minerals are also present, with various amounts of finely divided silica. Pure kaolinites are not plastic, but ball clays are, and this property makes them much in demand for making pottery, chinaware, floor and wall tile, and other ceramic products. Plasticity is needed so that clay can be molded into desired shapes and will hold those shapes while it is being fired. Reserves of ball clays in the United States are on the order of 100 million tons, mostly in Tennessee and Kentucky (Fig. 4-2). England has large deposits in Devonshire.

Clays of the kaolin group, both the high-quality kaolins and ball clays, are widely used in making firebrick, insulating brick, and other products (refractories) that must have high resistance to heat. "Fireclays" is a general term applied to clays used for such purposes. Clays underlying some coal seams have good refractory properties and are mined at numerous localities. The midwestern states are major sources of refractory clays.

Clay minerals of the smectite group form an important class of clay materials. They are marketed partly as "bentonites," partly as "fuller's earth," but the distinction between the two both in mineral composition

Figure 4-2 Part of a clay mine near Mayfield, Kentucky. The clay beds in the middle ground are overlain by beds of sand and gravel forming the wall of the pit in the background. Four distinct beds are being mined, the highest by means of the power shovel. The clay beds range from 1.25 to 4 feet in thickness.

and in use is rather vague. They have many uses. The main uses are in refining, filtering, clarifying, and decolorizing oils; in mud used in drilling wells for petroleum and natural gas; in pelletizing iron ore concentrates; and as a bonding agent for foundry sands.

There are several important types of bentonites. Sodium bentonites, mined mostly in Wyoming and adjacent states, have a remarkable capacity for absorbing water, swelling to as much as 15 times original volume and forming a gel-like mass. In drilling muds they act to seal cavities and fractures in the drill hole walls. They serve as a viscous suspension medium for ground barite, a heavy mineral that is added to the mud to increase its density so that the mud displaces more of the weight of the string of drill pipe. In foundries, sodium bentonite is used to bond the sand molds made for casting metals. Its bonding properties are also responsible for its use in pelletizing the concentrates made from taconite, which is low-grade iron ore. Such concentrates are very fine grained and, until they are pelletized, cannot be used as feed in the furnaces in which pig iron is made.

It is quite impossible to describe here all the special types of clays now recognized in industry, or their many uses. Enough has been said to indicate that in the clays man has found a wonderfully versatile group of materials. The large size and wide distribution of resources of many of the clays ensure that they will be available in adequate amounts for the foreseeable future.

OTHER CERAMIC MINERALS

A ceramic product is a solid composed of materials that have been subjected to heat (fired) above 875°F. The most familiar examples are cement, pottery, porcelain, chinaware, earthenware, wall tile, and glass, but many other products are included in the ceramic group. Refractory brick (a refractory is any heat-resistant material) is important for linings of fireplaces, chimneys, and metallurgical furnaces. Several dozen different minerals are used in various ceramic products. Besides cement and clays, which we have already discussed, the most important are feldspars, nepheline syenite, and silica minerals.

The feldspar minerals are aluminosilicates of sodium, potassium, and calcium. Potassium-rich feldspars are used in making pottery, porcelain, and wall tile. The feldspar serves chiefly as a flux, causing fusion of ceramic mixtures at temperatures below the melting points of clays and other constituents, reacting with them, and binding them with vitreous material. In glassmaking, sodium-rich feldspar and the related mineral, nepheline, are preferred. There are many glasses with special compositions, but for centuries most glass was made from silica sand, lime, and soda ash (sodium carbonate). Soda feldspar, with or without nepheline, is now added to glass as a source of alumina, which improves the strength of the glass, its resistance to thermal shock, its workability during forming, and its resistance to "blooming"—the frosting that develops on the outside of glass when it is exposed to weather.

Feldspars are the most abundant group of minerals in the rocks of the earth's crust, but deposits from which high-purity feldspar can be obtained are not widely distributed. Until World War II, potash feldspar was obtained almost entirely from pegmatites (Fig. 4-3), exceptionally coarse-grained rocks of granitic composition from which, after blasting, feldspar could be sorted out by hand. Current practice involves mass mining, crushing, and separation of feldspar from associated minerals by froth flotation. Similar methods are used to recover soda-rich feldspar for glassmaking. Most U.S. production of soda-rich feldspar is from deposits in the Spruce Pine district of North Carolina. Most of the potash feldspar is produced in Connecticut and Georgia. The United States is self-sufficient in feldspar production, and reserves of feldspar-bearing rocks are very large. Consumption is about 700,000 tons per year.

Quartz sands for glassmaking are one of many types of industrial sands, including the common sands used in construction. Glass sands are quartz sands of high purity; they provide most of the silica that is the main

Figure 4-3 Part of the Alto Feio pegmatite, Paraíba, Brazil, consisting of large crystals of feldspar (light gray) partly separated by masses of quartz (white). The man's hand is near the upper end of a large crystal of feldspar. Other crystals of feldspar in the wall are of equal size.

constituent of many types of glass. The best glass sands are very low in iron (less than 0.015 percent), since the iron oxides in which iron usually occurs in sand are powerful and undesirable coloring agents. Low-iron sands command a premium price.

MINERAL ABRASIVES

Cutting, grinding, and polishing are important processes in many industries, and a variety of abrasive materials is in use. Mineral abrasives have been in use for thousands of years, and they are still used in large amounts. Quartz sands are used in sand blasting and in common sandpaper. They are also used in cutting some of the softer stones such as limestone and serpentine. Emery, a natural mixture of corundum (natural alumina), spinel (magnesium aluminate), and magnetite (magnetic iron oxide), was employed in ancient times and is still used in making special abrasive papers and cloth for grinding stone or metal surfaces.

The premium abrasive is the diamond, prized because it is the hardest of all known substances and can be used for cutting, grinding, and polishing of the hardest stones and metals. Special bits composed of metal set with diamonds are the major tool for drilling hard rocks and therefore play an important role in mineral exploration. Metal disks with diamonds set in their rims are used to cut and slice rocks and other hard, brittle materials. Pencils with diamond tips are used to mark glass and other hard materials. These are only a few of the hundreds of applications of diamonds in industry.

The primary sources of diamonds are curious volcanic pipes that are found in certain places in Africa and were first recognized in the famous Kimberley district of South Africa (Fig. 4-4). Subsequently they have been found in Tanganyika, West Africa, Namibia, the United States, the Soviet Union, and most recently in Australia. Ratios of diamond to rock in the pipes that have been mined range from 1 to 10 million to 1 to 17 million, depending on the proportions of gem diamonds and diamonds of industrial grade. Erosion of diamond-bearing pipes feeds the diamonds into river channels, in the gravels and coarse sands of which the diamonds may become concentrated as placer deposits. Such deposits are important sources of diamonds in Sierra Leone, Zaire, Angola, South Africa, and Brazil. The pipes of the Kimberley district shed diamonds into the Orange River system, down which they were carried westward to the Atlantic ocean. Further concentration by waves led to formation of beach placer deposits adjacent to the Orange River mouth. They have been richly productive of diamonds. Diamonds can now be synthesized, at high temperatures and very high pressures, and synthetic diamonds fill many of the requirements for industrial grades. The world is thus relieved of much

Figure 4-4 Top: The "Big Hole," the open pit in the upper part of the famous Kimberley diamond pipe, South Africa. Bottom: A vertical section of the pipe, mined to a depth of 3520 feet. Open-pit mining was followed by underground mining by means of a shaft and crosscuts to the pipe at various levels.

of its dependence on diamond deposits for its supply of an essential substance. Gem diamonds, however, still come only from the natural deposits.

There are other mineral abrasives. Crushed garnet is used in grinding and surfacing glass and in making special grades of sandpaper and abrasive cloth. These are used mainly in woodworking but are also used in finishing leather, hard rubber, plastics, and the softer metals. Next to the diamond, the hardest known mineral is corundum, the natural form of aluminum oxide. Deposits of corundum are very uncommon and have never yielded adequate supplies for abrasive purposes. Today the place of corundum has been taken by synthetic alumina, supplemented by synthetic silicon carbide, tungsten carbide, and boron carbide, this last equal in hardness to the softer grades of diamond. While silica sands continue to be important owing to their cheapness and availability in large amounts, synthetic abrasives have largely taken over the markets for the very hard abrasive materials.

THE FERTILIZER MINERALS

Since World War II there has been strong concern with the pressure of world population on available food supplies. Improved health care has reduced infant mortality and has increased average life-spans. Populations have grown rapidly, particularly in the less-developed countries, and further growth is predicted (Fig. 4-5). In sub-Saharan Africa alone, population has grown since 1960 from 210 million to 293 million people. Food requirements of the world have grown to unprecedented levels.

The problem has stimulated a worldwide drive to increase agricultural production. There has been substantial success in some countries. In the United States the drive has been so successful that it has created a severe problem of overproduction. This is due in no small part to heavy use of

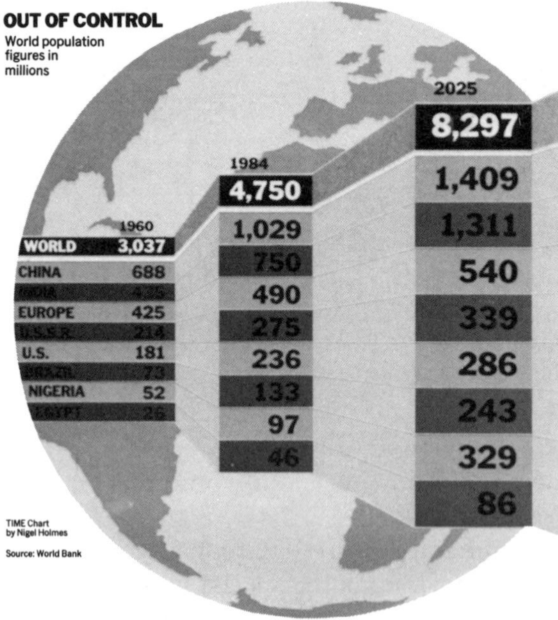

Figure 4-5 The population explosion. Copyright © *Time*, Aug. 6, 1984, by permission, all rights reserved.

manufactured fertilizers containing the three essential nutrients—nitrogen, phosphorus, and potash. As indicated in Table 4-2, use of fertilizer materials has grown at exponential rates since World War II, both in the United States and in the rest of the world.

For millennia, nitrogen has been supplied to soils by application of manure and other organic wastes, or fish meal, supplemented by nitrogen fixed by soil bacteria. Those sources of nitrogen, however, have never been sufficient to supply optimum amounts of nitrogen to the various crops of the world. The grain crops, in particular, deplete the nitrogen in soils, and wastes left over after those crops are harvested are not sufficient to restore the nitrogen. However, during the first half of the nineteenth century, as the role of nitrogen in plant growth became clear, it was realized that nitrogen could be added to soils in the form of soluble compounds such as ammonia, ammonium sulfate, and sodium nitrate. Around 1840, large deposits of sodium nitrate discovered in the Atacama desert of northern Chile began to be mined. After 1880 these were exploited on a large scale, and for 40 years gave Chile a very profitable monopoly of world nitrate supply to industry and agriculture. However, by the early 1920s the monopoly had been broken owing to the development and application of processes for producing synthetic compounds of nitrogen, most notably the Faber process for the manufacture of synthetic ammonia. For this process nitrogen is taken from the air, and hydrogen is supplied from water, natural gas, oil, or other sources. In the United States, the prices of nitrogen compounds are closely tied to the price of natural gas, since this is the principal source of hydrogen. Synthetic ammonia and ammonium sulfate and nitrates manufactured from

TABLE 4-2 Consumption of Fertilizer Components (Thousands of Tons)[a]

	1950	1960	1970	1980	1984[b]
Nitrogen in nitrogen compounds					
United States	1,500[b]	4,059	10,961	17,236	14,870
World	4,934	14,184	42,747	78,086	88,000
Phosphate rock					
United States	9,610	17,091	27,163	44,963	43,882
World	25,142	45,550[b]	93,365	150,904	157,629
Potash (K_2O)					
United States	1,411	2,336	4,728	6,999	6,724
World	4,500[b]	10,000	20,013	31,702	30,754

[a]Total consumption includes consumption for nonfertilizer uses: nitrogen, about 20 percent; phosphate, about 10 percent; potash, about 5 percent.

[b]Estimated.

Source: U.S. Bureau of Mines, *Mineral Commodity Summaries.*

ammonia now supply most of the nitrogen used in the world. Ammonia can be applied directly to soils, or it can be used to produce soluble nitrogen compounds such as ammonium nitrate and ammonium phosphate. The world is no longer dependent on mineral nitrates or natural organic sources for its nitrogen supply.

Phosphate can also be supplied in limited amounts from organic residues and fish meal, but the main supply today is from deposits of phosphate rock that have been discovered in various parts of the world. The most widespread deposits are of sedimentary origin, formed by precipitation of phosphorus from seawater in the form of the mineral apatite. Apatite in these deposits is a combination of calcium, phosphorus, oxygen, and carbon dioxide, with minor amounts of fluorine and other elements, including tiny percentages of uranium. The deposits have formed in the geologic past at certain places along the margins of continents. Besides apatite, phosphate rocks contain calcite (calcium carbonate), clay, and sand in various proportions, and the phosphate rocks vary accordingly in phosphate content. The richest deposits consist of rock containing 30 percent phosphorus pentoxide (P_2O_5) or more, but leaner rocks, if unconsolidated, can be processed to remove impurities. Average grades of phosphate rock mined in Florida are only 4.5 to 8.0 percent P_2O_5, but the crude ore is processed to yield concentrates ranging from 30 to 33 percent P_2O_5. At Lee Creek, North Carolina, the rock as mined contains 15 percent P_2O_5 and is readily processed to a material grading 30 percent P_2O_5. Hard phosphate rocks containing much carbonate cannot be used as sources of phosphate. As mentioned earlier, the phosphatic limestone formation of central Tennessee contains only 17 to 23 percent calcium phosphate. Over sizable areas, however, the rock has been exposed to leaching by downward-moving groundwaters, and the carbonate has been removed. The residual deposits thus formed have been an important source of phosphate for the agricultural areas of the midcontinent.

Sedimentary phosphate deposits are known on every continent except Antarctica. In North America the most important deposits are those of Florida, North Carolina, and a belt extending from northeastern Utah through eastern Idaho, western Wyoming, and western Montana. Very large deposits are found abroad in a belt extending across northern Morocco into Algeria and Tunisia. There are also deposits in Jordan and Israel. There are many other deposits, at present of lesser importance, in Europe, South America, Asia, and Australia. Offshore deposits are now under investigation.

Another very important type of deposit consists of apatite-bearing rocks that have formed from molten material in the throats of certain ancient volcanoes. Prior to 1948 such deposits were thought to be very rare, but many have been discovered during the intervening years, in Africa, South America, and Canada. The enormous deposits of the Kola Peninsula, in northern Russia, consist of nepheline syenite, a silicate rock, with 16 to 18 percent P_2O_5 (in apatite). The rock is processed to obtain a concentrate with 39 percent P_2O_5. At Palabora, in South Africa, an apatite-bearing pyroxenite (another silicate rock) yields a concentrate with about 36 percent P_2O_5. Other deposits of the same type, like those of Sekulu, Uganda, and Araxà, Brazil, were originally apatite-bearing carbonate rocks from which the carbonate has been dissolved out by percolating groundwaters. The residues are minable, phosphate-rich soils. Such deposits are especially important to countries that have no sedimentary phosphate deposits.

The droppings of birds (guano) are rich in phosphate. At certain places in the world where birds nest in great numbers, accumulations of guano are formed and can be harvested. On Ocean and Nauru islands in the South Pacific, phosphate leached from guano has moved downward and has replaced underlying limestone. The resulting materials are exceptionally rich in phosphate. The deposits of Nauru have brought great wealth to that tiny nation.

Production of phosphate and the reserve base of phosphate rock in various countries are shown in Table 4-3. In the early 1970s (Emigh, 1972), there was great concern over the adequacy of world phosphate resources, but as the table shows, they are enormous relative even to the greatly

TABLE 4-3 Production of Phosphate Rock in 1983 and 1984 and Estimates of the Reserve Base (in Thousands of Tons)

	Production		Reserve Base
	1983	1984[a]	
United States	47,126	54,013	5,952,800
Israel	3,273	3,300	99,000
Jordan	5,235	5,500	562,000
Morocco	22,162	23,148	23,150,000
Senegal	1,377	2,200	143,000
South Africa	3,022	2,865	2,866,000
Togo	2,294	2,200	55,100
Tunisia	6,530	6,170	132,000
Other market economy countries	13,221	13,200	3,274,000
China	13,779	14,330	231,500
Soviet Union	29,983	30,313	1,433,000
Other central economy countries	794	827	358,000
World totals	148,796	158,066	38,256,400

[a]Estimated.

Source: U.S. Bureau of Mines, *Mineral Commodity Summaries.*

increased world demand for phosphate. However, the reserve base includes material that is subeconomic at present, and some increase in the price of phosphate will take place as higher-grade deposits are depleted.

Phosphate rocks can be ground and applied directly to soils, but the apatite in them is relatively insoluble and becomes available to plants only very slowly. Phosphate is therefore treated with sulfuric acid or nitric acid to convert the apatite to more soluble compounds. Phosphate rocks can also be treated in electric furnaces to recover phosphorus in elemental form. The phosphorus is then treated to convert it into phosphoric acid, which is used partly in producing fertilizer, partly in producing various industrial chemicals. A by-product of treatment of phosphate in American plants is fluorine, formerly a toxic waste but now an important source of refrigerants (freon) and other fluorine chemicals.

The record 53.6 million tons of phosphate rock produced by the United States in 1981 had a value (f.o.b. mine) of about $1.4 billion. About 25 percent of the rock produced was exported directly, and about 45 percent was processed into fertilizers and chemicals for export. U.S. reserves of phosphate rock are large. Remaining phosphate rock (beneficiated to commercial grades) recoverable at $30 or less per ton, hence economic at present prices, has recently been estimated by the U.S. Bureau of Mines at 1.3 billion tons, with another 2.4 billion tons available at costs up to $40 per ton. Despite the large reserve base, the industry is faced with rising costs as the more cheaply minable deposits become depleted. In Florida, which has the largest reserves, phosphate mining is faced with competition from other uses of land and with rising costs of environmental protection. The United States has been the world leader in phosphate rock production but may not be able to maintain its competitive position in world markets.

Potash deposits are quite different in origin from phosphate deposits. All the world's potash deposits have formed by evaporation of seawater or the waters of saline lakes. As indicated in Table 4-4, the principal constituents dissolved in seawater are chloride and sodium, the constituents of common salt. Potassium amounts to only 0.038 percent of seawater. Potassium salts (the chloride and sulfates) are highly soluble, much more so than gypsum (hydrated calcium sulfate) or common salt, and they are only precipitated in the final stages of evaporation. In the geologic past, the advanced evaporation necessary for deposition of potash salts from seawater has taken place only under special conditions, where an arm of the sea has been largely cut off from the main body of the ocean. An arid climate in the area surrounding the arm is an essential condition. In this situation, evaporation from the arm exceeds the inflow from streams; more water moves into the basin from the ocean and is evaporated. The waters of the basin become more and more saline. Gypsum is deposited first, then halite. If the process goes on long enough, potassium salts are

TABLE 4-4 Average Composition of Seawater

	Percentage of Total Seawater	Percentage of Total Dissolved Constituents[a]
Water	96.552	
Chloride	1.898	55.05
Bromine	0.007	0.19
Sulfate	0.265	7.68
Bicarbonate	0.014	0.41
Fluorine	< 0.001	< 0.01
Boric acid	0.003	0.07
Magnesium	0.127	3.69
Calcium	0.040	1.16
Strontium	0.001	0.03
Potassium	0.038	1.10
Sodium	1.056	30.61
Totals	100.000	99.99

[a]Major constituents. Tiny percentages of many other constituents make up the remaining 0.01 percent of dissolved matter.

Source: B. Mason, *Principles of Geochemistry*, 1958, p. 186.

finally precipitated. The various salts are deposited as successive layers on the floor of the basin. The world's largest deposits of gypsum, common salt, and potash salts have formed in this manner.

Mining of potash deposits began in the Stassfurt region of Germany around the middle of the nineteenth century, and for some decades the Stassfurt mines were the sole source of potash supply. Later in the century, deposits were discovered in Alsace (now in northeastern France). Production and prices of potash were controlled by a cartel of Stassfurt and Alsace producers until World War I. During the war, the United States was cut off from the European sources, and a frantic search for domestic deposits ensued. This led ultimately to development of the deposits of Searles Lake, California, and later to the discovery of deposits of the Carlsbad area, New Mexico. In the 1950s additional deposits were found under the Paradox Valley of eastern Utah.

The deposits at Searles Lake have a unique history. During the glacial epoch, for considerable periods of time, the Searles Lake basin was the ultimate sink for the Owens River system that drained the eastern slopes of much of the Sierra Nevada. Evaporation of the waters of Searles Lake continued for long periods, salts from the river system were concentrated in saline brines, and finally the lake dried up completely. The result of all this is the thick deposit of salts, about 30 square miles in surface area, from which the brines are pumped. In contrast, the deposits of Carlsbad, New Mexico, are of marine origin, like those of Stassfurt and Alsace.

They are layers of rock rich in potash salts and are mined underground by conventional methods. The rock deposits of the Paradox Valley, Utah, are also marine deposits, but, as indicated in Chapter 2, are mined by means of wells. Potash is also recovered as a by-product of salt from the brines of Great Salt Lake. The brines of the Dead Sea (actually a salt lake) are a major source of potash and are being worked by both Israel and Jordan.

Since World War II, major deposits of potash salts have been discovered in Saskatchewan and New Brunswick in Canada and in England, Thailand, Ethiopia, Gabon, and Brazil. Russia and Poland have large reserves. All are bedded deposits of marine origin. Those of Saskatchewan are by far the largest in the world.

Data for U.S. and world production and the reserve base are given in Table 4-5. The great magnitude of the reserve base gives assurance that there will be no shortage of potash for a long time to come.

At present, about one-third of the U.S. potash supply comes from domestic mines, and almost all the remainder comes from Saskatchewan. The cost of potash for fertilizer is important to the American farmer. Prices for potash nearly doubled from 1977 to 1981 owing to heavy demand. They declined sharply during 1982 and were still low in mid-1985.

TABLE 4-5 United States and World Production of Potash in 1983 and 1984 and Estimates of the Reserve Base (in Thousands of Tons)

	Production		Reserve Base
	1983	1984[a]	
United States	1,575	1,764	397,000
Canada	6,838	7,385	10,692,000
France	1,694	1,764	55,100
West Germany	2,666	2,700	661,000
Israel	1,102	1,320	661,000[b]
Italy	138	143	44,000
Jordan	187	496	661,000[b]
Spain	724	550	55,100
United Kingdom	333	353	33,000
Other market economy countries	24	22	496,000
Developing countries	—	—	132,000
East Germany	3,781	3,802	1,102,000
Soviet Union	10,251	10,472	4,189,000
Other centrally planned economies	28	28	22,000
World totals	29,341	30,841	19,200,000

[a]Estimated.

[b]Dead Sea reserve base divided equally between Israel and Jordan.

Source: U.S. Bureau of Mines, *Mineral Commodity Summaries.*

SALINE MINERALS

Gypsum and potash are members of the group of saline minerals extracted from the salt deposits and salt lakes of the world. The group also includes halite (common salt), trona (sodium carbonate), sodium sulfate, boron salts, magnesium compounds, and lithium salts. Bromine is recovered from wells that tap saline brines in Arkansas and Michigan. The various saline minerals are all used in significant amounts in chemical industry (Table 4-6), but common salt is the most important of them all. It is essential to human life and is one of the most basic raw materials of chemical industry. Extensive marine deposits of saline minerals are known on all continents. Reserves of common salt in the deposits are so large as to be virtually inexhaustible. Deposits of gypsum are less common, but still, as we have seen, they contain very large reserves. Marine deposits also contain large reserves of magnesium compounds.

Since the waters of the oceans are very uniform in composition and have been about the same in composition for hundreds of millions of years, marine salt deposits of various ages are much the same in composition the world over, except that many of them lack the potassium and magnesium salts because of incomplete evaporation. Salt deposits formed in inland lakes (Fig. 4-6), however, are much more varied in composition, reflecting differences in the salts supplied from individual drainage basins. Thus the waters of the Caspian Sea, which is really just a huge salt lake, are much richer in sulfate than ocean water. Thick deposits of sodium sulfate have formed in the Karaboghas Gulf, a shallow arm of the sea on its eastern side. In the deposits of various salt lakes we find, in addition to sodium chloride, gypsum, potash salts, and magnesium salts, trona, sodium sulfate, salts of boron, and sometimes lithium

TABLE 4-6 U.S. Production of Certain Saline Minerals, 1983 and 1984 (in Thousands of Tons)

	1983	1984[a]
Salt	32,973	38,750
Sodium carbonate	8,467	8,350
Sodium sulfate	423	530
Boron[b]	637	680
Bromine	185	170
Magnesium compounds	625	650

[a]Estimated.

[b]Boric oxide content of boron minerals and compounds.

Source: U.S. Bureau of Mines, *Mineral Commodity Summaries.*

Figure 4-6 Salt deposits extending across the floor of Death Valley, California, as seen from Dante's View.

salts. The Green River Basin in Wyoming, one of the lake basins in which the great oil shale deposits of the Green River formation were formed, contains thick deposits of trona (sodium carbonate) interbedded with oil shales. They are the world's largest trona deposits. Boron salts were deposited in large amounts in many lakes of Nevada and eastern California. The region of the lakes has been one of much volcanic activity in recent and glacial times, and the boron has been contributed to the lakes from volcanic emanations. Lithium salts may also be present in the brines of interior basins. Those at Silver Peak, Nevada, are an important domestic source of lithium, and far larger reserves of lithium in brine have been proved in the salt deposits of the Atacama desert in northern Chile. The brine of the Dead Sea has been estimated to contain 1 billion tons of bromine. It has become clear that the deposits and brines of salt lakes are among the world's great mineral resources. There are still many saline lake systems that have not been completely explored.

SULFUR

Sulfur is one of the most important, perhaps *the* most important, of all inorganic industrial chemicals. Most industrial products require the use of sulfur in one form or another during their manufacture. Most of the sulfur used is not visible to the consumer, because sulfur is employed largely in processing other materials. About 80 percent of the nearly 14 million tons consumed annually in the United States is converted to sulfuric acid, and 75 percent of that, or 60 percent of total consumption, is used in production of agricultural fertilizers. The rest of the sulfur consumed finds industrial application in the manufacture of tires, paper, plastics, textiles, explosives, and a host of other things.

Sulfur comes from three major sources: deposits of native sulfur, sulfur in petroleum and natural gas, and deposits of sulfides of the metals. Fortunately, taken together, the three sources are widely distributed over the earth, and sulfur is available from them in enormous amounts. Man has only begun to tap the sulfur resources of the earth.

Native sulfur deposits are of two principal kinds—volcanic deposits and deposits formed from gypsum or anhydrite (anhydrous calcium sulfate). Volcanic deposits of native sulfur are numerous in Japan, New Zealand, Central and South America, and Italy—all regions where there has been active volcanism in recent times. The sulfur has formed by oxidation of hydrogen sulfide gas emitted by volcanoes. Volcanic deposits have been worked in Japan and elsewhere in the past but are no longer important sources of production. Resources in them are very large, but costs of producing sulfur from the deposits are too high. On the other hand, native sulfur deposits derived from gypsum or anhydrite supply about one-half of the total sulfur consumed by U.S. industry today and are important sources of sulfur in other countries. We have seen that gypsum deposits formed by evaporation of seawater are very extensive and are found in a number of countries. Sulfur deposits formed from the evaporite deposits, however, are not common. The most important are along the Gulf Coast of Louisiana, Texas, and Mexico and in Sicily, West Texas, Poland, the Ukraine, and Iran.

The native sulfur deposits of Sicily have been worked since the 15th century, but the origin of deposits of this type has only become apparent during the past 40 years. Their origin is a curious one. The deposits have formed from beds of gypsum and anhydrite. Certain bacteria, as part of their metabolism, reduce the sulfate of gypsum or anhydrite to sulfide, and hydrogen sulfide is formed. Oxidation of the sulfide produces native sulfur. Petroleum must be present, because the bacteria require it as a source of energy. Carbon dioxide is produced during the bacterial processes, and this combines with the calcium released by decomposition of gypsum or anhydrite to form calcite—that is, limestone. Crystalline sulfur, sometimes in handsome transparent or translucent crystals, is deposited in cavities in the limestone. The native sulfur deposits of West Texas and the salt domes of the Gulf Coast of the United States and Mexico have a similar origin.

Metal sulfides are found in large deposits in many parts of the world and have been important sources of sulfur for nearly 150 years. The most abundant and most important are the sulfides of iron—pyrite, marcasite, and pyrrhotite—but sulfur is also obtained from sulfides of zinc, copper, lead, and other metals. When sulfides are roasted in air, the sulfur is converted to sulfur dioxide, which can be converted to liquid sulfur dioxide or to sulfuric acid. About 40 percent of total world production of sulfur is from sulfide sources.

Sulfur is present as hydrogen sulfide in natural gas in certain fields and may form as much as 50 percent of the gas produced. Recovery of the sulfur began in the United States in 1944 and has grown enormously here and in Canada. Sulfur is also recovered from crude oils of certain fields and from the tar sands of Alberta. Sulfur from oil and gas now supplies about one-third of the annual U.S. consumption. Much of the sulfur potentially available from crude oil and natural gas elsewhere in the world, however, is lost to the atmosphere, wasted, and a source of air pollution. Sulfur in coal is a very large potential source, but it is not being recovered at present, although a process for recovering it has been developed by the U.S. Bureau of Mines. Sulfur from coal may be important in the future.

There are no estimates of world reserves and resources of sulfur, but they are in the billions of tons. There is no fear of shortage, even if the present waste of sulfur continues.

FLUORSPAR

Fluorspar is the commercial name for the mineral fluorite, a compound of calcium and fluorine. In some deposits it occurs as handsome cubic crystals, colorless, green, blue, violet, or yellow, often with attractive internal color banding that appears in slices of the crystals. Although it is not very hard and fractures easily, vases, cups, figurines, and table tops were carved from it in ancient times. Today it is important because it is the major source of fluorine chemicals that are of great variety and have almost innumerable uses in refrigerants, insecticides, preservatives, ceramic additives, opaque enamels, glassware, and other products. It is used in many industrial processes, as a flux in the manufacture of steel and aluminum, in making abrasives, in making Fiberglas insulation, in glass etching and polishing, and in electroplating, for example. There is a large and growing family of fluorine-bearing organic compounds. Much of the fluorspar consumed is first converted into hydrofluoric acid, which, like sulfuric acid, is a basic industrial chemical.

Fluorspar is thus an essential industrial mineral; there are no substitutes for it in most of its applications. At one time self-sufficient in the mineral, the United States now imports more than 80 percent of its supply, mainly from Mexico and South Africa. Mexico has substantial reserves, but the world's largest reserves are those discovered in recent years in South Africa, which has become an important supplier to the world. In the

United States, with the possible exception of Alaska, prospects for discovery of new major fluorite deposits are poor. The principal U.S. reserves of fluorine are actually in the phosphate rocks of Florida and the Rocky Mountains. Typical phosphate rock contains about 3.5 percent fluorine. Some of this is being recovered as fluosilicic acid, which contributes about 6 percent of the U.S. supply of fluorine. If all the recoverable fluorine were actually produced, it could contribute about 40 percent of annual fluorine consumption. Reserves of recoverable fluorine in phosphate rocks of the United States are in the tens of millions of tons. World reserves of fluorspar have been estimated at about 118 million tons, but estimates are very rough.

TALC AND PYROPHYLLITE

Talc, in the form of talcum powder, is a familiar article in the American household. Soft, unctuous, and chemically inert, it is used both to prevent or soothe minor irritations of the skin and as a component of cosmetic preparations. However, more than 90 percent of the roughly 1 million tons of talc produced and consumed annually in the United States is used for other purposes. About 50 percent is used for fillers and extenders in paint, paper, plastics, and rubber, not just because it is relatively cheap, but because it improves the properties of those products and does not react chemically with substances with which it is mixed. Another 30 percent is used in various kinds of ceramic products including dinnerware, pottery, and wall tile. It promotes translucency and improves resistance to crazing, the development of fine cracks. The rest of the talc produced finds a variety of uses, for talc is a very versatile industrial material. Most talc is used in the ground state, but block talc can be machined into a variety of shapes, including the pencils that are uniquely suited to marking hot metals and making the lines that the tailor uses to indicate alterations needed in our clothes. Talc is harmless to fabrics and easily removed by brushing.

Talc is a hydrous magnesium silicate, whereas pyrophyllite is a hydrous aluminum silicate, but the two minerals have the same atomic structure and are so similar in physical properties that they cannot be distinguished without x-ray or chemical analysis. The two minerals are interchangeable in some of their uses, but about half the roughly 200,000 tons of pyrophyllite produced in the United States is used in making various refractory products, and another 20 percent is used as an inert vehicle for insecticides. In 1981 talc was produced in 11 states, but most of the production came from Vermont, New York, Texas, and Montana. Pyrophyllite deposits are much less common than talc deposits, and in the United States pyrophyllite was mined in 1981 in only two states—North Carolina and

California. Reserves and resources of talc are thought to be large, but there are no firm figures either for the United States or for the rest of the world. The largest reserves of pyrophyllite are apparently in Japan. There is no concern about shortages of these two useful minerals in the near future, but longterm availability (see Chapter 6) is uncertain.

ASBESTOS AND SOME OTHER NONMETALLIC MINERALS

No nonmetallic mineral material has attracted so much unfavorable attention in recent years as have the asbestos minerals. Yet they are a remarkable group of substances. They have high resistance to heat, chemical reagents, and electric currents. They occur in finely fibrous, soft crystals; these can be spun into thread and woven into cloth, or formed into board, pipe, or other shapes. The resulting products are fireproof and highly resistant to friction. No other substance, synthetic or natural, has properties matching those of asbestos. Thousands of uses for it have been found in industry. Unfortunately, the fibrous habit that contributes much to its usefulness is also responsible for its behavior as a carcinogen. It cannot at present be replaced in some of its applications, especially those in friction-resistant materials, but uses that lead to dispersal of the fibers in the air are now being reduced or entirely eliminated. Since 1979 the use of asbestos in the United States has been reduced from 564,000 tons per year to 210,000 tons.

There are actually three asbestos minerals that enter into commerce. The most important is chrysotile asbestos, a hydrous magnesium silicate. The best deposits of chrysotile are in Canada, Zimbabwe, and the Soviet Union; they yield long-fiber asbestos, the most valuable grade. In the United States, asbestos deposits occur in California and Vermont, but they yield only short-fiber asbestos. The other types of asbestos, crocidolite (blue asbestos) and amosite, are produced only in South Africa.

There are many other nonmetallic minerals that play various roles in the processes and products involved in the complex industries of the modern world. There is graphite, natural carbon, used in pencil leads, metallurgical crucibles, brushes for electric motors, and special lubricants. There is the curious family of zeolite minerals. The arrangements of atoms in zeolites are such that their crystals have holes of various sizes. They can be used in refining petroleum, because the holes trap undesirable components. There are the aluminum silicate minerals, used in making crucibles and other ceramics that must be resistant to heat and shock. There are the titanium minerals that yield titanium dioxide, which has a remarkable ability to scatter light. It is used to improve the covering power of paint, so that a single coat will often suffice rather than the two or three that were formerly required. There is diatomite, a rock made up of

Figure 4-7 Diatomaceous earth from Lompoc, California, as seen under the scanning electron microscope. The earth is made of fragments of the siliceous tests of different types of diatoms. Magnified 320 times. Photograph by E. D. Glover.

trillions of the tiny siliceous skeletons of diatoms, one-celled plants that build microscopic structures of incredible complexity and beauty (Fig. 4-7). Diatomite is minutely porous and is a marvelous filtering material, but it has many other uses. There are the mineral materials used in making rock wool, a product that leaped into prominence when the energy crunch began. There are still others; our list is far from complete. The minerals that have been discussed, however, indicate the range and complexity of the use of nonmetallic minerals in industry and the range of their availability to man. We will go no farther here.

SUMMARY: THE NONMETALLIC MINERALS

The broad array of nonmetallic minerals used in industry has been indicated in this rather brief review along with the reasons for their importance as industrial raw materials. The usefulness of these materials has been markedly expanded through research, which has led to a better understanding of their chemical compositions and the reasons for their chemical and physical behavior. Through the development of better processing methods, the quality of many nonmetallic mineral products has been improved. Trial-and-error methods formerly employed in the utilization of many nonmetallic minerals have given way to processes that are closely controlled by reference to clearly defined properties of the minerals. In use, many of the nonmetallic minerals compete with one another; this has provided a stimulus to improvement of processed materials and has assured the consumer of adequate supplies at reasonable cost.

For most of the nonmetallic minerals there is no threat of shortage for many decades to come, and prospects for discovering new deposits of nonmetals and for converting them to reserves are good. From the standpoint of geographic distribution of reserves and resources, we can recognize three general classes of nonmetallic mineral commodities. There are those like common clay, stone, and cement materials, which are very widely distributed over the world and are produced in many countries. There are others, like phosphate, potash, sodium carbonate, and titanium minerals, that are less widely distributed but are still available from a number of sources through international trade. Finally, there are those such as asbestos and fluorspar, for which the world is currently dependent on sources in just a few countries. This situation could change in the future, but the degree of change is unpredictable, because exploration of the nonmetallic mineral resources of the world is still far from complete. There have been many changes in the geographic pattern of availability just since World War II. The case of potash is an example. Prior to 1945, major potash deposits were known only in Germany, France, the Soviet Union, Poland, Spain, and the United States. Since 1945, deposits have been discovered in Great Britain, Thailand, Ethiopia, Brazil, the Congo (Brazzaville), Saskatchewan, and New Brunswick, far more than doubling the world's known reserves. In addition, the magnitude of reserves of potash in the Dead Sea has been recognized. The new discoveries not only have broadened the distribution of reserves but have relieved the world of fear of shortages of potash. The magnitude of reserves and resources of the other saline minerals so vital to world chemical industry has become evident since World War II, but there are regions of saline lakes that are still only partially explored, notably in southern Africa, east central Africa, Asia, and the Andean region of South America. Given the record of mineral exploration during the past 40 years and the knowledge of how much of the earth's crust lies open to further exploration, only a confirmed pessimist can view the future of nonmetallic minerals darkly.

REFERENCES AND SOURCES OF INFORMATION ON NONMETALLIC MINERALS

Alexandersson, G., and Klevebring, B.-J., 1978, *World Resources: Energy and Minerals.* Walter de Gruyter, Berlin, 248 pp.

Bates, R. L., 1960, *Geology of the Industrial Rocks and Minerals.* Harper and Brothers, New York, 441 pp.

Brobst, D. A., and Pratt, W. P., 1973, United States mineral resources. U.S. Geological Survey, Professional Paper 820, 722 pp.

Emigh, G. D., 1972, World phosphate reserves—are there really enough? *Engineering and Mining Journal,* Vol. 173, No. 4, pp. 90–95.

Le Fond, S., ed., 1975, *Industrial Minerals and Rocks,* 4th ed. American Institute of Mining, Metallurgical, and Petroleum Engineers, New York, 1360 pp.

Mason B., 1958, *Principles of Geochemistry.* Wiley, New York, 310 pp.

U.S. Bureau of Mines, 1985, *Mineral Facts and Problems,* in press. U.S. Government Printing Office, Washington, D.C.

U.S. Bureau of Mines, annual publication, *Mineral Commodity Summaries.* U.S. Government Printing Office, Washington, D.C.

5 Mineral Raw Materials

THE METALS

INTRODUCTION

The earliest mineral materials used by man were undoubtedly stone, clay, flint, and other nonmetallic substances. Gold and copper appear to have been discovered around 15,000 B.C., but these and other metals were widely used in the Middle East only after 5000 B.C. Besides copper and gold, tin, lead, silver, mercury, and small amounts of antimony, arsenic, and bismuth came into use. All these metals occurred either as native metals or as minerals of simple composition from which metals could be extracted in primitive charcoal furnaces. Charcoal served both to provide the necessary heat and to reduce the metals to the elemental form. Once the metals had been recovered, they were easily worked and fabricated. For casting, remelting in the charcoal furnace was also easy. By 3000 B.C. tin, copper, gold, silver, and lead were in widespread use. By adding tin to copper, bronze was produced, and it became the prime metal for implements and weapons. It was tougher, harder, and more durable than copper. The smelting of iron ores to produce the metal began at some unknown time prior to 1300 B.C. As iron became available it partly supplanted copper and bronze, especially in weapons, but copper and bronze remained the principal metals in use until the development of the iron cannon in the Middle Ages. Brass, the alloy of copper and zinc, was in use late in the pre-Christian era.

The metallurgy of ancient times was very primitive, and its products were very uneven in quality. The compositions of metallic ores were only crudely understood, and impurities not only were common in the metals produced but they varied in amounts. Metallurgical procedures could not be closely controlled. Progress in metallurgy was made over the centuries, but it was very slow. Modern metallurgy began only in the 18th century. Until 1735, when cobalt was discovered, only 11-metals were known. Between 1735 and 1835, however, there was rapid improvement in analytical methods. The mineral and chemical compositions of ores could now be investigated. Twenty-six more metals were discovered, including the very important metals manganese, chromium, aluminum, nickel, tungsten, calcium, potassium, sodium, and magnesium. The stage was thus set for rapid advances that led in the nineteenth and twentieth centuries to development of the complex metallurgical industries of today. Methods of extracting the newly discovered metals were devised. A fantastic array of alloys of the metals, each with its special properties, came into being. Today every metal present in the rocks of the crust finds some application in industry. The usefulness of the metals to the human race is thereby enormously increased.

In this book we have not tried to cover all the nonmetallic minerals, nor will we try to examine all the metals. We shall, however, consider the major metals used in industry today, their uses and importance, the size and distribution of major reserves and resources, and the problems of availability that metals present. One important difference between metallic and nonmetallic mineral commodities should be pointed out. Many of the nonmetallic materials, such as sand, gravel, stone, clays, salt, and gypsum, are bulk materials that have relatively low unit values. Values at minehead are measured in dollars or tens of dollars per ton. Costs of transport are a substantial factor in costs to the consumer. These materials command internal markets, and domestic products are relatively invulnerable to competition from imports. The metals, however, have high unit values measured in hundreds or thousands of dollars per ton. Transport costs are not an important element in costs to the consumer. The metals are therefore heavily involved in international markets and international trade. Domestic metal-producing industries are highly vulnerable to competition from imported metals. This is reflected in problems that are discussed in the following sections on the metals.

IRON AND STEEL

Iron and steel play a unique role in the industry and civilization of today. Each year more than 100 million tons of steel are consumed in the United States, more than the total of the tonnages of all the other metals. There are three basic reasons. One is that iron occurs in large deposits of minerals, mainly oxides of iron, from which it can be extracted easily and cheaply. The second is that iron and the steels made from it are remarkable structural materials—hard, tough, and structurally strong. The third is that iron and steel are easy to work; they can be cast, forged, machined, rolled, and otherwise fabricated into any form that is desired, from needles to sheets to rails and enormous girders. Man has found no other metal that can be adapted to such an extraordinary range of uses.

As noted earlier, iron was in use by 1300 B.C. By the time America was discovered, nearly all European nations had developed methods of recovering iron from its ores and making steel, and the metals were being used for household articles, tools, implements, and weapons. The first product of the smelting of iron ore was pig iron, produced by heating iron ores with charcoal or, later on, with coke produced from coal. The carbon of charcoal or coke combined with the oxygen of the iron minerals in the ores to produce carbon dioxide, which passed off as a gas. The iron was left behind as the metal. Pig iron could be melted and cast into various shapes. It could also be converted into wrought iron by heating it in a furnace and reworking it; most of the carbon and other impurities present in pig iron were removed in the process. Cast iron tended to be brittle, but wrought iron was ductile and malleable, hence readily worked into various shapes. Treating wrought iron with carbon gave steel, but for a long time steel could be made only in small batches—a hundred pounds at most. The Bessemer converter, introduced in 1856, revolutionized steelmaking, because it would handle batches of 25 tons or more of pig iron. For the first time in history, large-scale production of steel became possible.

The technology of iron and steel has undergone a steady evolution since 1856. Various kinds of cast and wrought iron are made and account for about 15 percent of total iron and steel production. The size of blast furnaces for making pig iron has been enormously increased; furnace outputs range up to 10,000 tons per day. Most of the world's steel is still made by a two-stage process—iron ore to pig iron, pig iron to steel—but processes for direct conversion of iron ore to steel have been developed. In steelmaking, the Bessemer process gave way long ago to the more efficient open hearth process, capable of producing batches of 200 to 300 tons of steel in 6 to 8 hours. The open hearth process has in turn given way to the basic oxygen process, which is now the dominant steelmaking process in the United States.

In this process, a pear-shaped converter lined with refractory brick is charged with molten iron and high-grade scrap iron. A pipe is pushed down into the converter from the top, and pure oxygen is blown at high velocity through the molten metal to oxidize the impurities. Fluxes (fluorspar and silica) are added to form a slag. Manganese is added to scavenge oxygen and sulfur, taking them into the slag. Up to 300 tons of steel can be produced in about 45 minutes. The product of the furnace is carbon steel, containing 0.05 to 1.25 percent carbon. The molten steel is transferred into refractory-lined ladles for pouring into molds or for casting into continuous strands, which are then cut into various shapes for rolling. Rolling produces steel sheet or plate, girders, rails, rods for drawing into wire, and other shapes. Steel can also be poured directly into molds of various shapes to give products requiring a minimum of machining and finishing. Forging is done by hammering or pressing the hot metal. Alloy steels are made separately in electric furnaces.

A description of all the uses of steel would make a book in itself; even a listing would take many pages. Tools, implements, machines, structural materials, household articles, and appliances in almost infinite variety involve the use of steel. Articles made from steel are among the most familiar in our lives.

A modern steel plant is a very large and complex assemblage of facilities, including yards for receiving iron ore and coke, blast furnaces, steel-making furnaces, casting shops, rolling mills, foundries, and electric furnaces, together with necessary offices for management and administration. Coke may be brought into the plant from outside or it may be made from coke ovens at the plant. The total plant represents an investment of hundreds of millions of dollars, and as technology changes, large sums of money must be spent for replacing obsolete equipment. The world steel industry is growing, changing, and highly competitive. Plants in the United States and Europe that dominated the world steel industry after World War II must now compete with more modern plants built in Japan, Brazil, South Korea, and elsewhere. These plants not only are more efficient but have the advantage of lower labor costs. Steel industries are very important to national economies, not only as sources of iron and steel for internal consumption but also as sources of employment and of income from exports of steel and steel products. Steel industries in some countries are therefore supported by government subsidies in one form or another.

In this context, the U.S. steel industry has not fared well in recent years. At the close of World War II, the industry was the largest and strongest in the world. It was the key component in "the arsenal of democracy." In the 1970s, however, deterioration set in owing to several factors—obsolescence of plants, rising labor costs, decline in internal markets due to changes in the automobile industry, and loss of export markets to foreign producers. After World War II, U.S. steel production increased irregularly to a peak of nearly 150 million tons in 1973, but by 1981 it had declined to 120 million tons, and net imports of major iron and steel products had risen to 20 million tons. The Soviet Union became the world's leading producer of steel in 1971, and in 1982 Japan's production of steel exceeded that of the United States for the first time. The recession of 1982 brought general depression to world steel industries, but the American steel industry was by far the most severely affected.

Production of raw steel dropped to 74.6 million tons, a decline of 38 percent from 1981. A recovery to 93 million tons is indicated for 1984, but that is far below the average annual production for the period 1971 to 1981. Some major plants in the United States have been closed. A major reorganization of the U.S. steel industry appears to be under way, but to what degree the industry will recover in the future is uncertain at this time.

The enormous capital investment required for a fully integrated iron and steel plant has encouraged the construction of "minimills" in the United States and certain developing countries. Minimills are small plants in which steel is made from scrap pig iron and scrap steel. Such plants are viable in special markets but obviously are dependent on the major plants ultimately for furnace feed. Nonetheless, their importance is growing, and they are becoming a significant factor in the steel industry of the United States. Low unit costs of labor, location near specific markets, and adaptability to changes in markets give them advantages over the major U.S. steel plants and over competitive imports from other countries.

There is no lack of iron ores in the world. Large deposits of ore are found on every continent of the world except Antarctica. Locations of the principal reserves and major sources of current production are indicated in Table 5-1, but there are deposits in many other countries. Iron ore deposits have formed by a number of geologic processes, but much more than half of the world reserves and other identified resources are found in bedded sedimentary deposits. The most important deposits of this class are banded iron formations of Precambrian age, most of them formed around 2200 million years ago (Fig. 5-1). The formations underlie large areas in the Lake Superior region, eastern Quebec and Labrador, Russia, Western Australia, India, Venezuela, Brazil, Gabon, Manchuria, and South Africa. They are called banded iron formations because typically they consist of layers rich in iron oxides or iron carbonate alternating with layers rich in silica (Fig. 5-1). They were formed by chemical precipitation from water, probably in shallow marine basins but possibly in enormous lakes. The deposits of the iron formations are of two general kinds. One kind consists of unaltered iron formations, the taconites, the best of which contain 25 to 40 percent iron. The other type consists of enriched portions of the iron formations, in which the iron content has been increased either by removal of the original silica or by replacement of the silica by iron introduced by percolating solutions. In either case enrichment is the work of circulating waters. The ores of enriched deposits contain 50 percent to as much as 70 percent iron. Enriched deposits have yielded enormous amounts of ore—more than 3 billion tons in the Lake Superior region alone. The large reserves of Brazil and Western Australia are in enriched deposits containing 50 percent or more iron. The deposit

TABLE 5-1 World Production of Iron Ore and the Reserve Base (in Millions of Tons)

	Production		
	1983	1984[a]	Reserve Base
United States	42.1	57.1	27,800
Australia	81.5	94.1	36,960
Brazil	98.1	112.0	17,470
Canada	37.0	40.2	28,100
France	17.6	16.8	2,460
India	42.8	44.8	7,950
Liberia	16.5	16.8	1,800
South Africa	18.3	23.5	10,400
Sweden	14.6	16.8	5,150
Venezuela	10.8	13.4	2,240
Other market economy countries	64.1	67.2	13,600
Mainland China	78.4[a]	79.5	10,080
Soviet Union	270.0	271.0	66,080
Other communist countries	25.5	25.8	1,000
Totals	817.3	879.0	231,090

[a]Estimated.

Source: U.S. Bureau of Mines, *Mineral Commodity Summaries.*

Figure 5-1 Banded iron formation (taconite) in the wall of the Minntac open pit, Mesabi Range, Minnesota. Dark layers are rich in magnetite (magnetic iron oxide); white layers are silica.

of Carajas, Brazil, found in the remote country south of the Amazon River mouth in 1967, has proved reserves of 15.7 billion tons of ore averaging 66.7 percent iron. It is the largest high-grade iron ore deposit in the world. From Carajas and other enriched deposits in the state of Minas Gerais, Brazil now produces more iron ore than any other country in the Free World. Reserves of high-grade ore in various deposits in Western Australia are reported as 23.6 billion tons. The Soviet Union has very large reserves. Those in South Africa in Precambrian iron formations have been estimated at 5 to 10 billion tons.

Although Precambrian deposits contain the largest reserves and resources of iron, sedimentary iron formations of younger age in Europe, Alabama, and Newfoundland have been important sources of iron ore. Those of western Europe, together with the coal of that region, were the basis of the Industrial Revolution. Of other types of iron ores, the most important currently are the high-grade magnetite deposits of the Kiruna-Gällivare region of northern Sweden. These are related, in a manner not yet firmly established, to volcanic activity, either surficial or subterranean. They have been an important source of high-grade ore to western Europe for many decades and are important to the economy of Sweden.

In the United States by far the largest iron ore deposits are in the Mesabi Iron Range of northern Minnesota, the Marquette Range of northern Michigan, and the Gogebic Range of northern Wisconsin. For decades production from the ranges was from the high-grade ores enriched by leaching of silica from banded iron formation. Some of the ore bodies were of enormous size (Fig. 5-2). By the 1950s, however, most of the

Figure 5-2 Part of the open pit of the Hull-Rust-Mahoning mine, Mesabi Range, Minnesota, which yielded over 500 million tons of high-grade iron ore and was called America's stockpile of iron ore for World War II. A covered conveyor belt was used to transport ore from pit bottom to rim. The giant pit dwarfs the heavy equipment (power shovel and large trucks) used in mining; they are only pinpoints in the center of the photograph.

high-grade ores had been mined out, and production shifted to the taconites. The taconites are mined in huge open pits, some of them miles in length. Taconite is crushed and processed to remove the siliceous impurities. The resulting concentrates, with 66 to 70 percent iron, are formed into pellets to make them suitable for furnace feed. The pelletized concentrates are a superior product owing to their high iron content, their uniform composition, and their physical form. The processes of concentration and pelletizing, however, require extra energy, and with the rise in costs of fuel there has been difficulty in competing with high-grade imported ores. Prospects for the taconite mining industry are not as bright as they were 10 years ago. Iron ore production has declined from 98 million tons in 1973 to a low of 42 million tons in 1983. The United States relies increasingly on imported ores as well as on imported iron and steel.

THE FERROALLOY METALS

A modern steel industry does not manufacture just iron and simple steel. It produces a broad range of alloy steels created by adding one or more other metals in carefully controlled amounts. The metals are added for a variety of purposes—to remove impurities, improve hardenability, improve corrosion resistance, produce alloys with special magnetic or electrical properties, improve resistance to high temperatures, improve machinability, increase the strength of rolled products, control grain size, and improve toughness and weldability. Although sulfur, selenium, and phosphorus are added to certain steels to improve machinability, principal additives are metals, and these are known as the ferroalloy metals. The most important are manganese, silicon, chromium, nickel, vanadium, molybdenum, tungsten, cobalt, and titanium. It is worth noting that prior to 1800, none of these metals was used in steelmaking except accidentally as unrecognized impurities that might or might not improve the properties of steel. The deliberate, controlled use of ferroalloy metals is the hallmark of the modern age of steel; actually it is the age of alloy steels. Without the alloy metals, the steel industry as we know it today would simply not exist. The availability of the ferroalloy metals is therefore of intense concern to every steel-making country in the world. It is a particular problem for the United States, because only three of the metals can be obtained in adequate amounts from domestic ores.

The ferroalloy metals are not used solely in making alloys with iron and steel. They enter into many alloys that contain no iron, and they enter into innumerable compounds used in industry. The pure metals also have their uses. It is impossible to indicate fully, within the scope of this book, the role of the ferroalloy metals and their compounds in modern civilization, but something of the range of uses will be examined in discussing the individual metals.

We turn now to the question of availability of this remarkable group of elements, examining the nature, size, and distribution of reserves and resources, and the important sources of current production. The ferroalloy metals present some special problems of supply. A note of caution is necessary in considering estimates of reserves and identified resources (the reserve base) of the metals. Data for the ferroalloy metals are incomplete and uneven in quality. Excellent estimates are available for some individual deposits but not for others. In general, therefore, the estimates should be taken to indicate only the order of magnitude of the world's reserve base.

Manganese

Manganese is used mainly to scavenge sulfur and oxygen from steel, but it is also useful as an alloying element, imparting strength, hardness, and hardenability to steel. Up to 14 pounds of manganese is consumed for every ton of steel produced. No substitute for it has ever been found. Manganese ores for steelmaking are in large part smelted with iron to produce various grades of ferromanganese containing 74 to 90 percent manganese, or to produce spiegeleisen, with 16 to 23 percent manganese. One alloy or the other is then fed to steel-making furnaces in carefully controlled amounts. Some manganese is supplied in the form of silicomanganese, a name applied to alloys of silicon and manganese containing 65 to 68 percent manganese. There are many useful manganese chemicals. Manganese oxides are essential components of standard dry batteries.

Manganese ores as marketed commonly contain 46 to 50 percent of the metal in the form of oxides, and large reserves of ores that can be concentrated to such grades are found in Australia, Brazil, South Africa, Gabon, and India (Table 5-2). The Soviet Union has large deposits of lower-grade ores that are beneficiated to marketable grade; Russian production is currently the largest in the world. The only sizable manganese deposits in North America are in Mexico. They yield manganese carbonate ores of low grades, but roasting the ore to drive off carbon dioxide gives a product containing about 42 percent manganese.

As indicated in Table 5-2, the United States produces no manganese ore at the present time. Small high-grade deposits, chiefly in Georgia, Arkansas, Montana, Nevada, and Arizona, were worked and depleted in the past. Intensive searches for manganese in the United States during the two world wars failed to discover large, high-grade deposits. U.S. identified resources of manganese are substantial but too low in grade to be workable except at prices 5 to 20 times the present price of manganese ore, according to a recent study (Kilgore and Thomas, 1982). The United States is dependent on imports, and so are the nations of western Europe. The problem of manganese supply is thus not one of limited reserves but of their very restricted geographic distribution.

TABLE 5-2 Major World Sources and Reserve Base of Manganese Ore (in Thousands of Tons)

	Mine Production			
	1983	1984	Reserve Base	Percent Mn in Ore
Australia	1,491	1,800	470,000	35
Brazil	2,300[a]	2,300	142,750	27–46
Gabon	2,047	2,300	242,500	44
Ghana	~500	no data	20,000	30
Hungary	~170	no data	215,000	20.5
India	1,455	1,400	106,000	<38–46
Mexico	~900	no data	29,000[b]	27
South Africa	3,181	2,700	396,000[c]	>40
Soviet Union	11,500[a]	11,600	2,015,000	8.5–24.5
China	1,760[a]	1,800	734,000	6.0–50.0

[a]Estimated.

[b]Incompletely evaluated. Much larger reserves may be present.

[c]Ore with less than 40 percent Mn estimated at 13.5 billion tons.

Sources: Production data from U.S. Bureau of Mines, *Mineral Commodity Summaries;* data for reserve base from DeYoung et al. (1984).

The manganese supply problem of the United States has changed since World War II. At that time the United States was the world's principal producer of ferromanganese, and its supply problem was simply that of obtaining manganese ores. More than half the total supply of manganese is now obtained in the form of ferromanganese manufactured in other countries. The U.S. ferromanganese industry has correspondingly declined. South Africa contributes 43 percent of the total supply of the alloy as well as 33 percent of the manganese ore still used directly in the United States.

Silicon

In metallurgy, the most important use of silicon is for the production of various grades of ferrosilicon, alloys of iron and silicon in which the silicon content ranges from 25 to 95 percent. They are used in the steel industry to remove oxygen from the molten metal and as a constituent of alloy cast irons and alloy steels. The second most important use is in the production of silicon metal, which is used mainly in alloys with aluminum but is also used in alloys with copper and nickel, in the production of silicon chemicals, and silicon carbide abrasives. Silicon of high purity is critical to the manufacture of semiconductors for use in computers and related devices. It is an essential component of photovoltaic cells, which convert solar energy to electrical energy.

TABLE 5-3 World Production of Silicon (in Thousands of Short Tons)

	Production	
	1983	1984[a]
United States	333	440
Brazil	140	150
Canada	102	100
France	167	170
Italy	63	60
Japan	128	130
Norway	367	380
South Africa	106	110
Spain	72	70
Other market economy countries	339	350
Soviet Union	549	550
Other centrally planned economies	430	440
Totals	2796	2950

[a]Estimated.

Source: U.S. Bureau of Mines, *Mineral Commodity Summaries.*

The major sources of silicon are high-purity sands and quartzites, which are ancient sands that have been solidified by heat and pressure in the earth's crust. Deposits are available at a number of places in the United States and the rest of the world. U.S. and world reserves are so large that no one has bothered to estimate them. As we have seen, high-purity silica sands are used for glassmaking and other purposes; no separate figures are available for sands used as sources of silicon and ferrosilicon. Production data for silicon are given in Table 5-3. Although U.S. reserves of silicon are very large, for years about one-sixth of the U.S. supply of silicon has come from abroad. The production of silicon metal is energy-intensive, and the availability of cheap hydroelectric power has led to sizable production in Canada and Norway, the principal sources of U.S. imports.

Chromium

Chromium metal is a latecomer to the industrial world. The element was not discovered until 1797, and until the early 1900s it was used mostly in the form of chromium chemicals. However, its use in metallurgy has grown steadily during the 20th century and at present accounts for more than half the total chromium consumed in the United States. The rest is used in chromium chemicals and in refractory products (bricks, e.g.) for metallurgical furnaces.

The world's supply of chromium metal and compounds is obtained entirely from deposits of the mineral chromite, which is a complex compound of chromium, iron, magnesium, and aluminum with oxygen. Its black color gives no hint of the bright, almost silvery metal that is obtained from it. The chromium content of chromite, generally expressed as its content of chromic oxide (Cr_2O_3), varies from deposit to deposit, from about 22 percent to as much as 60 percent. The contents of the other metals also vary. In the past, the variations in composition were strongly reflected in patterns of use. Chromite high in chromium and low in iron was preferred for metallurgical use, chromite high in iron was used as a source of chemicals, and chromite high in aluminum was used for brick and other refractory products for metallurgical furnaces. Advances in technology have made the three grades of chromite interchangeable to a considerable extent, except that chromite high in aluminum is less desirable for chemical or metallurgical purposes, being relatively low in chromium. The change is important, because it has made it possible to use South African ore, formerly used only for chemical purposes, to make ferrochromium. South Africa now rivals the Soviet Union as a source of chromite.

Most of the chromite used in metallurgy is smelted and converted to ferrochromium, a group of alloys of iron and chromium containing 50 to 75 percent of the latter metal. Ferrochromium is then added to steel in controlled amounts. The most important chromium alloys are the family of stainless steels, which have 12 to 30 percent chromium and have high resistance to rusting and corrosion, but chromium is also used in many alloys containing other metals. High-speed tool steels, for example, contain chromium, tungsten, vanadium, and sometimes molybdenum. Chromium is also added to certain cast irons. Coating steel with chromium protects it from rusting and gives the bright appearance attractive in such items as the trim on automobiles. In metallurgy it is used to enhance such properties as hardenability, strength, and resistance to corrosion and abrasion. The varied uses of chromium make it one of the essential elements in modern industry.

Deposits of chromite are found on all continents, but most deposits are small. The United States has hundreds of deposits, in the Appalachians, the Coast Ranges of California, the Sierra Nevada, the mountains of western Washington and Oregon, and Alaska. Numerous deposits have been worked on a small scale from time to time, mostly at times when prices of chromite have been abnormally high. Remaining identified resources are mostly low in grade. They cannot compete with large, high-grade deposits abroad (Lemons et al., 1982). The only sizable identified domestic resources of chromite (about 10 million tons) are in the Stillwater Complex of southwestern Montana. They are not competitive with foreign ores, and no chromite is being produced in the United States at present.

TABLE 5-4 World Production and Reserve Base of Chromite (in Thousands of Tons)

	Production		
	1983	1984[a]	Reserve Base
Brazil	310	350	10,000
Finland	375	400	32,000
India	400	450	66,000
Philippines	365	400	32,000
South Africa	2460	2500	6,300,000
Turkey	440	450	80,000
Zimbabwe	475	500	830,000
Other market economies	351	400	25,000
Albania	990	1000	22,000
Soviet Union	2700	2700	142,000
Other centrally planned economies	55	60	4,000
World totals	8921	9210	7,540,000[b]

[a]Estimated.

[b]Rounded.

Source: U.S. Bureau of Mines, *Mineral Commodity Summaries.*

World production and reserve bases of chromite in various countries are shown in Table 5-4. Although there is currently important production from the Soviet Union, Turkey, the Philippines, and elsewhere, there is a remarkable concentration of reserves in South Africa and Zimbabwe—95 percent of the total world reserve base. Those two countries will ultimately be the sources of most of the world's chromium supply.

The world's economic chromite deposits are all contained in bodies of igneous rock formed in the crust by crystallization of molten magma. Most of the deposits are lenses or layers of limited extent. In contrast, the major deposits of the Great Dyke of Zimbabwe and the even larger deposits of the Bushveld Complex of South Africa are layers that extend for miles in outcrop. The Steelpoort chromite seam (Fig. 5-3), one of a series of chromite deposits in the Bushveld Complex, has been traced for about 40 miles. It probably contains more chromite than all the deposits of the rest of the world outside South Africa and Zimbabwe.

In U.S. industry, there is a parallel between manganese and chromium. For many decades the United States was the world's leading producer of chromium metal and ferrochromium, importing mainly chromite ores. Now most ferrochromium is produced in the chromite-producing countries, particularly the Soviet Union, South Africa, and Zimbabwe. Imports of ferrochromium now account for about half of U.S. imports of chromium-bearing materials. The remainder is chromite used largely in making refractories and chemicals.

Figure 5-3 The dark layer is the Steelpoort chromite seam described in the text. At the extreme right, a few feet above the Steelpoort seam, is the thinner Leader seam; erosion has removed most of it here.

Nickel

Nickel is another very useful metal. It finds extensive use in the form of corrosion-resistant alloys in equipment for chemical manufacturing plants and petroleum refineries. Nickel is one of the constituents of stainless steels, but there are many other alloys with steel and other metals. In many of these it serves to improve toughness, strength, hardness, and resistance to heat and corrosion. Special alloys having good strength at high temperatures are used for aircraft engines, superchargers, and jet engines. Nickel-bearing alloys have many other uses in airframes, motor vehicles, electrical and other machinery, household appliances, structural elements in buildings, and armor plate for warships and tanks. It is used in nickel-cadmium batteries and in many compounds employed in chemical industry. Like chromium it is a versatile industrial material.

Nickel is obtained almost entirely from two kinds of deposits, laterite deposits and sulfide deposits. Laterite deposits are found in certain mountain belts, such as the Appalachians; the belt extending across Europe through Albania, Greece, Turkey, and Iran; the Ural Mountains in the Soviet Union, and the belt extending from Guatemala across Cuba to the Dominican Republic. Another belt along the southwest margin of the Pacific Ocean includes the great deposits of New Caledonia, productive for more than 100 years. Important deposits are now being developed in Brazil. Scattered along the belts are bodies of rock called peridotite, which consists of iron-magnesium silicates containing a few tenths of a percent of nickel. Peridotite is not ore; costs of extraction of the nickel would be far too high. Some of the peridotite bodies, however, occur in humid tropical regions and have been subjected to the leaching action of rainwaters percolating downward from the surface. In this situation, iron in the silicates is oxidized and converted to iron oxides. Silica, magnesium, and nickel are leached out, carried away in part, but in part reprecipitated lower down in the peridotite body. The final result is an upper zone greatly enriched in iron (potential iron ore), a blanketlike deposit. Nickel is enriched both in the lower part of the blanket and in less weathered peridotite beneath it. The nickel-enriched material (1 to as much as 3 percent nickel) is ore. Thus arise the lateritic iron and nickel deposits of the world. They contain billions of tons of material with 40 to 50 percent iron. They are not mined for iron at present, because they contain small amounts of chromium, nickel, and cobalt that make them less desirable than ores that lack those elements. They are, however, important resources of iron for the future. The nickel-rich materials, on the contrary, are important sources of nickel supply today.

The other important class of nickel deposits consists of concentrations of sulfides of iron, copper, and nickel in certain bodies of igneous rocks. Economic deposits of this class are relatively rare. By far the largest reserves of nickel in ores of this class are in the Sudbury district in Ontario and the Thompson district in Manitoba, but the Soviet Union and Australia have substantial reserves in sulfide ores.

From 1875 to 1905, the laterite deposits of New Caledonia were the principal world source of nickel and were the basis of a monopoly of world nickel supply. By 1905, however, the great sulfide deposits of the Sudbury district had been developed, and Canada became the world's dominant producer, remaining so until the 1960s. Since that time production from other countries has increased, as shown in Table 5-5. In 1982 and 1983, Russian production exceeded that of Canada. Since 1960 production from laterite deposits has increased and now accounts for about 40 percent of world annual supply.

The United States has no major economic nickel deposits. The only nickel mine in the country, in a lateritic deposit at Riddle, Oregon, has produced only about 8 percent of annual U.S. requirements for new metal. A body of igneous rock northeast of Duluth, Minnesota, contains sulfide deposits estimated to have about 13 million tons of nickel, with some copper, but the ore is low in grade. Extensive exploration and testing of the deposits have been done, but whether and when the deposits can be mined economically is uncertain at this time. Smaller subeconomic resources are known in the Stillwater Complex of Montana.

The United States is heavily dependent on imports for adequate supplies of nickel; so are its allies in western Europe. The United States can no longer depend on Canada for most of its supply, as it did in the past,

TABLE 5-5 World Mine Production and Reserve Base of Nickel (in Tons)

	Production		Reserve Base
	1983	1984[a]	
United States	—	7,500	2,800,000[b]
Australia	99,200	110,000	5,300,000
Botswana	19,300	20,000	500,000
Canada	134,300	148,000	14,800,000
Indonesia	51,400	52,000	5,800,000
New Caledonia	69,400	45,000	17,000,000
Philippines	20,900	10,000	5,100,000
South Africa	22,600	23,000	2,900,000
Other market economy countries	80,972	85,000	19,900,000
Cuba	41,170	42,000	25,000,000
Soviet Union	187,000	190,000	8,100,000
Other centrally planned economies	33,100	35,000	4,000,000
World totals (may be rounded)	759,342	768,000	111,000,000

[a]Estimated.

[b]Mostly uneconomic material; see text.

Source: U.S. Bureau of Mines, *Mineral Commodity Summaries.*

but the increase in the number of sources of nickel since World War II enhances the prospects for uninterrupted availability of the metal.

Vanadium

The principal use of vanadium is in alloy steels to improve strength. U.S. annual consumption has ranged from 6500 to 10,000 tons in recent years. Sources and resources of vanadium are not easy to define. Nearly half the world's reserve base is in South Africa, in deposits in the upper part of the Bushveld Complex. Like the chromite deposits, the vanadium deposits are layers (Fig. 5-4) that can be followed for tens of miles. Large reserves of vanadium are attributed to the Soviet Union but are not easily verified, although Russian annual production must certainly be based on substantial reserves. There are large deposits in Australia. U.S. reserves include vanadium in uranium-vanadium deposits of the Colorado Plateau area and vanadium present in phosphate rock in the Rocky Mountain belt; some vanadium is recovered annually there as a by-product of phosphate production. However, U.S. domestic production comes partly from processing petroleum residues, ash from coal-burning utility plants, and imported iron slags, for none of which are reserve figures available. Reliance on imports is not especially significant, since production from domestic sources could be increased if necessary, and vanadium can be

Figure 5-4 The black layer covering part of the floor of the gulch, together with the thick black layer above it at the right, forms the 8-foot-thick Main Magnetite Seam of the Eastern Bushveld Complex, South Africa. The seam consists of vanadium-bearing iron-titanium oxide. The continuation of the seam can be seen in the left bank, resting on its floor of lighter-colored rock.

replaced by molybdenum and other metals to a considerable extent. In recent years, reliance on imports has ranged between 25 and 40 percent of annual consumption.

Molybdenum

About 75 percent of the molybdenum produced in the United States is consumed by the steel industry in making alloy steels and alloy cast irons to which it imparts toughness, hardenability, strength, and resistance to corrosion. Such alloys are used in transportation equipment, industrial machinery, tubing for oil and gas pipelines, equipment for oil and gas refineries, boilers, condensers, and other industrial equipment.

U.S. and world production and reserve base of molybdenum are shown in Table 5-6. For many decades the United States has been the world's leading producer of molybdenum and an important supplier to other countries. U.S. reserves are large, and additional deposits have been discovered in recent years in the western states and Alaska. One of these, the Quartz Hill deposit in southeastern Alaska, may be the world's largest deposit of molybdenum. The major molybdenum deposits belong in the class of porphyry deposits, bodies of granitic rock through which metal sulfides are disseminated. Large-scale, mass mining of the deposits is the rule. Some porphyry deposits, like the great ore bodies of the Climax

TABLE 5-6 World Mine Production and Reserve Base of Molybdenum (in Tons)

	Production		Reserve Base
	1983	1984[a]	
United States	16,975	49,500	5,900,000
Canada (shipping)	11,550	9,050	1,000,000
Chile	17,550	17,650	2,700,000
Mexico	5,850	6,650	250,000
Peru	2,900	3,500	250,000
Other market economy countries	590	650	290,000
Centrally planned economies	14,350	15,000	2,585,000
Total (may be rounded)	69,000	102,000	12,975,000

[a]Estimated.

Source: U.S. Bureau of Mines, *Mineral Commodity Summaries.*

and Henderson mines in Colorado, yield only molybdenum, with a little tungsten. Others, like those at Bingham, Utah, are mined mainly for copper but yield molybdenum as a by-product. Certain copper deposits in British Columbia, Chile, Peru, and Mexico are of this type. World reserves of molybdenum are adequate to meet requirements well into the next century, and additions to reserves through future exploration are very likely.

Tungsten

Only a few thousand tons of tungsten a year is used in the United States. It is, however, a most essential metal for cutting tools of various kinds. It is an essential metal for defense industries. Tungsten-bearing steels are used in high-speed drilling, cutting, and shaping, because they have great hardness at high temperatures and are resistant to wear and corrosion. Tungsten carbides are widely used in machine tools and in drill bits used in mining.

The U.S. reserve base of tungsten consists mostly of low-grade ores, and production from them is very sensitive to fluctuations in world price. During the period 1969 to 1982, world price, adjusted for inflation, ranged from $24 to $85 per short ton unit, one short ton unit being equal to 20 pounds of tungsten trioxide or 16 pounds of tungsten. U.S. production during the same period ranged from 1740 to 4075 tons per year, whereas consumption ranged from 6624 to 11,828 tons. In recent years the United States has relied on imports for 40 to 58 percent of its supply. China is not only the largest producer of tungsten but is thought to hold the world's largest reserves. Substantial amounts of tungsten are recovered as scrap, and secondary tungsten has amounted to as much as one-third of the U.S. annual supply.

The world tungsten reserve base may be large enough to meet demand for many decades, even though world consumption of tungsten has been rising rather steadily. However, the data on which the tungsten reserve base is calculated are uneven in quality, so the estimates are not firm. Offsetting the uncertainty, however, is the likelihood that future exploration will add substantially to the reserve base. The United States is likely to rely heavily on imports, because domestic production adequate to satisfy domestic demand can only be achieved at high cost.

Cobalt

Only 34 thousand tons of cobalt is used annually in the world, and less than 8000 tons has been used annually in the United States in recent years. It is, however, an essential component of an important group of alloys to which it imparts hardness, toughness, and resistance to abrasion, corrosion, and heat. It is a component of high-speed steels used in machining iron, steel, and nonferrous metals. Its most critical uses are in alloys used in jet engines, because they resist high temperatures, but alloys of cobalt with aluminum, nickel, or rare earths make strong permanent magnets for use in motors and other electrical devices. It would be very difficult to replace.

Most of the world's cobalt is a by-product obtained from ores of nickel and copper. Cobalt is obtained from both laterite and sulfide ores of nickel, but about 65 percent of world production is a by-product of copper ores mined in Zambia and Zaire. No cobalt is mined in the United States; resources are substantial but are low-grade and metallurgically difficult. Cobalt cannot currently be produced from them at acceptable costs.

THE NONFERROUS METALS

The nonferrous metals are those that are used in metallurgy for the most part independently of steel. The most important members of the group are aluminum, copper, lead, zinc, tin, antimony, bismuth, arsenic, magnesium, and mercury. All these metals, except for aluminum and magnesium, occur in minerals from which they can be recovered by simple metallurgical methods and have therefore been in use since ancient times. In modern times their usefulness has been extended by the development of an array of alloys having a wide range of properties and diversified applications in industry. In addition, each of them is used in manufacturing compounds that have various uses in chemical industry.

Copper

As indicated earlier, copper has a long history of service to man, in tools, implements, weapons, and products of the arts and crafts. In many of its ancient uses it has been supplanted by other metals, but decrease in these uses has been more than offset by the large requirements for copper arising from development of the electrical industry. Copper is needed for wiring, armatures, and coils of motors and generators, switches, bus bars, transformers, industrial controls, and other electrical and electronic devices. Its key property is its high electrical conductivity, but its ductility, malleability, and resistance to corrosion are also essential properties. In construction, it is the best metal for roofing and plumbing. It enters into numerous alloys such as brass and bronze. It has wide applications in all the transportation industries.

Copper deposits have formed in the crust by a number of different geologic processes, but there are three major geologic types of deposits—porphyry, "stratabound," and volcanogenic. Porphyry copper deposits were mentioned in connection with molybdenum. Typically these are bodies of granitic rock that have been shattered and veined or impregnated with sulfides of iron, copper, and sometimes molybdenum and other metals. Grades of deposits currently being mined range from 0.35 percent to 3.0 percent copper. Tonnages of ore in various deposits range from tens of millions of tons to more than a billion tons. There are numerous deposits in the great mountain belt that extends from Alaska and British Columbia down through our western states, through Mexico and Central America, and, as the Andean ranges, along the western side of South America. Deposits of this type are now known in mountain belts of other continents. The porphyry copper deposits are among the great mineral treasures of the world. They lend themselves to highly efficient mass mining and recovery of copper from the ores. "Stratabound" deposits are beds of sedimentary rock that contain enough copper to be mined economically. This group also includes some very large deposits, the greatest of which are the deposits of Zambia and Zaire. A deposit of this type near White Pine, in northern Michigan, has been worked for 30 years. Volcanogenic deposits, the so-called "massive sulfide" deposits, have formed as a consequence of submarine volcanic activity. They are distributed widely over the earth in volcanic belts of many ages. They are not as large as deposits of the other two groups, but some of them have been richly productive of copper as well as zinc, lead, silver, and gold (Fig. 5-5). To the geologist they are especially exciting, because they illustrate so well the importance of full understanding of the processes by which mineral deposits have formed in the crust of the earth. Certain deposits of the volcanogenic class have been known for a very long time,

Figure 5-5 The dark material, consisting of layers rich in zinc sulfide, is part of the highly productive volcanogenic massive sulfide deposit of Kidd Creek, Ontario. The zinc-rich zone is underlain by a zone rich in iron-copper sulfide (not shown).

but their origin was not really understood until the 1950s and 1960s. Recognition that they are directly related to submarine volcanic processes has led to a highly successful search for other deposits in the volcanic belts of the world.

The story of exploration and development of copper deposits since World War II is one of remarkable success, both in discovering major deposits and in improving methods of mining and processing copper ores. In 1933, world copper reserves were estimated at about 100 million tons of contained metal. By 1964, reserves were estimated at 212 million tons. Since then about 135 million tons of copper has been produced from the world's copper mines, yet reserves are now estimated at over 500 million tons, and prospects for discovery of additional deposits are very bright. Since 1964, the real price of copper has actually decreased, although lower grades of ore are now being mined.

Table 5-7 shows data for U.S. and world production and reserve base of copper. For more than 80 years, the United States was the world's leading producer of copper. In recent years, however, the industry has fallen on hard times and is struggling to maintain its competitive position. U.S. reliance on imports has increased, averaging about 15 percent in recent years. Chile has replaced the United States as the world's principal producer of copper. There are various reasons for the state of the domestic copper industry—declining grades of ore, high labor costs, high costs of pollution control, and competition with subsidized foreign production. The future of the domestic copper mining and smelting industry is in doubt.

TABLE 5-7 World Mine Production and Reserve Base of Copper (in Thousands of Tons)

	Production		
	1983	1984[a]	Reserve Base
United States	1144	1157	99,200
Australia	282	276	17,600
Canada	689	689	35,300
Chile	1386	1378	106,900
Peru	355	408	35,300
Philippines	301	276	19,800
Zaire	590	579	33,100
Zambia	599	595	37,500
Poland	419	419	14,300
Soviet Union	1102[a]	1102	39,700
Other countries	2000	2072	121,200
World totals	8867	8951	559,900

[a]Estimated.

Source: U.S. Bureau of Mines, *Mineral Commodity Summaries.*

Aluminum

Aluminum is both the newcomer and the upstart among the nonferrous metals. It forms 8 percent of the earth's crust and is thus the most abundant of the true metals, but the ancients did not know of its existence. It was identified only in 1825, and only in 1885 was a process developed by which it could be extracted commercially from ores. Since 1895, however, aluminum production has steadily increased, and it has displaced copper as the second most important metal in tonnage used annually, both in the United States and in the rest of the world. It has very desirable properties: it is light in weight, resistant to corrosion, and pound for pound is as good an electrical conductor as copper. Pure aluminum is very soft, but combined with other metals it forms alloys that are much harder and tougher than the pure metal. Aluminum has replaced both iron and copper in many of their important uses. Power transmission lines are now made mostly of aluminum. It finds extensive use in the building industries, in aircraft and automobiles, in packaging (it is non-toxic), in miscellaneous consumer goods, and in chemicals.

Most of the aluminum in the crustal rocks is bound up in silicates, from which it can be extracted only at high cost. The world's aluminum supply comes almost entirely from deposits of bauxite, which consists of hydrated oxides of aluminum. Bauxite is formed where bodies of silicate rocks are subjected to intensive weathering in humid tropical climates.

The conditions required are the same as those that produce the lateritic iron and nickel ores; in fact, bauxites are simply aluminum-rich laterites. In forming bauxite, nature has performed for us the energy-consuming task of separating aluminum oxide from silica. Like the lateritic iron ore deposits, bauxite deposits are generally blanketlike. Most of them can be mined cheaply in large-scale open-pit operations.

The raw bauxite must first be heated (calcined) to drive off water. It is then treated in huge vats with a strong, hot solution of caustic soda. The aluminum oxide in the ore is dissolved, the solution drained off, and the aluminum precipitated as pure aluminum hydroxide. This is in turn calcined to give pure aluminum oxide, alumina. The alumina is then reduced to the metal in huge pots containing a molten bath of sodium aluminum fluoride. Carbon, supplied mainly in the form of coke, removes the oxygen; molten aluminum is left. Electrical energy drives the process. From 5 to 6 tons of bauxite are required for each ton of aluminum metal produced. To reduce alumina to the metal, 1300 pounds of coke and 14,000 to 16,000 kilowatt hours of electrical energy per ton of metal are needed. The total energy required to produce a ton of aluminum metal from bauxite is 244 million Btu, the energy equivalent of about 17 tons of good bituminous coal. In contrast, the energy required to produce a ton of steel from iron ore is about 25 million Btu. The availability of relatively cheap electrical power is therefore of critical importance to the location of an aluminum-smelting industry. If ore can be brought to the smelter by water transport, it is far cheaper to ship ore than to supply energy to mining sites and smelt the ores where they are mined. This explains why Canada and Norway are important producers of aluminum; both have cheap hydroelectric power close to seaports. In the United States, there is a concentration of aluminum smelters in the region of the Tennessee Valley Authority and in the Pacific Northwest, where power is supplied by the Bonneville Power Administration.

The only sizable bauxite deposits in the United States are near Little Rock, Arkansas. They have been substantially depleted and now furnish only about 5 percent of U.S. bauxite requirements. The rest of U.S. supply of aluminum raw materials comes from imports. For many decades the United States has been the world's leading producer of aluminum, but the structure of U.S. industry is changing. Until the 1960s, raw material consisted solely of bauxite, which was converted to alumina and then to aluminum in American plants. Since then, however, imports of alumina have increased, because more and more alumina is being produced in the bauxite-producing countries. As aluminum smelters are built in the bauxite-producing countries, the United States is becoming an importer of the metal.

The aluminum industry is a critical component of the defense industries of the United States, and there has been much concern over the increased

TABLE 5-8 World Production of Aluminum (in Thousands of Tons)

	1983	1984[a]
United States	3,696	4,442
Australia	524	783
Canada	1,203	1,356
West Germany	805	827
Norway	789	860
China	420[a]	419
Soviet Union	2,200	2,315
Other[b]	5,647	5,721
Totals	15,284	16,723

[a]Estimated.

[b]Other major producers include Japan, Italy, France, and Indonesia.

Source: U.S. Bureau of Mines, *Mineral Commodity Summaries*.

dependence of the country on foreign sources, not only for aluminum raw materials, but even for aluminum metal. The United States has very large resources of aluminum-bearing, nonbauxitic materials, and processes for extracting alumina from various materials have been developed. The kaolin deposits of Georgia and South Carolina, for example, are a huge resource of aluminum. A recent study (Peterson et al., 1981), however, indicates that significant production of alumina from these materials could only be achieved at twice the present cost of producing alumina from bauxite. The alternate materials, including kaolin, are mostly silicate materials, and much of the added cost would be the energy cost of separating alumina from silica.

The energy cost of producing aluminum highlights the importance of recycling the metal. In effect, 90 percent of the energy required to produce the metal originally is also recycled. At present about 17 percent of the U.S. supply of aluminum consists of aluminum cans and other scrap. This is an important contribution.

Table 5-8 gives data for production of aluminum, and Table 5-9 gives data for production and the reserve base of bauxite. World reserves are enormous and adequate for many decades to come. Comparison of the two tables indicates that whereas 70 percent of world supply of aluminum metal is produced by industrialized countries, bauxite reserves and production are mainly in underdeveloped countries. This is an unstable situation, and as time goes on, aluminum production is gradually shifting to the countries that have large bauxite reserves.

TABLE 5-9 World Production and Reserve Base of Bauxite (in Thousands of Dry Tons)

	Production		
	1983	1984[a]	Bauxite Reserve Base
United States	748	981	44,000
Australia	17,006	27,558	5,070,000
Brazil	7,716	8,267	2,500,000
Greece	3,197	3,527	720,000
Guinea	12,213	13,228	6,500,000
Guyana	1,974	2,094	992,000
India	2,120	2,094	1,320,000
Jamaica	8,047	7,606	2,200,000
Suriname	1,929	1,874	660,000
Hungary	3,215	3,200	330,000
Yugoslavia	3,858	3,968	440,000
Soviet Union	5,071	5,071	330,000
Other countries	6,698	385	3,400,000
Totals	73,792	86,853	23,506,000

[a]Estimated.

Source: U.S. Bureau of Mines, *Mineral Commodity Summaries.*

Lead

More than half the lead consumed annually in the United States goes into storage batteries. The remainder has a variety of applications in construction, ammunition, electrical devices, gasoline additives, ceramics, glass, and type metal. The Romans used it for plumbing, leading to much speculation as to its effect on their life-spans. Its use in paints has been discontinued in the United States owing to toxicity problems, and its use in tetraethyl lead, the antiknock additive for gasolines, is being phased out owing to pollution problems. Substitution of plastics has reduced or eliminated some of the uses of lead, and consumption of lead in the United States has been slowly declining since 1973. However, the United States still uses about a million tons of lead per year. There is no substitute for it in storage batteries. More than 90 percent of the lead in storage batteries is recycled, and because of this and the recovery of other scrap, scrap constitutes about half of the U.S. annual supply.

As indicated in Table 5-10, the U.S. and world reserve bases are large. Prospects for additions to reserves are good. Major discoveries of lead deposits have been made in Alaska and western Canada in recent years. Lead deposits are rather widely scattered over the world. There was significant production of lead from 42 countries in 1980. Half the world pro-

TABLE 5-10 World Mine Production and Reserve Base of Lead (in Thousands of Tons)

	Production		
	1983	1984[a]	Reserve Base
United States	495	375	26,760
Australia	526	474	30,860
Canada	278	331	18,740
Mexico	201	209	4,410
Morocco	112	110	2,200
Peru	226	220	3,310
South Africa	88	88	5,510
Yugoslavia	132	110	5,510
Other market economy countries	661	683	17,640
Centrally planned economies	973	981	30,860
Totals	3692	3581	145,800

[a]Estimated.

Source: U.S. Bureau of Mines, *Mineral Commodity Summaries.*

duction in that year, however, came from four countries—Canada, the United States, the Soviet Union, and Australia—and those countries held nearly 60 percent of the world reserve base.

Zinc

The largest single use of zinc is in galvanizing steel—that is, in coating sheet steel with a thin film of zinc to keep the steel from rusting. Zinc also has important uses in brass and a variety of other alloys, and zinc compounds are important in the chemical, paint, and rubber industries. However, zinc must compete with other metals and plastics, and U.S. consumption of the metal appears to have stabilized at an annual rate of about 1 million tons. Worldwide, however, zinc production and consumption have been rising rather steadily. Production of 1983 was nearly double that of 1964. Only about 10 percent of the new zinc metal used can be recycled, so most of the annual supply is from newly mined zinc.

Table 5-11 indicates the most important Free World sources of world zinc supply and gives estimates of the reserve base. Zinc, like lead, is mined in many countries. In 1980, 35 countries produced 20,000 tons or more of the metal. At the 1984 rate of production, world reserves are equivalent to about 46 years' supply. Additional deposits will certainly be found, and there are large resources of subeconomic materials from which zinc could be made available at somewhat higher prices. It is generally true that deposits of lead and zinc do not lend themselves to easy

TABLE 5-11 World Mine Production and Reserve Base of Zinc (in Thousands of Tons)

	Production		
	1983	1984[a]	Reserve Base
United States	303	292	58,400
Australia	766	711	43,000
Canada	1180	1306	61,700
Mexico	283	309	8,800
Peru	610	623	13,200
Other market economy countries[b]	2132	2150	99,200
Centrally planned economies[c]	1612	1609	35,300
Totals	6886	7000	319,600

[a]Estimated.

[b]Ireland, Spain, Sweden, and Japan are important market economy producers.

[c]Production is mostly from the Soviet Union, Poland, and China.

Source: U.S. Bureau of Mines, *Mineral Commodity Summaries.*

or rapid delineation of large tonnages of ores. The reserves in a deposit or group of deposits are proved progressively over a period of years. In this respect there is marked analogy between reserves of zinc and proved reserves of petroleum and natural gas.

Tin

Although tin is still used to a minor extent in bronze, the principal application of tin in ancient times, its modern uses are based mainly on its low melting point (232°C), its resistance to corrosion, its malleability, its lack of toxicity, and its high electrical conductivity. Its most essential use is actually in solders in electrical circuits, but it is still used to coat sheet steel to prevent rusting. Solder and tinplate account for about 60 percent of U.S. consumption of tin. As a component of alloys it has a number of uses in machinery, including use in engine bearings. For decades tin cans (made of thin sheet steel coated with tin) were essential to the food storage industry, but aluminum and plastic-coated cans have largely supplanted them. During the past 20 years its use in the United States has been slowly declining, but no substitute for it is likely in the field of alloys with low melting points, and it continues to be an essential metal.

Table 5-12 gives data for tin production and the tin reserve base. Over 90 percent of world production comes from deposits in a belt extending from eastern Siberia through mainland China and southeast Asia to Indonesia. The belt contains 70 percent of the world reserve base. The only sizable reserves in the western hemisphere are in Bolivia and Brazil. The

TABLE 5-12 World Mine Production and Reserve Base of Tin (in Tons)

	Production[a]		
	1983	1984[b]	Reserve Base
United States	Negligible	Negligible	44,000[c]
Australia	10,700	9,900	198,000
Bolivia	26,900	27,600	154,000
Brazil	13,200	15,400	77,000
Burma	1,800	1,800	11,000
Indonesia	19,800	28,700	750,000
Malaysia	46,300	44,100	1,224,000
Nigeria	1,100	1,300	22,000
Thailand	22,000	22,000	298,000
United Kingdom	4,500	4,400	99,000
Zaire	3,500	3,300	22,000
China	16,500	17,600	88,000
Soviet Union	40,800	38,600	88,000
Other countries	16,800	15,400	319,000
Totals	223,900	230,100	3,394,000

[a]Production figures rounded to nearest 100 tons, reserve base figures to nearest thousand tons.

[b]Estimated.

[c]The U.S. reserve base consists of uneconomic material.

Source: U.S. Bureau of Mines, *Mineral Commodity Summaries.*

United States has no tin reserves as such, although small amounts of tin are produced annually as a by-product of molybdenum production.

Mercury

Mercury has a peculiar combination of physical and chemical properties that make it almost uniquely useful in a number of applications in chemical and electrical industries. Its liquidity at room temperature together with its high electrical conductivity makes it very useful in electrical switches and other electrical apparatus, as well as in thermometers. In the past its chemical properties caused it to be used in the manufacture of alkalis and chlorine, but problems of toxic pollution have led to curtailment of such uses and its use as a fungicide in paints. Total world production of mercury in 1984 is estimated at about 71,000 tons, mostly from the United States, Algeria, Spain, and the Soviet Union. U.S. consumption of mercury has declined in recent years; in 1984 consumption was about 2050 tons.

More than half the known world reserves of mercury are in three countries—Spain, Italy, and the Soviet Union. The largest of all deposits are those of Almadén, in Spain, which have been worked almost continuously since Carthaginian and Roman times and are still far from exhausted. In recent years about half the U.S. supply has come from mines in California and Nevada. Canada and Mexico also have substantial reserves. Production from western hemisphere sources over any period of time is very difficult to maintain, because the price of mercury has been as volatile as the element itself, ranging (in constant 1978 dollars) from $135 to $986 per 76-pound flask during the period 1958 to 1979. Despite the wild fluctuations in price and the consequent difficulty in estimating what deposits actually contain reserves, needed supplies of mercury have been forthcoming for centuries and seem likely to continue to be available.

Magnesium

Commercial production of magnesium metal began at about the same time as production of aluminum. Magnesium forms about 2 percent of the earth's crust. Resources of raw materials range from magnesium in seawater to deposits of magnesium carbonate on land, and these resources are for all practical purposes inexhaustible. The use of magnesium, however, has grown only very slowly, and world annual demand is less than 300,000 tons. It is the lightest of the common metals, a distinct advantage for use in transportation equipment, but it is brittle at room temperatures and has only limited structural strength. It lacks the versatility of aluminum and is half again as expensive to produce. Much of the metal is used in alloys with aluminum and other metals. The United States produces about 40 percent of world supply of the metal and exports part of its output to other countries.

Far more important than its use as a metal is the use of magnesium in nonmetallic compounds. Magnesium oxide is an important material for linings of metallurgical furnaces, insulators, and other refractory products. Magnesium chemicals are used in many industries. Annual U.S. consumption of magnesium in various compounds has ranged in recent years from 550,000 to 1 million tons.

Other Nonferrous Metals

Besides the nonferrous metals discussed above, there are a number of others that are used in various amounts in industry today. Lithium and beryllium are the lightest of the metals, but they are available in more limited amounts than the metals described above and are much more expensive to produce. In January 1984 lithium metal was quoted at about $21 per pound, and beryllium was quoted at $241 per pound. As metals, lithium and beryllium are used chiefly in special alloys with aluminum or copper. Lithium is also a component of many industrial chemicals.

Major reserves of lithium are known in the United States, Zaire, and Chile. Reserves are large relative to current demand, but if thermonuclear fusion ever becomes a viable source of power, lithium is likely to be a major component of the fuel, and demand will greatly increase. Beryllium resources are rather widely scattered over the world, but the only large reserve is in a deposit at Spor Mountain, Utah.

Of the other nonferrous metals, antimony, arsenic, and bismuth are the most important, naming them in declining order of tonnages consumed annually. They are used in various alloys, but they also enter into various industrial chemicals, and such uses account for most of the consumption of arsenic. All these metals are produced largely as by-products from ores mined chiefly for copper, lead, zinc, silver, tungsten, or molybdenum. Shortages of supply are not anticipated.

THE PRECIOUS METALS

Two metals, gold and silver, have had a unique role in history. Both were among the earliest discovered metals, and both became highly prized for their beauty and rarity compared to other metals known in ancient times. Both were used for many centuries not only in the arts and crafts, but in coins and bars that were tangible items of wealth and the principal medium of exchange. They made up the bulk of the storied treasures of princes and potentates and were hoarded by all who could afford them as protection against the vicissitudes of fortune.

The quest for gold has led men to the corners of the earth and in some countries has influenced the course of history. It led to the conquest and plunder of the ancient civilizations of Central and South America. In 1848 it set off the rush to California that triggered the development of our western states. The rush to the Klondike set the stage for modern development of America's orphan child, Alaska. The discovery of gold on the Witwatersrand in South Africa's Transvaal disturbed a simple agricultural society and set the stage for the tragedy of the Boer War. Only Antarctica has never seen the hardships, triumphs, and historical dramas of man's continuing search for silver and gold.

The quest goes on, because man has never found enough of the two metals to satisfy more than a limited group of desires and needs. For most metals the standard unit is the ton, and prices are set in terms of dollars per ton or pound. For gold and silver, ounces* are nearly always the units for discussion, and prices are in terms of dollars per ounce. Gold and silver are restricted to uses consistent with such prices. It is unfortunate that this must be so, because both metals have properties

*The troy ounce, equal to 1.097 avoirdupois ounces, is used for precious metals.

that qualify them for uses in industry far beyond the present ones. Both gold and silver are excellent conductors of electricity and heat. Both are extremely malleable and ductile and are readily cast or otherwise formed into any desired shape, including the incredibly thin gold leaf that can be laid over other metals. They are soft, but they can be hardened by alloying them with small amounts of copper. Both metals are very beautiful; their soft, warm colors are unmatched by those of other metals. Silver tarnishes, as every householder knows, in contact with the sulfur of the atmosphere, and it dissolves in nitric acid. Gold, however, does not tarnish, and it can be dissolved only in royal water, aqua regia, a mixture of concentrated nitric and hydrochloric acids. Gold shares this property with the more lately recognized metals of the platinum group, and together they are known as the noble metals.

Both gold and silver are widely used at present, as in the past, for jewelry and other items of exquisite workmanship. They are now too valuable and scarce to be used much for coinage. Maple leaves and Krugerrands are investment items, not coins for circulation, but the major national treasuries of the world hold stocks of gold as hedges against inflation, and both gold and silver are hoarded in large amounts in the form of coins and bars, partly for speculation, partly as protection against economic or political instability. Beyond this, however, both gold and silver have important applications in electrical and electronic devices. Silver salts are the basis for the photographic industry, and photochemicals account for about 40 percent of the silver consumed annually in the world. Electrical and electronic products consume about 25 percent, silverware and jewelry about 15 percent, and alloys and miscellaneous applications the rest. No one really knows how much silver is tied up in jewelry, sterling, and hoarded coins and bars, but when speculation drives up the price of silver, jewelry and coins pour into the marketplace. That is what happened in 1979 and early 1980, to the dismay of those who attempted to corner the supply of silver.

Gold likewise has some important industrial uses, but the great bulk of world production still finds other applications. Pretorius (1981) reports that 1765 metric tons (about 57 million troy ounces) of gold were sold in the world in 1979. Private bullion purchases accounted for 20 percent, official coins for 16 percent, and medallions for 2 percent. About 42 percent was used for jewelry, 5 percent for dentistry, 5 percent for electronics, and 4 percent for miscellaneous purposes.

Silver is found mostly in deposits formed by hot solutions that have circulated along fractures in the crust in regions of former volcanic activity. A great series of such deposits studs the volcanic belt that extends from Nevada and western Utah southward into Mexico, Central America, and the Andean region of South America. Some of the silver mines in the belt are legendary—the Comstock Lode, the treasure from which helped finance the Civil War; Pachuca-Real del Monte, worked almost continuously since the time of the Spanish Conquest in the 1500s but worked still earlier by Indians from which the Spaniards wrested the secret of the deposits; and Potosí, Bolivia's silver treasure chest. All these and many other deposits have been worked mainly for their silver content, although gold has also been won from them. Extremely rich gold deposits have been found in the same belt. The Mother Lode of California and Goldfield, Nevada, with its curtains of gold along fractures in the rocks, are famous examples. The heyday of those fabulous deposits is over. Most of the silver produced in the world today is from deposits in which the metal occurs with base metals—lead, zinc, or copper in various proportions. The highly productive silver deposits of the Coeur d'Alene district of Idaho, certain mines in Mexico, and the mines of Peru are in this class. The most productive silver mine in North America at present, however, is the deposit at Kidd Creek, Ontario, a massive zinc-copper sulfide deposit in an ancient volcanic belt. Silver there is an important by-product.

Principal producers of silver are indicated in Table 5-13, with data on the world reserve base. The silver supply situation is always a confusing one. In 1983, estimated U.S. consumption of silver was 197 million ounces, of which 42 million ounces was supplied from domestic mines and about 35 million ounces from secondary metal (old scrap). An additional 32 million ounces was recovered as new scrap, scrap generated during 1983

TABLE 5-13 World Mine Production and Reserve Base of Silver (in Millions of Troy Ounces)

	Production, 1983	Reserve Base
United States	32.1	1820
Canada	33.3	1600
Mexico	49.4	1060
Peru	47.9	610
Bolivia	6.1	—[a]
Chile	9.6	—[a]
South Africa	5.5	—[a]
Japan	8.9	—[a]
Australia	25.4	—[a]
Other market economy countries	41.2	1345
Poland	24.7	—[b]
Soviet Union	46.0	—[b]
Other central economy countries	12.1	2000
Totals	342.2	8435

[a]Included under other market economy countries.

[b]Included under other central economy countries.

Source: U.S. Bureau of Mines, *Mineral Commodity Summaries.*

fabrication of silver. At the same time, net imports were 161 million ounces. Import reliance was estimated at 61 percent. If the reader is unable to make sense of these figures (new metal plus old scrap plus new scrap equals 109 million ounces), it is because changes in industry stocks and Treasury stocks are taken into account in calculating import reliance. Evaluating figures for production and imports and exports of silver is confusing enough. Still worse is the problem of estimating how much silver is available in private hands. One estimate is that privately owned silver in the United States in 1983 was about 2.4 billion ounces, roughly 12 years' U.S. consumption at the 1981 level. There is no estimate for the world as a whole, but the amount must be many times the U.S. total. Estimates of silver hoarded in India alone range from 2 billion ounces to nearly 5 billion ounces.

Gold deposits are widely scattered over the world. Historically, world gold production has come mostly from deposits of two kinds—vein (lode) deposits and placer deposits. Like silver, gold has been deposited by heated waters moving along fractures in the crust; the veins are fillings of the fractures and consist largely of quartz. The placer deposits are generally gravels and coarse sands containing gold eroded from the veins and washed into streams (Fig. 5-6). Owing to its high specific gravity, gold becomes concentrated in the coarser stream deposits. In California and many other places the placer deposits were found first, by panning stream gravels for gold. The lode deposits were located as the gold in the gravels was traced upstream to its sources. Thus the gold deposits of California were traced back to their sources in the Mother Lode and other vein systems of the state. In Victoria, Australia, gold-bearing gravels were followed downstream under lava flows that had buried the lower parts of river channels.

In 1887, the largest placer deposits in the world were found on the Witwatersrand of South Africa. The deposits formed about 2 billion years ago and were buried beneath thousands of feet of lavas and sediments. The deposits extend for miles (Fig. 5-6) and go far below the surface. In places the mines extend to depths of more than 12,000 feet; they are the world's deepest mines. The deposits are by far the most important gold deposits in the world. It is estimated that they have already produced between 40 and 50 percent of all the gold produced during historic times. They currently yield about 50 percent of the annual world supply and still contain about 50 percent of world gold reserves. The second largest reserves, perhaps 15 percent of the world total, are thought to be in the Soviet Union, which now produces about 25 percent of the world supply. Production from the South African deposits is expected to decline during the remainder of the century, and the Soviet Union is likely to become the world's leading gold producer (Pretorius, 1981).

In the United States, the rise in the price of gold that followed deval-

Figure 5-6 Top: Hydraulic monitor washing gold-bearing gravels into the sluiceway in the middle and right foreground. Hunker Creek, Yukon. Photograph by L. H. Green. Bottom: The waste dumps from the deep underground gold mines extend many miles along the Witwatersrand, near Johannesburg, South Africa.

uation of the dollar in 1971 has led to a resurgence of interest in gold deposits. Certain low-grade gold deposits have become economically workable, and two types of gold deposits previously unrecognized have been discovered in California and Nevada in the same mountain belts in which the bonanza deposits were found in earlier times. Total world gold reserves are estimated at about 1.3 billion ounces at a price of $500 per ounce. An additional 2 billion ounces is estimated to be available at $600 per ounce. The U.S. share of world production is now about 3 percent; U.S. reserves are estimated at about 11 percent of the world total. Besides reserves in the ground, stocks of gold above ground are estimated at about 2.1 billion ounces. Gold is so valuable that a high percentage of it is recycled. It has been said that most of the gold mined in historic times is still available.

The platinum group metals, conveniently referred to as the PGM, are platinum, palladium, iridium, osmium, rhodium, and ruthenium. Although known as early as the 16th century, deposits of platinum were first reported from Colombia in 1748. In the early 19th century scientific investigations showed that crude platinum is typically a complex mixture of platinum with other metals that somewhat resemble it but that have certain distinct and valuable properties of their own. Important use of the metals developed in the present century, and the metals are now highly prized and rival gold in price. In the United States, more than half the annual consumption of PGM is in the automotive, petroleum-refining, and chemical industries, where they serve as catalysts, promoting chemical reactions of many kinds. Most of the rest is used in the ceramics, glass, and electrical industries and as a component of dental alloys. Platinum is also used in jewelry, but to a much lesser extent than gold and silver. Chemical inertness, conductivity, resistance to high temperatures, and catalytic properties are the main properties that make the PGM of great value in industry. Catalytic converters are an essential item in control of automobile emissions in the United States and are the largest single use of PGM. Palladium and rhodium are also used in emission control.

Of the various PGMs, platinum and palladium are by far the most important, accounting for about 90 percent of the annual world consumption of about 6.5 million ounces. About 95 percent of world reserves is shared equally by South Africa and the Soviet Union. South Africa is the principal producer of platinum, the Soviet Union of palladium. Canada ranks third as a producer but contributes only about 4 percent of world supply. Reserves in the United States are small, but sizable deposits of PGM have been identified in the Stillwater Complex of Montana. They are not yet being worked, but they constitute an important strategic reserve. World reserves of PGM are estimated at 1.2 billion ounces, enough for many decades, but in view of the importance of PGMs to all industrial countries, the concentration of reserves in just two countries of the world is a matter of some concern.

REFERENCES AND SOURCES OF INFORMATION ON METALS

Alexandersson, G., and Klevebring, B.-J., 1978, *World Resources: Energy and Minerals.* Walter de Gruyter, Berlin, 248 pp.

Brobst, D. A., and Pratt, W. P., eds., 1973, United States mineral resources. U.S. Geological Survey, Professional Paper 820, 722 pp.

DeYoung, J. H., Jr., Sutphin, D. M., and Cannon, W. F., 1984, International strategic minerals inventory summary report—manganese. U.S. Geological Survey, Circular 930-A, 22 pp.

DeYoung, J. H., Jr., Lee, M. P., and Lipin, B. R., 1984, International strategic minerals inventory summary report—chromium. U.S. Geological Survey Circular 930-B, 41 pp.

Kilgore, C., and Thomas, P. R., 1982, Manganese availability—domestic. U.S. Bureau of Mines Information Circular 8889, 13 pp.

Lemons, J. F., Jr., Boyle, E. H., Jr., and Kilgore, C. C., 1982. Chromium availability—domestic. U.S. Bureau of Mines Information Circular 8895, 14 pp.

Peterson, G. R., Davidoff, R. L., Bleiwas, D. I., and Fantel, R. J., 1981, Alumina availability—domestic. U.S. Bureau of Mines, Information Circular 8861, 23 pp.

Pretorius, D. A., 1981, Gold, gelt, gilt: Future supply and demand. *Economic Geology,* Vol. 76, pp. 2032–2042.

Ross, J. R., and Travis, G. A., 1981, The nickel sulfide deposits of Western Australia in global perspective. *Economic Geology,* Vol. 76, pp. 1291–1329.

Rosta, J., 1983, Silver. *Engineering and Mining Journal,* Vol. 184, No. 3, pp. 61–64.

Skinner, B. J., 1976, *Earth Resources.* Prentice-Hall, Englewood Cliffs, N.J., 150 pp.

U.S. Bureau of Mines, annual issues, *Mineral Commodity Summaries.* U.S. Government Printing Office, Washington, D.C.

U.S. Bureau of Mines, *Mineral Facts and Problems.* Issued at 5-year intervals. The 1985 edition is in press. U.S. Government Printing Office, Washington, D.C.

6 Outlook for World Supplies of the Nonfuel Minerals

In Chapter 3, we examined the outlook for supplies of energy minerals. We turn now to the question of availability of nonfuel minerals in the future. This book is mainly concerned with the mineral problems of the United States and especially with supply of minerals to meet our national needs. In succeeding chapters, therefore, we will examine our present mineral position, the prospects for future production from domestic mines, and the factors that will influence domestic mineral production. However, some of the nonfuel materials we consume must be had from sources abroad, so our problems of mineral supply cannot be considered solely in terms of availability from domestic resources. They must be examined against the broad background of the availability of nonfuel minerals from the resources of the world, and that is the subject of the present chapter.

Since World War II there have been many expressions of opinion regarding future availability of nonfuel minerals, from warnings of imminent exhaustion to predictions of unlimited supplies. The result has been uncertainty and confusion on the part of policy makers and the general public. Some general reasons for this were discussed in Chapter 2, but at this point we must inquire more specifically into the future availability of nonfuel mineral commodities. The discussion will not lead to firm forecasts for all of them, but it will serve to indicate the range of availability of various minerals, the problems involved in forecasting availability, and the sources of uncertainty about future supplies of the nonfuel minerals.

First we must define what is meant by "availability" and "adequacy." Availability over any period of time is defined in terms of (1) the amounts of extractable mineral raw materials present in identified resources of the earth's crust, and (2) the rates at which they can be economically extracted. Adequacy is defined as the ratio between annual supply and annual demand for minerals. Without indulging in economic semantics, we will define demand in terms of actual consumption of minerals. We really have no choice.

As a basis for forecasting future availability and adequacy we have three kinds of information: estimates of the reserve base for each mineral or metal; statistics of world mineral production, which give us a measure of both past and present mineral production and consumption* and indicate their trends; and information on anticipated changes in patterns of mineral use. Let us start with the reserve base, recalling that for any mineral commodity the reserve base includes measured and indicated reserves, marginal reserves, and subeconomic resources that are somewhere near the boundary between reserves and other resources. The proportions of these various components in the reserve base estimate of a mineral are generally not indicated. Estimates of the reserve base are much better for some mineral commodities than for others for which information is less accurate and less complete. The estimate of the reserve base for copper, for example, is certainly more accurate than that for tungsten. In Table 6-1, for each of the commodities listed, the 1985 world reserve base as currently estimated is given, and in the next column the 1980 production is given. If we divide the reserve base by the figure for production, we get the figure in the final column, the reserve base/production (RB/P) index for the mineral. The RB/P index tells how many years' supply, at the 1980 production rate, is represented in the reserve base. The year 1980 is selected because it is the latest "normal" year—that is, not a year of recession in mineral industry.

The RB/P index is only a starting point for a forecast of availability. It does not tell us at what rates a mineral commodity can be produced in the future, nor does it tell us what prices per unit of production will be necessary to make the various components of the reserve base available. The problems involved are illustrated in a U.S. Bureau of Mines study of availability of phosphate from the reserve base of the United States (Fantel et al., 1983). The reserve base in 1981 was estimated at 6.4 billion metric tons of potentially recoverable phosphate. Of this, only about 1.3 billion metric tons was classified as true reserves—that is, phosphate recoverable at a price of $30 per metric ton of product. The average price of phosphate in 1981 was $27 per metric ton. At a price of $45 per metric ton, an additional 2.4 billion tons could become available. The remaining 2.7 billion metric tons of the reserve base could become available only at

*There are no separate data for annual world consumption of minerals. However, over any period of a few years, world consumption is approximately equal to world production.

TABLE 6-1 World Production, Reserve Base, and Reserve Base/Production Index for Some Important Mineral Commodities

	Reserve Base (Thousands of Tons)[a]	Production 1980 (Thousands of Tons)[a]	Reserve Base/ Production Index (RB/P)
Salt	∞	181,600	∞
Magnesium	∞	351	∞
Cement	Very large	978,000	Very large
Lime	Very large	131,623	Very large
Gypsum	Very large	78,290	Very large
Clays (common)[b]	Very large	449,000	Very large
Sodium carbonate	>43,200,000	8,459	>5,107
Sodium sulfate	5,100,000	2,169	2,351
Chromite	7,540,000	10,725	703
Potash	18,739,000	30,722	610
Vanadium	18,250	40	456
Manganese ore	12,000,000	29,000	414
Cobalt	9,200	33	279
Feldspar	>1,000,000	3,782	>264
Boron minerals	>300,000	1,175	>255
Phosphate rock	38,580,000	151,000	255
Bauxite	24,581,000	99,165	248
Ilmenite	905,000	3,979	227
Iron ore	231,056,000	1,090,432	212
Nickel	111,000	850	131
Fluorspar	645,000	5,000	129
Molybdenum	12,975	120	108
Antimony	5,175	74	70
Copper	562,000	8,421	67
Tungsten	3,813	60	63
Sulfur	2,976,000	56,900	52
Zinc	319,670	6,340	50
Talc and pyrophyllite	330,000	7,366	45
Mercury[c]	7,200	191	38
Lead	148,810	3,885	38
Barite	244,000	8,114	30
Asbestos	114,600	5,314	22
Tin	3,307	272	12

[a]Data from U.S. Bureau of Mines, *Mineral Commodity Summaries*, 1982 (production) and 1985 (reserve base).

[b]There are no data for world reserves of special types of clays.

[c]The standard unit for mercury is the flask, a flask being 76 lb. Numbers given are thousands of flasks.

prices in excess of $45 per metric ton and ranging to more than $100 per metric ton. U.S. production of phosphate in 1981 was 53.6 million metric tons. If we divide that figure into the reserve base of 6.4 billion metric tons, we get an RB/P index of 120, but if we divide 53.6 into the figure for reserves, 1.3 billion tons, we get a figure of 24 for the *reserve*/production index, which we will designate as R/P. The latter figure indicates how many years' supply of phosphate, at the 1981 production rate, is available at an average price of $30 per metric ton. The study suggests that both reserve base tonnages and RB/P indices must be substantially discounted in estimating supplies available at current prices.

There is still the problem of the rate at which phosphate can be produced economically from the reserve base. This is likewise a function of price. The U.S. Bureau of Mines analysis indicates that at $30 per metric ton, production from deposits currently being mined will decline to about 9 million metric tons per year in the year 2000. To maintain the 1981 production rate of 53.6 million metric tons through the year 2000, deposits that are currently nonproductive can be tapped, but a price of $45 (1981 dollars) per metric ton will be necessary to support the new operations. Considering the variables involved in these calculations, the figures developed in the study must be considered as approximations, but the principles involved are sound. Studies of the U.S. copper and uranium* mining industries (National Academy of Sciences, 1975) and the world gold industry (Pretorius, 1981) showed comparable relations between the reserve base, reserves, prices, and rates of production. We therefore repeat: RB/P indices are only a starting point in forecasting availability of mineral supplies. For realistic forecasts of availability we need analyses of production capability at various price levels. We have the necessary data for only a few nonfuel minerals and must use these, with some mental reservations, as representing the rest.

All analyses of production capability indicate that for any given level of reserves there will be a maximum rate of annual production that cannot be exceeded at acceptable costs. This relationship is implicit in M. K. Hubbert's analysis of U.S. petroleum supply (see Chapter 3) and is equally applicable to the nonfuel minerals. We can express the relationship as the ratio of *reserves* to maximum rate, or R/MR. There is no fixed value for this ratio, since reserves are a function of price and the ratio is influenced by other factors such as the characteristics of the deposits from which various nonfuel minerals are produced. From available information it seems to range from about 24 to 1 to about 40 to 1. In general, at any given time, R/MR will be significantly lower than RB/MR, the ratio of the *reserve base* to maximum rate of production, because the reserve base in-

*Although uranium is a fuel mineral, uranium mining operations are similar to nonfuel mineral operations.

cludes material that is not currently producible. From limited information it appears that RB/MR may range from about 60 to 1 to more than 100 to 1, the latter under adverse economic conditions.

From the foregoing discussion of the effect of prices on reserves and rates of production of mineral commodities, it is clear that availability of a mineral can be markedly influenced by price, but as pointed out in the chapter on energy minerals, rise in price restricts the use of a mineral. This is evident in Figure 6-1, which shows the relation between prices of metals and tonnages consumed in a typical year in the United States. The consumption of a metal available at a given price was roughly 100 times the consumption of a metal available at 10 times that price.

Let us return to the RB/P indices given in Table 6-1. From the discussion above it will be clear that we have only a very rough guide to possible rates of production of nonfuel minerals that can be achieved and maintained from the present reserve base and the total amounts that can ultimately be removed. The information we have, however, suggests that

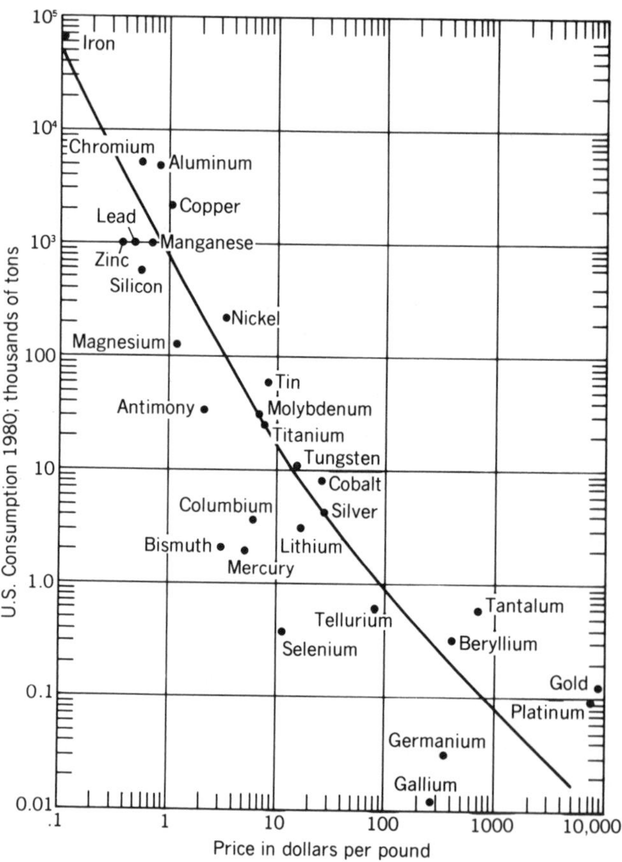

Figure 6-1 Relation between prices of metals and amounts consumed in the United States.

if the RB/P index for a mineral is lower than 50, present annual production is close to or actually at the maximum rate possible at present prices and that unless new *reserves* are added to the reserve base through discovery, technologic advances in extraction, or increases in price, production rates must shortly decline as the present reserve base is depleted. If the RB/P index is less than 25, the present rate of production is unlikely to be sustained regardless of price, without additions to reserves. On the other hand, if the RB/P index is above 50, production at present prices is possible for a time. On this rough basis we will appraise the future availability of nonfuel minerals at present rates of production and consumption.

Fortunately, the RB/P indices for some minerals (Table 6–1) are very large. The infinity indices for salt and magnesium indicate inexhaustible supplies. There are no estimates of reserve base for lime, cement, gypsum, or clays, but for each the base is known to be very large. The RB/P index for sodium carbonate is very large, and the present estimate of the reserve base is conservative. The RB/P indices for phosphate, sodium sulfate, potash, bauxite, boron minerals, feldspar, chromite, iron ore, and vanadium, measured against 1980 rates of production, seem comfortably large. The indices for fluorspar, nickel, manganese, cobalt, and molybdenum are less favorable, but they suggest supplies adequate for some decades. The base for fluorspar does not include large amounts of fluorine that are becoming available as a by-product of phosphate production.

With antimony, copper, and tungsten we move into the critical range. Longer-term availability of these metals will depend on success in adding to reserves. This will become apparent in subsequent paragraphs. For the remainder of the commodities on the list, with RB/P indices of 52 to 12, the need to add reserves to the reserve base is more urgent and more immediate.

The above discussion of future availability is based on 1980 production rates. But what if production of minerals continues to increase? Part of the concern over exhaustion of mineral resources stems from fears of rapid escalation in both world population and world per capita mineral consumption, the two combining to cause increasingly rapid growth in mineral production. In Figs. 6-2 and 6-3, per capita production of 22 mineral commodities in 1964 is compared with per capita production in 1980. Equating production with consumption, there has been an increase in per capita consumption of all but six commodities. The effect of this and of growth in world population is registered in the charts of Figs. 6-4 to 6-10, in which amounts of nonfuel minerals produced in each of the years from 1964 through 1984 are plotted. The charts show that there has been substantial growth in world mineral production during the period. Production of sulfur, soda ash, barite, cement, phosphate, potash, talc and pyrophyllite, aluminum, nickel, molybdenum, magnesium, and vanadium has doubled or nearly doubled. All the other commodities show

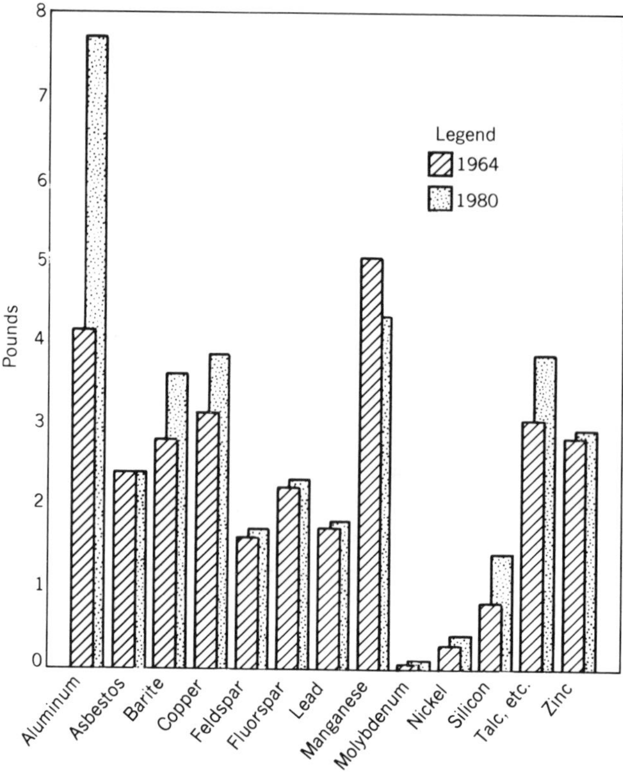

Figure 6-2 World per capita consumption of some important mineral commodities in 1964 compared with 1980. Prepared by D. E. Cameron.

increases since 1964. The curves indicate, however, that growth has not been uniform over the whole period. This fact is brought out in Table 6-2, in which annual growth rates from 1964 to 1973 are compared with annual growth rates for 1973 to 1980. Growth rates for all the minerals except barite, silicon, chromium, and tungsten are lower for the later period. We seem to see in this the impact of rising costs of energy on the mineral-based industries of the world. Future growth in mineral production is certainly suggested by trends since 1973, but at lower rates for most minerals than might have been forecast in that year.

On the basis of 1980 production rates we have a preliminary appraisal of future availability. However, a more realistic appraisal must take into account the consequences of future growth in mineral production in terms of total production of minerals and diminution of the 1985 reserve base. This means that future rates of growth in production (consumption) of various mineral commodities must be estimated. It is a difficult undertaking, because the rates will be affected by a number of variables—changes in the technology of mineral extraction, changes in patterns of use, world economic growth, population growth, energy costs, conser-

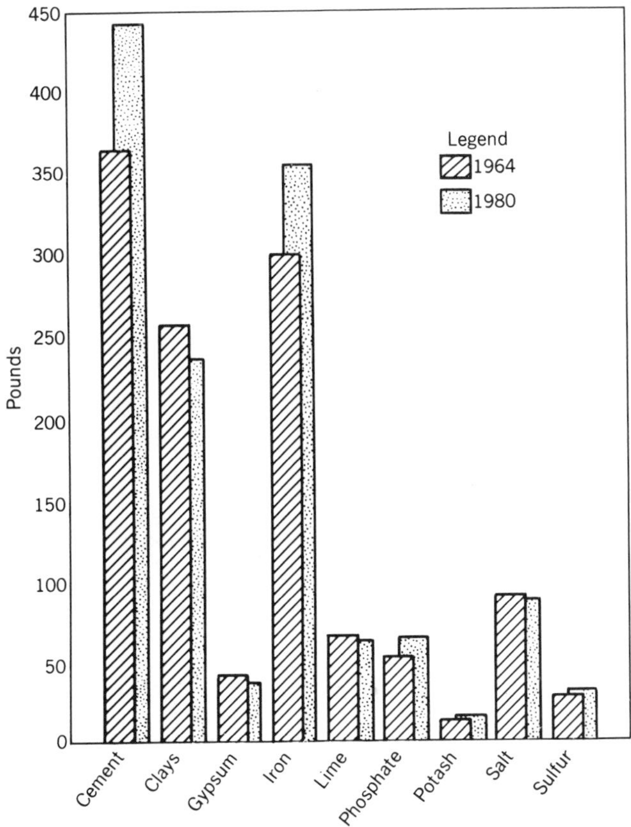

Figure 6-3 World per capita production of some important mineral commodities in 1964 compared with 1980. Prepared by D. E. Cameron.

vational measures, environmental constraints, and others. None of the variables can be quantified unless assumptions are made as to rates of change and degrees of influence of the variables on future rates of production. Nonetheless, estimates of growth rates have been made from time to time, especially for growth rates of U.S. consumption. The most recent are those of Malenbaum (1978), the U.S. Bureau of Mines (1980), Ridker and Watson (1980), and the Institute for Economic Analysis (Leontief et al., 1983). Their estimates for various mineral commodities for the period 1970 to 2000 are compared in Table 6-3. The most comprehensive data are those of the U.S. Bureau of Mines and the Institute for Economic Analysis. The forecasts use different assumptions and different methods of estimation; the calculated growth rates show a considerable range.

In the author's view, differences in estimates go back to an inability to quantify the variables that will determine future nonfuel mineral production. The basis for accurate forecasts of future consumption does not

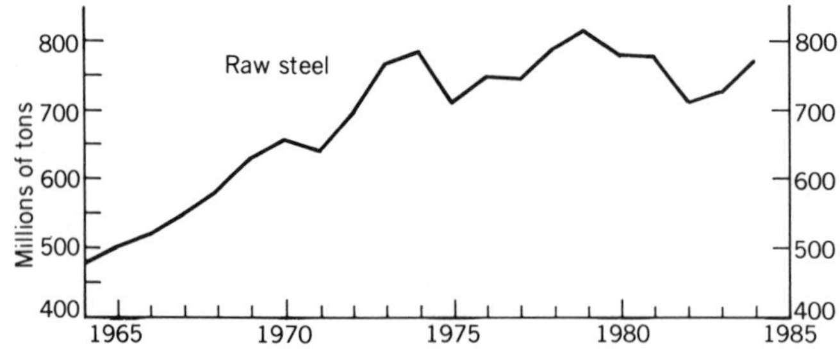

Figure 6-4 World production of raw steel. Data from U.S. Bureau of Mines, *Mineral Commodity Summaries* and *Minerals in the U.S. Economy*. The same publications are the sources of the data used in constructing the charts of Figs. 6-5 to 6-10.

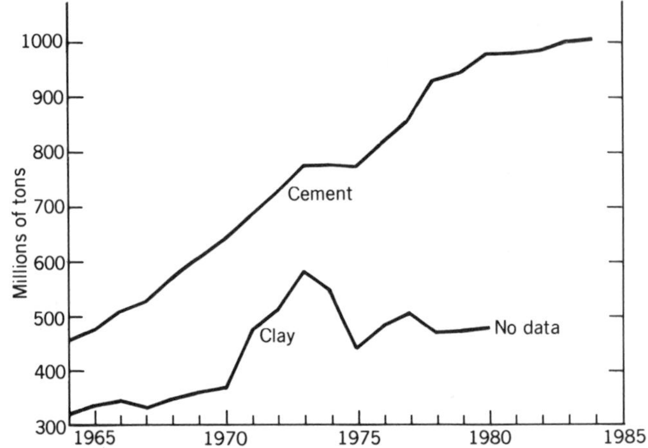

Figure 6-5 World production of cement and clay, 1964–1984.

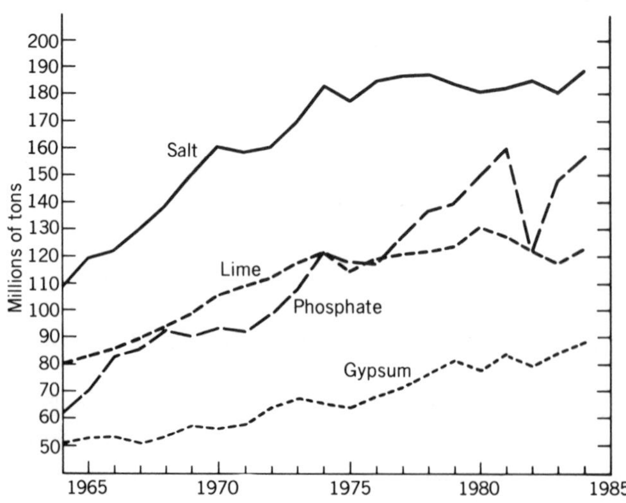

Figure 6-6 World production of salt, lime, phosphate, and gypsum, 1964–1984.

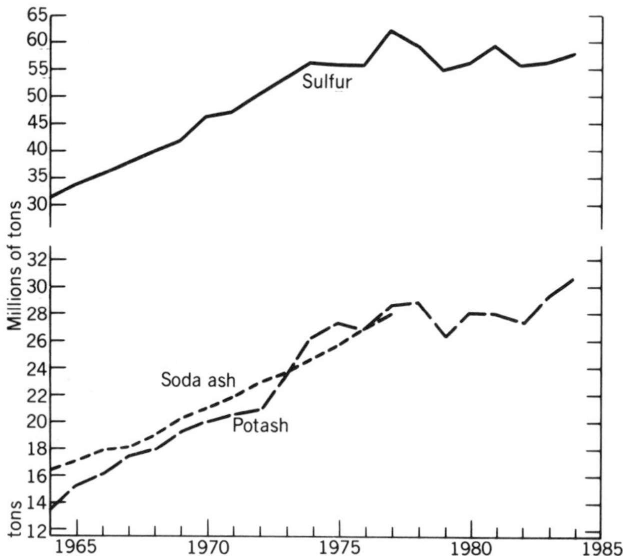

Figure 6-7 World production of sulfur, soda ash, and potash, 1964–1984. Data for soda ash for 1978–1984 are incomplete.

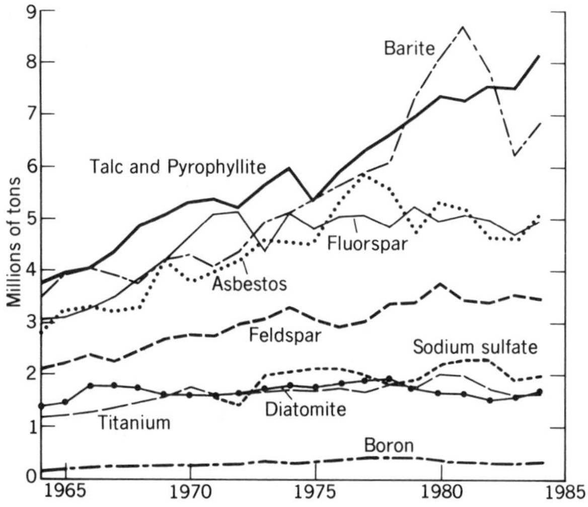

Figure 6-8 World production of barite, talc and pyrophyllite, asbestos, fluorspar, feldspar, sodium sulfate, titanium (titanium content of ilmenite and rutile used for nonmetallic purposes), diatomite, and boron (boron content of boron minerals and compounds), 1964–1984. There are no data for sodium sulfate for 1964–1970.

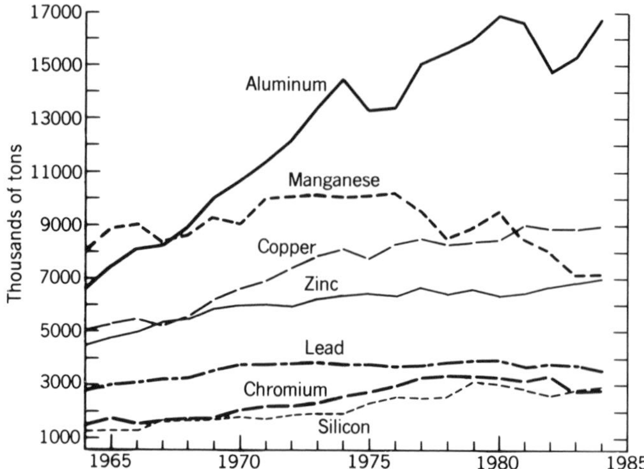

Figure 6-9 World production of aluminum, manganese, copper, zinc, lead, chromium, and silicon, 1964–1984.

exist. At any given time only a rough forecast of production over a period of years is possible. In this volume, therefore, no real forecast will be made. Instead we will simply answer a question, namely: What will happen to the world reserve base if the annual growth rates for 1973 to 1980 hold for the period 1981 to 2005. The choice of these growth rates is based on the fact that the graphs of Figs. 6-4 to 6-10 indicate a general slackening in world production of nonfuel minerals since the Arab oil embargo of 1973. Implicit in the choice of growth rates is that growth in world mineral production will resume as world recovery from the 1981–1983 recession

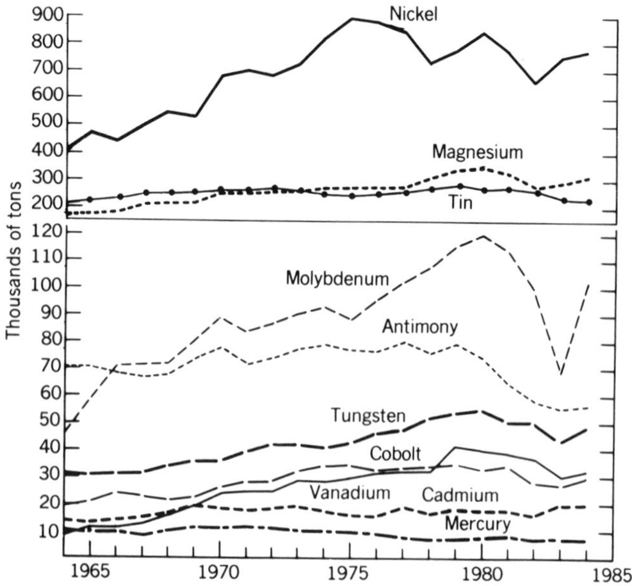

Figure 6-10 World production of nickel, magnesium, tin, molybdenum, antimony, tungsten, cobalt, vanadium, cadmium, and mercury, 1964–1984.

TABLE 6-2 Annual rates of Growth of World Production of Some Important Mineral Commodities

	Rate of growth[a]			Rate of growth[a]	
Commodity	1964–1973	1973–1980	Commodity	1964–1973	1973–1980
Aluminum	8.12%	3.45%	Magnesium	5.23%	4.32%
Antimony	1.06	−0.60	Manganese metal	3.36	−1.85
Asbestos	4.67	2.07	Molybdenum	7.74	4.2
Barite	3.96	7.33	Nickel	6.5	2.36
Boron	7.94	0.89	Phosphate	6.24	4.88
Cement	5.99	3.44	Potash	6.26	4.01
Chromium	5.21	5.17	Salt	5.13	0.91
Clays	6.72	−2.70	Silicon	5.17	6.45
Cobalt	5.96	0.87	Sodium sulfate	No data	1.22
Copper	4.87	1.01	Sulfur	6.08	.76
Diatomite	2.58	−0.60	Talc-pyrophyllite	4.66	3.82
Feldspar	4.26	3.12	Tin	2.21	0.48
Fluorspar	6.55	0.29	Tungsten	3.43	5.23
Gypsum	3.10	2.06	Vanadium	13.44	5.23
Ilmenite and rutile	4.27	2.91	Zinc	3.97	0.09
Lead	3.87	0.13	Raw steel	5.37	0.24
Bauxite	9.73	4.25	Mercury	0.88	−5.13
Lime	4.35	1.55			

[a]Calculated from data of U.S. Bureau of Mines, in *Mineral Commodity Summaries* and *Minerals in the U.S. Economy*.

TABLE 6-3 U.S. Nonfuel Minerals Consumption: Comparison of Alternative Projected Annual Rates of Growth (1970–2000)

	U.S. Bureau of Mines	W. Malenbaum[a]	Ridker and Watson[b]	Institute for Economic Analysis
Iron	1.18%	1.78%	1.87%	2.22%
Molybdenum	4.19	—	2.35	2.63
Nickel	3.15	2.00	4.05	2.66
Tungsten	3.90	2.30	2.07	2.95
Manganese	1.36	2.70	2.05	2.90
Chromium	2.70	1.20	2.40	3.14
Copper	2.70	1.90	2.30	3.05
Lead	0.90	—	4.40	4.80[b]
Zinc	1.15	2.00	2.60	1.46
Gold	−0.20	—	—	3.34
Silver	3.80	—	—	3.30
Aluminum	4.48	4.00	4.30	2.25
Mercury	−0.47	—	—	0.19
Vanadium	3.23	—	2.54	3.01
Platinum	3.80	2.60	—	3.20
Titanium	2.60	—	2.26	2.40
Tin	−0.33	0.90	2.20	0.79
Silicon	3.26	—	—	2.94
Fluorine	1.83	—	—	1.90
Potash	2.89	—	1.80	2.56
Soda ash	1.16	—	—	3.23
Boron	4.00	—	—	3.12
Phosphate rock	3.20	—	2.50	3.24
Sulfur	4.38	—	2.40	2.76
Chlorine	—	—	—	3.31
Magnesium	4.60	—	—	1.74

The table is taken from Leontief et al. (1983); their own estimates are given in the last column.

Note: Rates of growth computed on the basis of physical units.

[a]1975–2000.

[b]Includes primary as well as secondary demand.

Sources: U.S. Bureau of Mines: Bulletin 671, *Mineral Facts and Problems,* 1980 edition; W. Malenbaum (1977): *The Global 2000 Report to the President,* Vol. 2, 1980, pp. 206–207; R.G. Ridker and W.D. Watson: *To Choose a Future,* Johns Hopkins University Press, 1980, p. 153.

continues. The onset of recovery is clearly reflected in increase in production of most commodities in 1983 or 1984.

Table 6-4 shows the impact of production during 1981–2005 on the 1985 reserve base and on RB/P indices, assuming that the growth rates for 1973 to 1980 hold until 2005, except for barite, silicon, clays, and mercury (see notes in table). The results are mixed. For the first 14 commodities, from salt through chromite, the reserve bases remaining in 2005 are so large that adequate supplies for decades beyond 2005 can be safely assumed. The 1985 reserve base estimates for vanadium and feldspar are very conservative. For both these commodities the RB/P indices calculated for 2005 (remaining reserve base divided by production in 2005) are far above the critical level. The indices for bauxite and fluorspar are also comfortably large. Phosphate, antimony, ilmenite, and nickel are in the marginal range. For the remaining 11 commodities, twice as many as in 1980, RB/P indices are at or below the critical level. For those with indices of 35 to 21—copper, mercury, zinc, molybdenum, and sulfur—it is doubtful that the projected 2005 production rates could be achieved without substantial additions to *reserves* in the meanwhile. Even if 2005 production rates were reached, they could not be continued much beyond 2005. For the remaining commodities the prospects are still worse. Rates of production, if dependent on the 1985 reserve base, must begin to decline before the end of the century.

The above is a rough indication of availability. Given the inaccuracies of the reserve base and the uncertainties as to future growth rates, there is no point in attempting a more precise evaluation of future availability and adequacy of nonfuel mineral supplies. The general picture, however, is quite clear. For a number of the mineral commodities listed in Table 6-4, supplies are likely to be adequate well into the next century, or even beyond. For a considerable group of essential commodities, however, there is the prospect that supplies will be adequate only if substantial additions to reserves can be made, not only between now and the year 2005 but continuing in the years beyond. The critical group includes nearly half the metals listed in Table 6-4.

What are the chances of substantial additions to reserves and the reserve base in future years? For most of the mineral commodities, the chances appear very good. The record since 1947 is most encouraging. More than half present world reserves of iron ore, bauxite, copper, molybdenum, beryllium, chromite, vanadium, manganese, lithium, zinc, lead, sodium carbonate, phosphate, potash, and uranium have been discovered since 1947. Despite the prodigious production of minerals since 1947, world reserves of most minerals are at record high levels.

How long this can continue no one can foretell, but we know that exploration of the earth's crust is incomplete. It is reasonable to expect both that major discoveries will be made in years to come and that advances in the technology of mineral extraction and use will continue. We should not take Table 6-4 as a gloomy prediction of imminent failure of supplies of some essential mineral commodities for two reasons. One is that to do so would ignore the brilliant record of the mineral industry since World War II. The other is that gloomy predictions can be self-fulfilling, because they discourage efforts to overcome the problems that confront us. Rather than a prediction of disaster, Table 6-4 should be

TABLE 6-4 1981–2005 Consequences of World Production at 1973–1980 Annual Growth Rates

Commodity	Growth Rate 1973–1980 (Percent)	Production (Millions of Tons)[a]			Reserve Base, 2005 (Millions of Tons)	RB/P Index 2005
		1980[b]	2005	Total 1981–2005		
Salt	0.91	181.6	227.8	5,119.0	∞	∞
Magnesium metal	4.3	.35	1.0	15.8	∞	∞
Cement	3.44	978.0	2278.0	38,826.0	Very large	Very large
Clays (all types)	2.70[c]	479.0	786.0	15,649.0	Very large	Very large
Lime	1.55	131.6	193.3	4,042.	Very large	Very large
Silicon	6.45[d]	3.0	8.1	131.2	Very large	Very large
Diatomite	−0.60	1.65	1.49	38.2	Large	Large
Gypsum	2.06	78.3	130.0	2,579.	Large	Large
Sodium sulfate	1.22	2.17	2.9	63.8	5,036	1737
Potash	4.01	30.7	82.0	1,332.	17,407	212
Manganese ore	2.40	29.0	52.5	1,001.0	10,999	209
Cobalt	0.87	0.03	0.04	0.9	8.3	207
Iron ore	0.39	978.4	1077.0	25,722.	205,334	191
Chromite	5.17	10.73	37.8	575.0	6,965	184
Vanadium	5.23	0.04	0.143	2.1	16.2	113
Feldspar	3.12	3.78	8.2	144.	856	104
Bauxite	4.25	76.5	225.3	3,518	21,062	93
Fluorspar	0.29	5.0	5.8	130	515	89
Phosphate rock	4.88	150.1	494.0	7,450	31,130	63
Antimony	−0.60	0.07	0.06	1.2	3.6	60

Ilmenite	4.74	3.98	12.7	192.	713	56
Nickel	2.36	0.85	1.52	29.2	81.8	54
Copper	1.01	8.4	10.8	184.	378	35
Mercury	−5.13[e]	191.1[f]	151[f]	3,713[f]	3,487[f]	30
Zinc	0.09	6.3	6.5	160.	160	23
Molybdenum	4.20	0.12	0.33	5.4	7.6	23
Sulfur	0.76	56.1	67.8	1,551	1,425	21
Lead	0.13	3.9	4.0	98.9	50	13
Tungsten	5.23	0.06	0.21	3.1	0.7	3
Talc-pyrophyllite	3.82	7.4	18.8	311	19	1
Tin	0.48	0.27	None	3.3	0	0
Asbestos	2.07	5.3	None	115	0	0
Barite	7.33[d]	8.1	None	244	0	0

[a]Except mercury.

[b]Data from U.S. Bureau of Mines, *Mineral Commodity Summaries*, 1982.

[c]The negative growth rate of 1973–1980 is not likely to hold for 1981–2005. An average annual production of 500 million tons is assumed in calculating production during 1981–2005.

[d]These growth rates are unlikely to be maintained during 1981–2005. A growth rate of 4.00 percent is used here for both silicon and barite in calculating production during 1981–2005.

[e]Use of mercury during 1981–2005 is likely to decline more slowly than during 1973–1980. A growth rate of −2.00 percent is used here in calculating production during 1981–2005.

[f]Thousands of 76-pound flasks.

No data are given for sodium carbonate; data for world production and the world reserve base are incomplete. Known reserves are very large and will not be significantly depleted by 2005.

taken as a challenge and a warning that efforts to discover new deposits, to improve extractive processes, and to improve the technology of mineral use must not slacken in future years. More efficient methods of utilizing the world's energy resources must be developed. Otherwise rising costs of energy will set limits on both discovery and extraction of mineral raw materials. The social, economic, and political structure of the world must be such that it encourages the enormous endeavors that will be involved.

So much for the prospects for next 20 years, but what about prospects for the longer future? The crystal ball clouds over at this point. Who can foretell the technology and the mineral requirements even for 2050, much less for the centuries beyond? Who can foretell the future impact of rising costs of energy on the availability of nonfuel minerals to the world? At the present time man is making use of a very broad range of metals and nonmetallic materials present in the earth's crust. Some are common, but some are really very rare (Skinner, 1976), or rather we should say that deposits from which some mineral commodities must be obtained are rare. Those deposits cannot last forever. One can surely foresee that as time goes on, rising costs of the rarer minerals will stimulate efforts to adapt technology more and more toward use of the more abundant materials of the crust. In the variety of minerals being extracted and used, the century now drawing to a close may well be the golden age of mineral availability to man.

REFERENCES AND ADDITIONAL READING

Barney, G. O., 1980, *The Global Report to the President:* Vol. 1, *The Summary Report,* 360 pp.; Vol. 2, *The Technical Report,* 766 pp. Pergamon Press, New York.

Fantel, R. J., Sullivan, D. E., and Peterson, G. R., 1983, Phosphate rock availability—domestic. U.S. Bureau of Mines, Information Circular 8937, 57 pp.

Fischman, L. L., 1980, *World Mineral Needs and U.S. Supply Problems.* Resources for the Future, Washington, 535 pp.

Govett, G. J. S., and Govett, M. H., eds., 1976, *World Mineral Supplies: Assessment and Perspectives,* Elsevier, Amsterdam, 472 pp.

Lemons, J. F., Jr., Boyles, E. H., Jr., and Kilgore, C. C., 1982, Chromium availability—domestic. U.S. Bureau of Mines, Information Circular 8895, 14 pp.

Leontief, W., Koo, J. C. M., Nasar, S., and Sohn, I., 1983. *The Future of Nonfuel Minerals in the U.S. and World Economy.* Lexington Books, D.C. Heath, Lexington, Mass., 454 pp.

Malenbaum, W., 1978, *World Demand for Raw Materials in 1985 and 2000.* Mc-Graw Hill, New York, 126 pp.

Meadows, D. H., Meadows, D. L., Randers, J., and Behrens, W. W., 1972, *The Limits to Growth, A Report for the Club of Rome's Project on the Predicament of Mankind.* Potomac Associates–Universe Books, New York, 205 pp.

National Academy of Sciences, 1975, *Mineral Resources and the Environment.* Chapter VI, resources of copper. National Academy of Sciences, Washington, pp. 147–183.

National Academy of Sciences, 1975, *Mineral Resources and the Environment, Supplementary Report: Reserves and Resources of Uranium in the United States.* National Academy of Sciences, Washington, D.C., 236 pp.

Pretorius, D. A., 1981, Gold, geld, and gilt, future supply and demand. *Economic Geology,* Vol. 76, pp. 2032–2042.

Ridker, R. G., and Watson, W. D., 1980, *To Choose a Future.* Johns Hopkins University Press, Baltimore, 463 pp.

Skinner, B. J., 1976, A second iron age ahead? *American Scientist,* Vol. 64, No. 3, pp. 258–269.

Tilton, J. E., 1977, *The Future of Non-fuel Minerals.* Brookings Institution, Washington, D.C., 113 pp.

U.S. Bureau of Mines, annual volumes, *Mineral Commodity Summaries.* Department of the Interior, Washington, D.C.

U.S. Bureau of Mines, *Minerals in the U.S. Economy, 1964–1973, 1965–1974, 1966–1975, and 1968–1977.* U.S. Government Printing Office, Washington, D.C.

7 The Mineral Position of the United States

DEFINITION

We have examined the nature and distribution of important groups of mineral deposits, annual production and consumption, and reserves and resources available in the United States and in the world. It is time to assess the present mineral position of the United States. We have already done this in piecemeal fashion, but now we must take an overall view. We must begin by defining the term "mineral position." It is the relation between U.S. needs for minerals and its ability to supply those needs from its own resources. Having already discussed the energy position of the United States, we concern ourselves here with the nonfuel minerals.

THE HISTORICAL RECORD

The present mineral position of the United States is an outgrowth of the past. In colonial times, America's mineral needs were relatively few. Apart from stone and other local materials used for construction, minerals were supplied mainly by imports of finished goods. Largely cut off from imports during the Revolution, the country turned to its own resources. Iron mining and local coal mining were carried on in the Appalachian region, and modest amounts of certain other metals and nonmetallic minerals were produced. In the early decades of the nineteenth century, the pace of mineral development quickened. As the waves of settlers moved across the Midwest, the major lead and zinc deposits of southeastern Missouri, of the Tri-State district of Missouri–Kansas–Oklahoma, of eastern Tennessee, and of Wisconsin–Illinois were developed. Large deposits of many nonmetallic minerals were also discovered and brought into production.

In the 1840s copper deposits were discovered in the Keweenaw Peninsula of Upper Michigan, and in succeeding years the Keweenaw became one of the great copper-producing districts of the world. As midcentury approached, the discovery of gold in California in 1848 triggered the discovery and development of one major mining district after another in the western states. Cripple Creek, the Comstock Lode, Butte, the Mother Lode, Leadville, Coeur d'Alene, and Bisbee became names to conjure by, as they and other districts poured forth their mineral treasures.

Iron ore mining in the Lake Superior region also began in 1848; by the end of the century all the great iron ranges of Minnesota, Michigan, and Wisconsin were in production. Meanwhile the huge coal deposits of Pennsylvania and the midwestern and southeastern states were discovered. Coal and iron ore deposits of the Birmingham district of Alabama led to its development into the principal industrial area of the South. The petroleum industry began in Pennsylvania in the late 1850s. The mineral industry of the United States rapidly expanded.

When the 20th century began, the United States was already a major world producer and consumer of minerals, and its importance increased as the century wore on (Fig. 7-1). The value of nonfuel mineral production in constant 1967 dollars rose from $1.2 billion in 1900 to nearly $2.5 billion in 1929, responding to the burgeoning of industry in the United States and to economic development abroad.

Already in 1900 deficiencies in U.S. mineral resources had appeared,

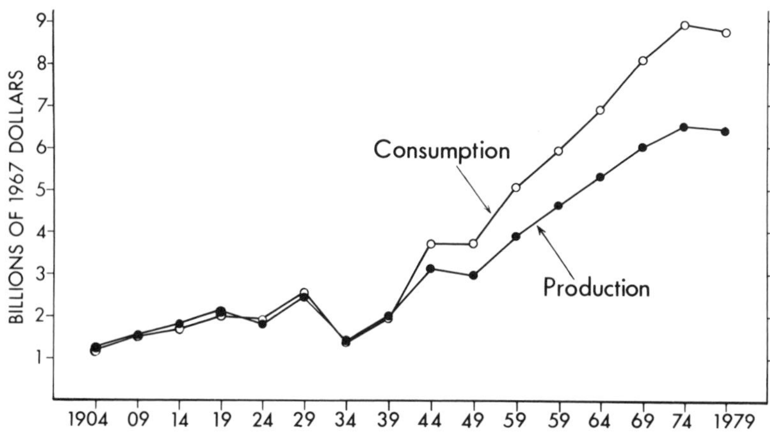

Figure 7-1 U.S. production and consumption of nonfuel minerals, 1900–1979. Each point on a curve is the average for the 5-year period ending with the year for which the point is plotted. Data from U.S. Bureau of Mines *Mineral Commodity Summaries* and *Minerals in the U.S. Economy* and from Spencer (1972). From Cameron (1982), by permission of Elsevier Publishing Company, Amsterdam.

and materials amounting in value to about 10 percent of United States production were being imported from various countries. Imports, however, were more than balanced in value by mineral exports. By 1929, the value of the country's imports had tripled, but so had the value of its exports. A balance was thus maintained, so that between 1900 and 1929 there was little change in the U.S. mineral position. If it did not improve, on balance it did not grow worse. This epoch of U.S. mineral history ended with the Great Depression of the 1930s. Both production and consumption declined and were not revived until World War II.

Changes in U.S. Mineral Position Since World War II

World War II was a turning point in the mineral history of the United States. As indicated in Fig. 7-1, from 1939 to 1974 U.S. consumption of nonfuel minerals increased more rapidly than production. The result was a large increase in the absolute amounts of imports and an increasing excess of imports over exports. Data that would make it possible to complete the chart through 1984 are not available, but the trend shown in Fig. 7-1 has continued. The U.S. Bureau of Mines estimates that in 1984 the value of net imports (total imports minus exports) of nonfuel minerals was $19 billion, approximately equal to 6.5 billion 1967 dollars. The gap between consumption and production has more than doubled since 1979.

Figure 7-1 is based on data for total production and total consumption of nonfuel minerals—that is, both primary (newly mined) and secondary (recycled) minerals. However, in assessing the mineral position of the United States it is necessary to examine the relation between U.S. primary production (mine production) and U.S. consumption of primary minerals.

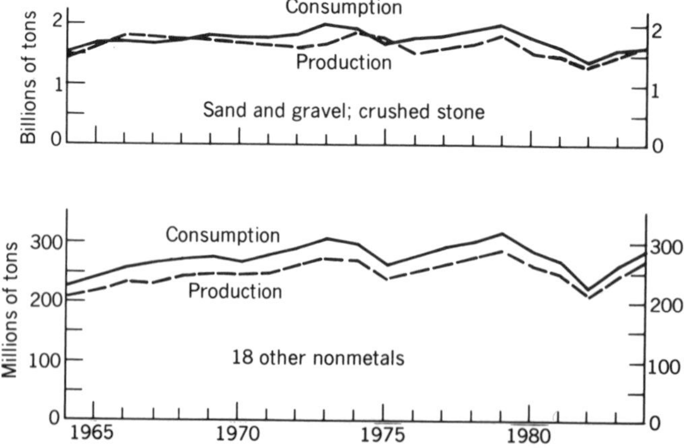

Figure 7-2 U.S. production and consumption of sand, gravel, and crushed stone, and 18 other nonmetals. Data from U.S. Bureau of Mines, *Mineral Commodity Summaries* and *Minerals in the U.S. Economy*.

This gives a true measure of the adequacy of mineral production from domestic mines. In Figs. 7-2 and 7-3, primary production and primary consumption from 1964 through 1984 are charted for sand and gravel plus crushed stone, for a group of 16 other nonmetallic mineral commodities, for iron (in terms of iron ore), and for 18 other metals. The mineral commodities charted account for approximately 99 percent of the total tonnages of metals and nonmetallic mineral commodities produced and consumed in the United States. For the nonmetallic group, on balance, the United States is close to being self-sufficient; there are very few serious deficiencies. The metals are another matter; primary production is far below primary consumption. Mine production of metals peaked in 1970 and has declined irregularly since. The metal mining industry has not shared in the general recovery from the recession of 1981–1982. For the nonmetallic commodities the curves indicate that the period of substantial growth in U.S. mineral production and consumption following World War II ended at about the time of the Arab oil embargo of 1973.

The United States has thus become significantly dependent on imports of certain minerals, and we must examine the nature of that dependence more fully. Charts published annually by the U.S. Bureau of Mines give a measure of dependence on imports of specific nonfuel minerals. The

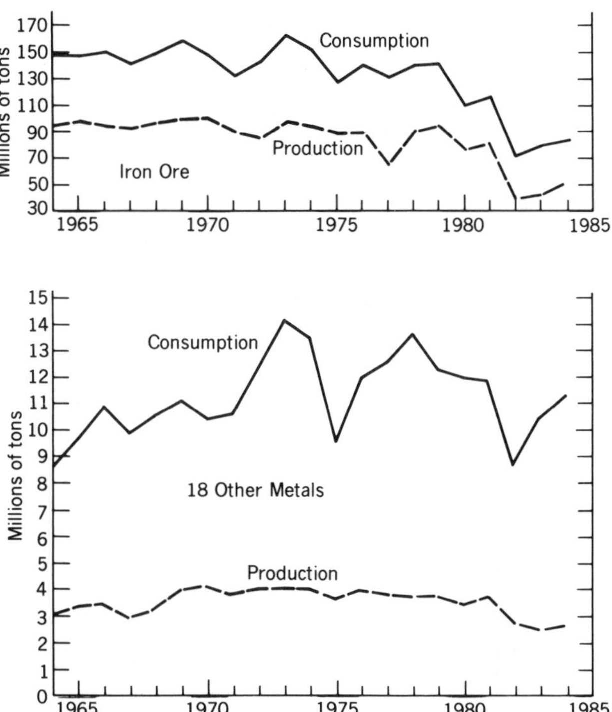

Figure 7-3 U.S. production and consumption of iron ore and 18 other metals. Data from U.S. Bureau of Mines, *Mineral Commodity Summaries* and *Minerals in the U.S. Economy*.

latest chart (Fig. 7-4) indicates that for 21 important mineral commodities dependence is 50 percent or more, and for another two dependence is 38 percent or more. There are a number of commodities not covered by the chart, but for most of them U.S. production is sufficient or nearly sufficient for domestic needs. On the right-hand side of the chart, principal sources of the various minerals are indicated. The names of the countries are in order of decreasing importance as sources of U.S. supply. Canada and Mexico contribute 25 percent or more of U.S. supplies of 21 of the 35 minerals, but for the remainder we are dependent on countries of South America and the eastern hemisphere, including China and the Soviet Union.

There are two common misconceptions of the nature of U.S. dependence on the mineral resources of other nations. The first is that the United States has built its industrial economy largely at the expense of the mineral

NET IMPORT RELIANCE: 1984
SELECTED NONFUEL MINERAL MATERIALS

U.S.A.

		MAJOR SOURCES
COLUMBIUM	100	Brazil, Canada, Thailand
MICA (sheet)	100	India, Belgium, France
STRONTIUM	100	Mexico, Spain
MANGANESE	99	So. Africa, France, Gabon, Brazil
BAUXITE & ALUMINA	96	Australia, Jamaica, Guinea, Suriname
COBALT	95	Zaire, Zambia, Canada, Japan
TANTALUM	94	Thailand, Malaysia, Brazil, Canada
FLUORSPAR	91	Mexico, So. Africa, China, Italy
PLATINUM GROUP	91	So. Africa, UK, USSR
CHROMIUM	82	So. Africa, Zimbabwe, USSR, Philippines
TIN	79	Thailand, Malaysia, Indonesia, Bolivia
ASBESTOS	75	Canada, So. Africa
NICKEL	74	Canada, Australia, Norway, Botswana
POTASH	74	Canada, Israel
TUNGSTEN	71	Canada, China, Bolivia
ZINC	67	Canada, Peru, Mexico, Australia
BARITE	64	China, Morocco, Chile, Peru
SILVER	61	Canada, Mexico, Peru, UK
MERCURY	60	Spain, Japan, Mexico, Turkey
CADMIUM	56	Canada, Australia, Mexico, Peru
SELENIUM	51	Canada, UK, Japan, Belg.-Lux.
VANADIUM	41	So. Africa, Canada, Finland
GYPSUM	38	Canada, Mexico, Spain
IRON & STEEL	23	Japan, EEC, Canada
COPPER	21	Chile, Canada, Mexico, Peru
SILICON	21	Canada, Brazil, Norway, Venezuela
IRON ORE	19	Canada, Venezuela, Liberia, Brazil
LEAD	18	Canada, Mexico, Australia, Peru
SULFUR	17	Canada, Mexico
GOLD	16	Canada, Switzerland, Uruguay
NITROGEN (fixed)	14	USSR, Canada, Mexico, Trinidad & Tobago
ALUMINUM	9	Canada, Ghana, Japan, Venezuela

Figure 7-4 U.S. net import reliance (imports minus exports) in 1984 for 32 mineral commodities. U.S. Bureau of Mines, courtesy of J.D. Morgan, Jr.

resources of the rest of the world. This is not true. As late as 1979 (and there has been little change since), the United States produced from its own resources 14.7 percent of total world production of nonfuel minerals and consumed 15.1 percent. In other words, mines and quarries *within the United States* furnished 97.4 percent of the tonnage of nonfuel minerals the nation consumed. About 80 percent of the world and U.S. tonnages, however, consisted of sand, gravel, and crushed stone. If we eliminate those three commodities from the totals, then U.S. production of other nonfuel minerals was 13 percent of world production, U.S. consumption was 15.2 percent, and the U.S. produced 85.5 percent of the nonfuel minerals it consumed. The remaining 14.5 percent was supplied by imports, but those imports amounted to only 3.7 percent of the production of nonfuel minerals (exclusive of sand, gravel, and crushed stone) in all the other countries of the world. Actually, the world's principal importers of nonfuel minerals are Japan (No. 1) and the nations of western Europe. As shown in Fig. 7-5, both are heavily dependent on imports for mineral supplies. In sharp contrast is the near self-sufficiency of the Soviet Union in minerals (Fig. 7-6).

The second misconception is that the United States draws most of its imports from the less developed nations of the world. Yet in 1979, of our total mineral imports, 38 percent came from less developed nations, 62 percent coming from other industrialized nations, including the Soviet Union. The view that the United States has been despoiling the mineral resources of the less developed countries is without foundation. It is interesting to note here that any attempt by the United States to restrict imports of minerals is strongly resisted by those less developed countries that have minerals available for export. Mineral exports are important sources of funds for these countries. For example, Chile, Zambia, and Zaire depend heavily on income from exports of copper. The economy of Jamaica depends on income from its bauxite deposits. Gabon derives essential income from exports of manganese ore. So it is for many less developed countries.

Successive net import reliance charts of the U.S. Bureau of Mines clearly portray the increase in U.S. dependence on imports since World War II. There are changes in the U.S. mineral position, however, that are not reflected in the charts. One is a change in our importance as a world producer and consumer of minerals. A complete picture of this change would require compiling data for U.S. and world production of all minerals. However, such data are lacking for some minerals, so the author has chosen to compile data for production and consumption of 18 important and representative commodities (Fig. 7-7) for which both world and U.S. data are available from 1930 through 1984. Figures for world consumption are not available, but as earlier pointed out, world consumption is approximately equal to world production. A single curve therefore suffices for both.

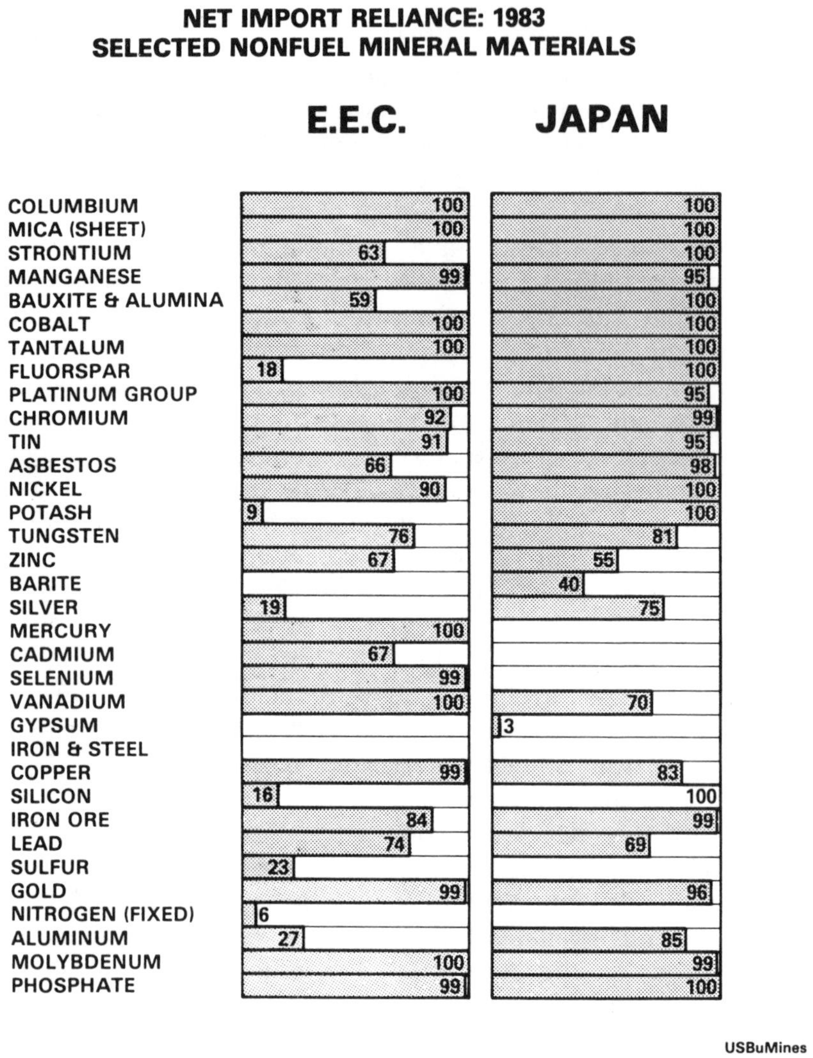

NET IMPORT RELIANCE: 1983
SELECTED NONFUEL MINERAL MATERIALS

	E.E.C.	JAPAN
COLUMBIUM	100	100
MICA (SHEET)	100	100
STRONTIUM	63	100
MANGANESE	99	95
BAUXITE & ALUMINA	59	100
COBALT	100	100
TANTALUM	100	100
FLUORSPAR	18	100
PLATINUM GROUP	100	95
CHROMIUM	92	99
TIN	91	95
ASBESTOS	66	98
NICKEL	90	100
POTASH	9	100
TUNGSTEN	76	81
ZINC	67	55
BARITE		40
SILVER	19	75
MERCURY	100	
CADMIUM	67	
SELENIUM	99	
VANADIUM	100	70
GYPSUM		3
IRON & STEEL		
COPPER	99	83
SILICON	16	100
IRON ORE	84	99
LEAD	74	69
SULFUR	23	
GOLD	99	96
NITROGEN (FIXED)	6	
ALUMINUM	27	85
MOLYBDENUM	100	99
PHOSPHATE	99	100

USBuMines

Figure 7-5 Net import reliance of the European Economic Community and Japan in 1983 for supplies of selected nonfuel mineral materials. U.S. Bureau of Mines, courtesy of J.D. Morgan, Jr.

The chart shows that from 1930 to 1945, the United States produced 42 to 58 percent of world supplies of the 18 minerals and consumed 39 to 56 percent. Since 1946, however, world production and consumption of the 18 minerals have increased at a phenomenal rate, doubling three times between 1947 and 1982. As we have seen, U.S. production and consumption also increased until the early 1970s, but at a much lower rate. By 1981, U.S. consumption was only about 12 percent of world production. The importance of the United States as a world producer and consumer of minerals has therefore greatly changed. In the late 1940s the country was the dominant producer and consumer of world minerals and played a dominant role in world mineral markets. That is no longer

U.S.S.R. NET IMPORTS OF SELECTED MINERALS AND METALS AS A PERCENT OF CONSUMPTION IN 1978*

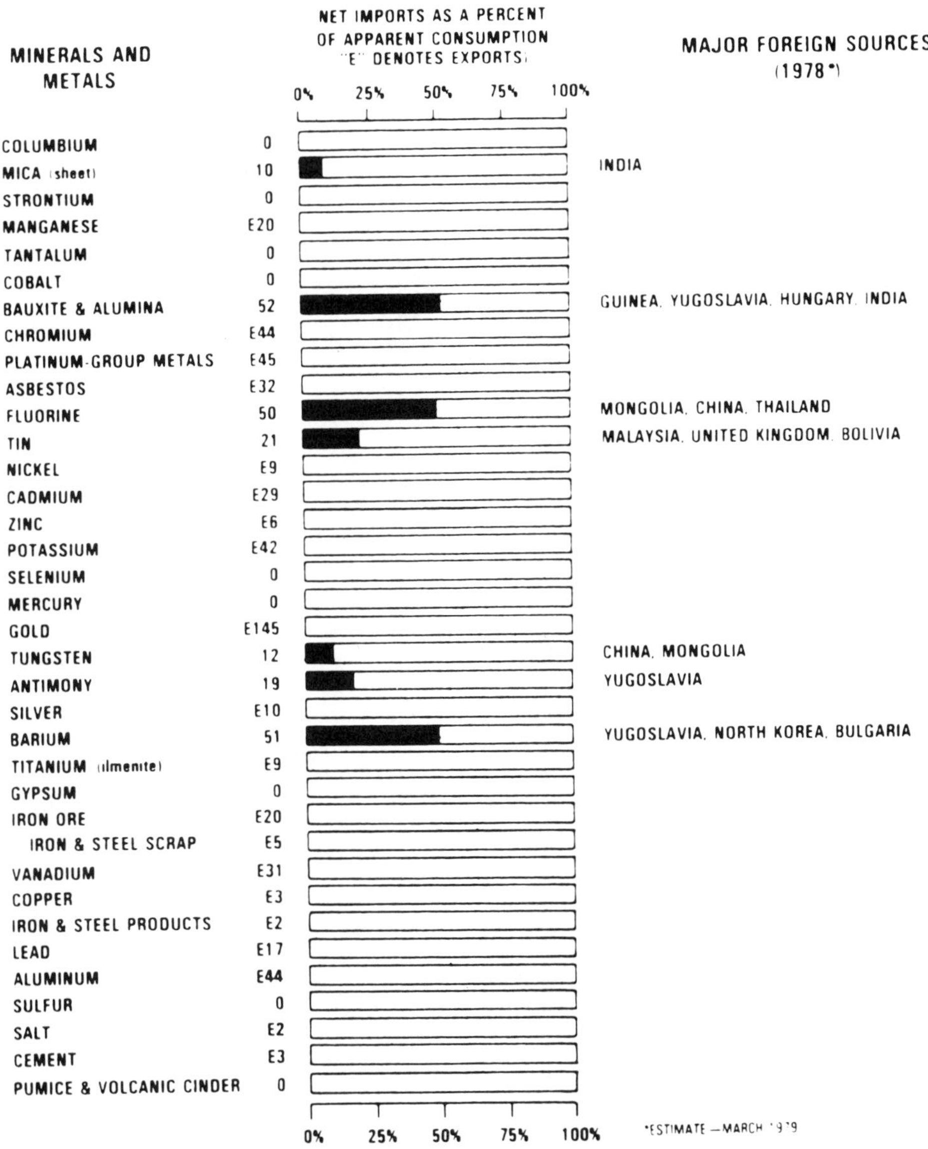

Figure 7-6 Net import reliance of the Soviet Union in 1978 for supplies of selected nonfuel minerals. U.S. Bureau of Mines, courtesy of J.D. Morgan, Jr.

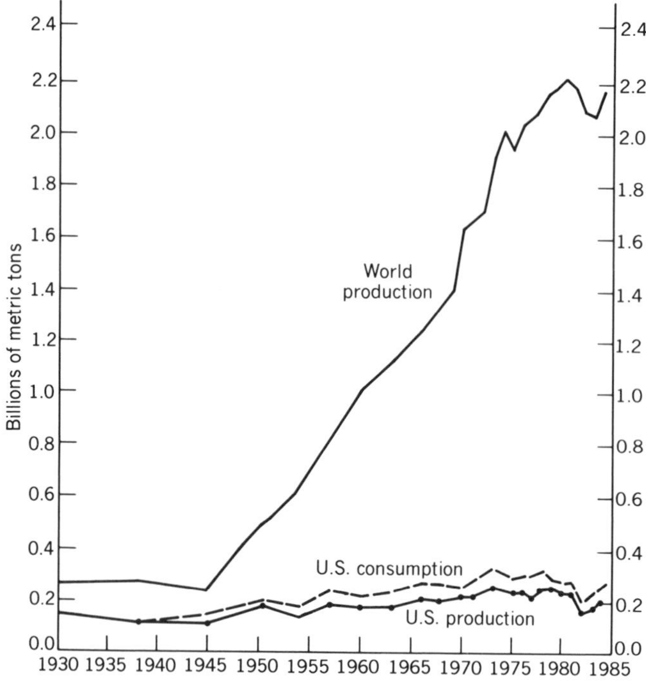

Figure 7-7 World consumption and U.S. production and consumption of 18 nonfuel minerals, 1930–1984: iron ore, bauxite, copper, lead, zinc, tungsten, chromium, nickel, molybdenum, manganese, tin, vanadium, fluorspar, phosphate, cement, gypsum, potash, and sulfur. Data from U.S. Bureau of Mines, *Minerals Yearbook, Mineral Commodity Summaries,* and *Minerals in the U.S. Economy.*

true, and the United States faces increasing competition both in selling mineral commodities and in acquiring supplies of minerals that cannot be obtained in adequate amounts from domestic mines.

Data for mineral imports and exports are one element of the mineral position of the United States, but there are two other important elements. One is the nation's capacity for smelting, refining, and other processing of mineral raw materials into the metals and other products that are needed by American industry. The other element is the capacity of American industry for converting those products into finished goods. The chart of Fig. 7-4 therefore only partly expresses the broad mineral position of the United States. Let us consider it more closely. It is based on two sets of data: (1) data for imports of metallic ores and concentrates, plus unprocessed nonmetallic minerals, and (2) data for imports of metals and semiprocessed nonmetallic minerals. In the early part of the period following World War II, such charts gave a fairly accurate picture of U.S. reliance on imports. Most imports were indeed metallic ores and concentrates plus nonmetallic raw materials. Reliance on imports was thus largely a matter of deficiencies in production of such materials from domestic mines and mills. American industry not only was very broad in scope but was vertically integrated. Given the necessary raw materials, the smelting, refining, and other processing plants, together with the factories needed for converting mineral-derived materials into consumer goods, were largely available within the United States. This structure was the product of a history of more than 100 years, a history of discovery of mineral deposits, development of mining operations, and building of smelters, refineries, and other processing plants. There followed the development of fabricating industries, in which mineral commodities were converted into useful forms. Minerals thus became the base of a broad industrial complex.

The process involved is clearly illustrated by the history of Butte, Montana—one of the world's most productive mining districts. Mining began at Butte in 1864, first for gold, then for silver. Copper ores were discovered in the early 1880s, and the first ore was mined in 1884. There was neither smelter nor refinery in Montana. During the next few years copper concentrates were hauled about 350 miles by oxcart to Corinne, Utah, then carried by rail to eastern ports for shipment to the smelters at Swansea, Wales. Some ore was later sent by rail to Portland, Oregon, thence by ship around Cape Horn, again to the Swansea smelters. There were now copper mines in the western United States but no smelting, refining, or fabricating industries.

This situation did not last long. A smelter and then a refinery were built in Montana, and shipments of ores and concentrates to Wales ceased. Not only that, but there developed over the years a whole structure of domestic fabricating industries based on the mines at Butte and those of other western districts. The copper ore that might have fed the growth of copper-based industries in England and Wales now fed the growth of American copper-based industries. In effect, Butte illustrates the fact that mineral industry tends to develop and flourish around major sources of mineral supply. As old sources are depleted, mineral-based industries survive for a time by importing ore and for a while longer by importing semifinished materials, but withering is inevitable. The problem is that such changes are slow and not easily perceived, either by the general public or by the makers of mineral policy. Just a few hundred jobs are lost at any one time, and the root causes go unperceived. There is none of the dramatic impact of the Arab oil embargo. This is what has happened at Butte, where the mines have finally closed, and both the great smelter at nearby Anaconda and the refinery at Great Falls have been abandoned.

The general history associated with Butte is being repeated in certain other mineral-based industries of the United States. Table 7-1 gives data for U.S. supply of chromite for 1948 and 1981. The chromium industry of the United States began late in the nineteenth century. It was based on chromite supplied from mines in Maryland and Pennsylvania. These

TABLE 7-1 U.S. Chromite Supply, 1948 Compared with 1981 (Short Tons)

		U.S. Imports		
	World Production	Chromite	Ferrochrome (Chromite Equivalent)	Ferrochrome (% of Total Supply)
1948	2,332,752	1,542,000	14,000	0.8
1981	10,200,000	760,000	730,000	49.0

and other domestic deposits were small and soon depleted; the United States became dependent on sources of chromite abroad. Until World War II, supplies of chromite were freely available, but as Table 7-1 shows, there has been a significant change since 1948. Smelters have been built in the producing countries, more are under way, and more and more of the U.S. supply is being imported as the alloy ferrochrome. The U.S. ferrochrome industry has been declining.

Table 7-2 gives data for manganese supply, again for 1948 and 1981. The story of manganese differs in detail from that of chromium, but the net result is the same. In Fig. 7-8, data for aluminum supply are given. Since 1966, an increasing proportion of our supply has been imported as alumina, the production of which from bauxite is the first link in the chain between mine and factory. Imports of aluminum metal appeared in the record for the first time in 1965. The zinc mining industry has a similar history. U.S. mine production of zinc declined from 611,000 tons in 1965 to 331,000 tons in 1982. Since 1968, primary smelter capacity has dropped 60 percent, and primary smelter production has dropped 75 percent.

As shown by Hewett (1929), over any extended period of time, as old mining districts decline and finally cease production, the smelting, refining, and other processing industries based on mining tend to be transferred to countries where new mineral districts are being discovered and developed. This is particularly true for the metals, because domestic mines and smelters must compete in international markets. The history of Butte is now being repeated in other parts of the world. Mineral districts that for a time were simply sources of ores and concentrates for the smelters,

TABLE 7-2 U.S. Manganese Supply, 1948 Compared with 1981 (Short Tons)

		U.S. Imports		
	World Production	Manganese Ore	Ferromanganese (Ore Equivalent)	Ferromanganese (% of Total Supply)
1948	4,305,600	1,256,597	200,000	13.7
1981	26,000,000	675,000	570,000	46.0

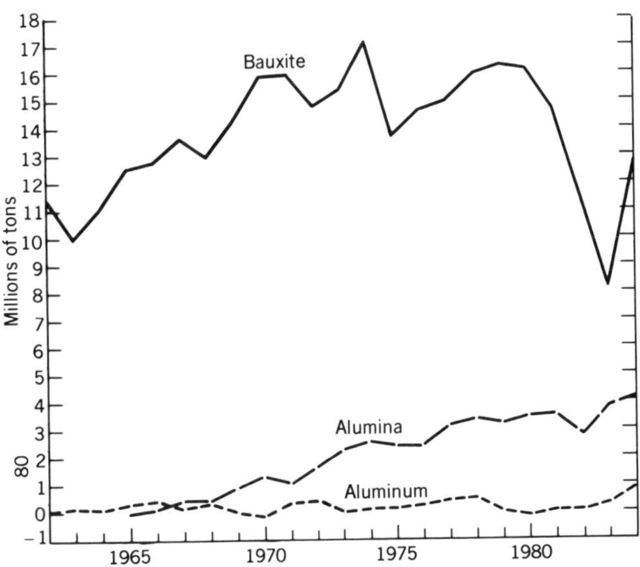

Figure 7-8 U.S. imports of bauxite, alumina, and aluminum, 1962–1984. Data from U.S. Bureau of Mines, *Mineral Commodity Summaries* and *Minerals in the U.S. Economy.*

refineries, and processing plants of the United States, western Europe, and Japan are now developing into complex undertakings of the kind exemplified by Butte at the height of its development. The end result is the transfer of important segments of mineral industry to mineral-producing countries. Transfer of smelting, refining, and processing capacity is actively encouraged, sometimes subsidized, by the countries in which important new mineral deposits are discovered. In the aluminum industry, most of the new alumina and aluminum plants are being built in countries such as Brazil, Venezuela, and Australia, which have the great bauxite reserves of the world. New ferrochrome plants in South Africa, Zimbabwe, and even Brazil and India are supplanting ferrochrome plants in the United States. The same thing is happening in the manganese industry and apparently in the copper industry. In the latter industry, however, part of the increase in U.S. metal imports results from the creation of a subsidized smelting and refining industry in Japan; this has partly supplanted U.S. smelting industry.

The consequences of industry transfer should be clearly understood, the more so because they are rarely given adequate consideration in the framing of U.S. mineral policy. When a mining district is shut down, there are not just loss of direct employment and loss of the wealth represented by the products of the mines. There is also loss of employment in the secondary and tertiary industries of supply and service to the mines, mills, and smelters. These effects are felt far beyond the boundaries of the mining district. The manufacturers of mining machinery in Milwaukee, Wisconsin, for example, have been strongly affected by the decline or cessation of activity in the copper mining districts of the West and in the iron ranges of Michigan and Minnesota. Federal, state, and local tax receipts decline.

Another change in the U.S. mineral position not reflected in charts such as Fig. 7-4 is an increase in the net amounts of minerals that are imported in the form of manufactured goods. The change is difficult to quantify, because we do not have data for recent years for the metal and mineral contents of imported manufactured goods. It is obvious, however, that the increase in recent years has been substantial, owing to rising imports of automobiles, tractors and other agricultural equipment, motorcycles, snowmobiles, sewing machines, radios, television sets, cameras, kitchen utensils, and innumerable other goods. These imports indicate that U.S. capacity for producing manufactured goods that are competitive in domestic markets has not kept pace with domestic demand.

The final important change in U.S. mineral and mineral-based industries is a consequence of the growth of multinational corporations. It is closely related to the others. What 35 years ago were vertically integrated U.S. industries are being progressively converted into international industries. The automobile industry is a conspicuous example. There used to be a sharp distinction between the U.S. automobile industry and the automobile industries of other countries. In May 1980, however, when layoffs in American plants brought a call for restrictions on imports of automobiles, the American public became aware that the distinction had become blurred, to say the least. The motors of American light trucks were being made at home, but the bodies were imported from Japan. One "American" car had parts imported from six countries. The same thing has happened in other industries. In the electronics industry, some parts of equipment are made in the United States, others in Japan or elsewhere, but the assembly may be done in Taiwan, Hong Kong, Singapore, or Mexico.

To conclude, in assessing the mineral position of the United States, it does not suffice to examine charts that show reliance on imports of mineral raw materials. The whole structure of American mineral-based industry must be taken into account, not only its capacity for mining, milling, smelting, and other processing of ores, but its capacity for converting mineral raw materials into finished goods. In the same way, the energy position of the nation depends not just on the amounts of fuel materials that it can produce, but on its capacity for producing useful energy from these materials. From the foregoing it will be evident that the broader mineral position of the United States has weakened significantly in recent years. There are many who see little prospect of reversing this trend, viewing it as firmly rooted in the economic evolution of the United States and in American relationships with other countries of the world. We return to this question in the final chapter.

FUTURE MINERAL SUPPLIES FROM DOMESTIC SOURCES

What will the future bring? In the previous chapter we discussed the availability and adequacy of world supplies of nonfuel minerals during the period 1985 to 2005. Here we examine, for the same period, the availability and adequacy of U.S. supplies from domestic sources. We start with similar data—rates of consumption, future rates of consumption as suggested by trends of the past 20 years, and data for the U.S. reserve base. There are the same pitfalls—weaknesses in reserve base data, uncertainties as to the impact of changes in technology of production and use, possible discovery of new resources, and rising costs of energy. There are additional pitfalls in the changes we have noted above in the nature of American mineral-based industry and the declining competitiveness of some segments of American mining industry. Under the circumstances, no firm forecast for all minerals is to be expected. Nonetheless, the effort will enable us to see something of the shape of America's mineral future.

As a basis for anticipating future demand, we need data for primary consumption of individual commodities. Since recycling of nonmetals is insignificant, primary consumption of nonmetallic minerals can be taken as roughly equal to total consumption. Consumption of sand and gravel plus crushed stone during 1964–1984 was shown in Fig. 7-2. Data for consumption of each of 18 other nonmetallic commodities are plotted in Figs. 7-9, 7-10, and 7-11. Figures 7-3, 7-12, and 7-13 give the necessary data for primary consumption of iron (in terms of iron ore) and 18 other metals. Ideally the graphs might be extended back to 1950, so as to show

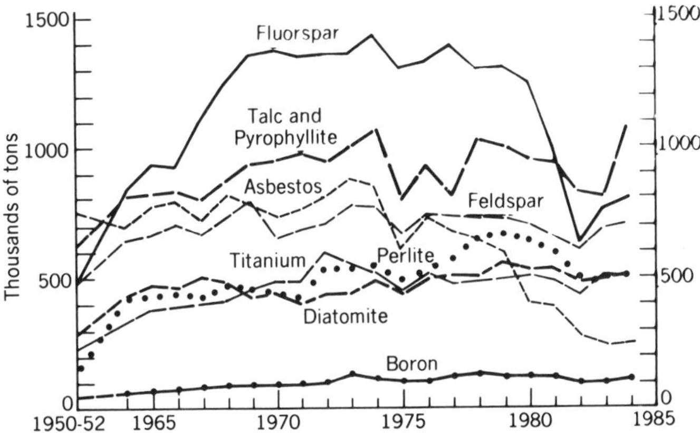

Figure 7-9 U.S. average annual consumption, 1950–1952, and annual consumption, 1964–1984, of fluorspar, talc and pyrophyllite, asbestos, feldspar, perlite, titanium (titanium content of rutile and ilmenite used for nonmetallic purposes), diatomite, and boron (boron content of boron minerals and compounds). Data used in constructing this chart and Figs. 7-10, 7-11, 7-12, and 7-13 from U.S. Bureau of Mines, *Mineral Commodity Summaries* and *Minerals in the U.S. Economy*.

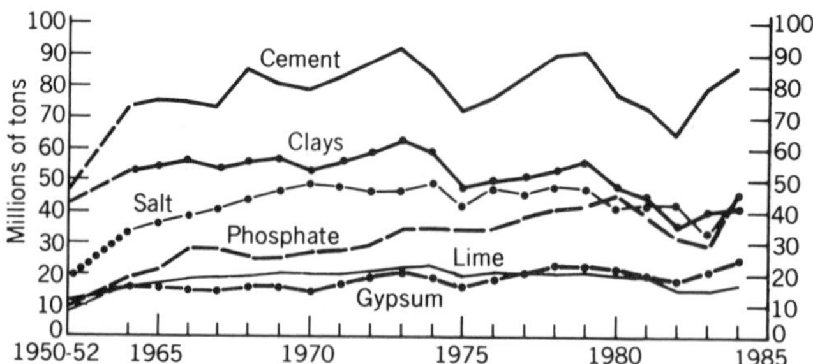

Figure 7-10 U.S. average annual consumption, 1950–1952, and annual consumption, 1964–1984, of cement, clays, salt, phosphate, lime, and gypsum.

the full pattern of growth since World War II. Unfortunately, the necessary data are not available for all commodities. However, at the left of each chart a figure is given for average consumption of each of a number of the minerals during 1950–1952. For any mineral, a comparison of this average with consumption in 1964 gives a measure of growth in consumption during the interval.

Most of the curves show irregular growth of consumption in the 1960s and early 1970s, a continuation of the trend that developed after World War II. For the 1970s, however, there is no consistent pattern. Consumption of some commodities peaked in 1973–1974, at about the time of the Arab oil embargo. Rising costs of energy have undoubtedly had an impact on mineral consumption. However, peaks of consumption for manganese, sulfur, barite, silicon, phosphate, potash, talc, soda ash, and perlite were reached only in the late 1970s or in 1980–1984. The recessions of 1975 and 1981–1982 caused drops in consumption. The effects of the 1981–1982 recession were especially severe. Some of the graphs reflect changes in patterns of use. Consumption of lead is declining as use of tetraethyl lead in gasoline is phased out. Silicon is being used in increased

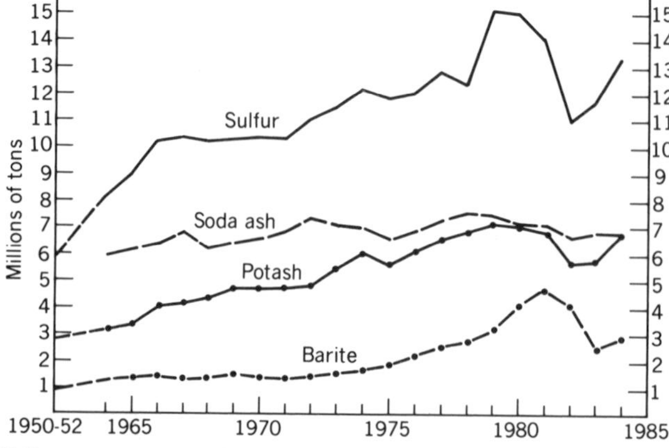

Figure 7-11 U.S. average annual consumption, 1950–1952, and annual consumption, 1964–1984, of sulfur, soda ash, potash, and barite.

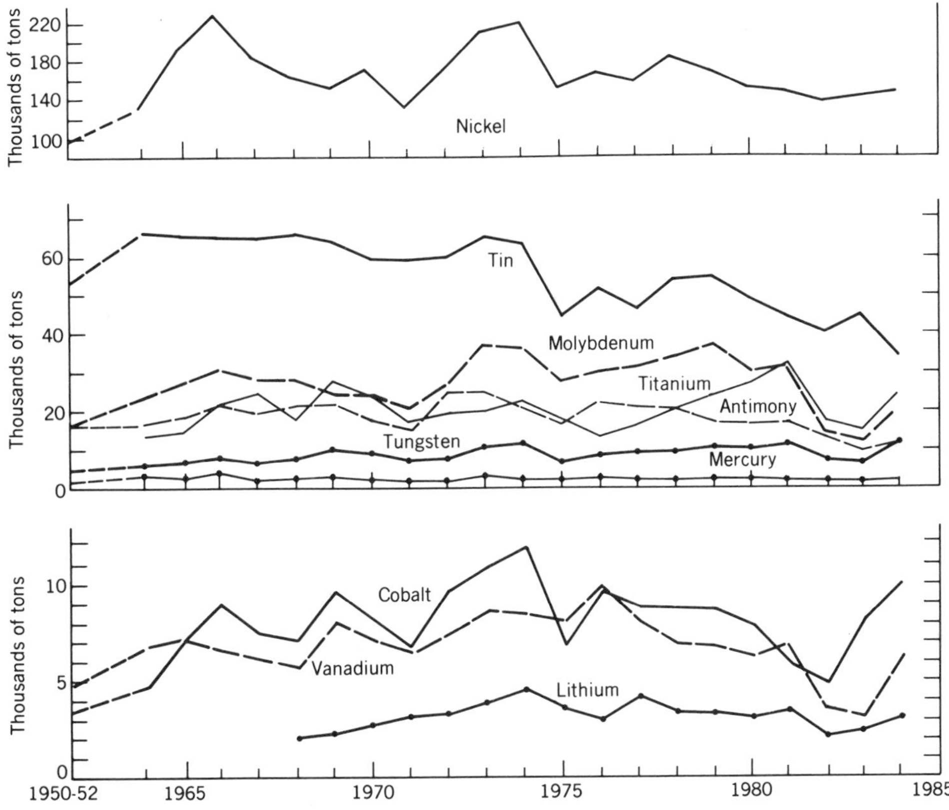

Figure 7-12 U.S. average annual consumption, 1950–1952, and annual consumption, 1964–1984, of primary nickel, tin, molybdenum, titanium, antimony, tungsten, mercury, cobalt, vanadium, and lithium.

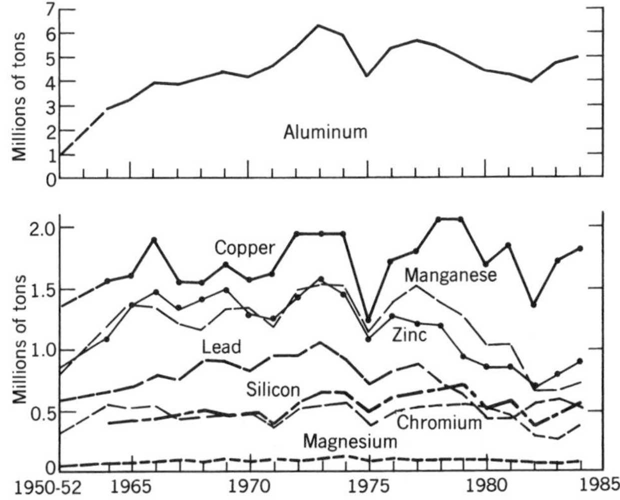

Figure 7-13 U.S. average annual consumption, 1950–1952, and annual consumption, 1964–1984, of primary aluminum, copper, manganese, zinc, lead, silicon, chromium, and magnesium.

amounts in steelmaking. Substitution of plastics has affected consumption of certain metals. Decreased use of asbestos and mercury reflects concern with their effects on health.

In appraising future world mineral supplies we made use of the reserve base/production index. For the United States we make use of an index that is the ratio of the reserve base to primary consumption, or RB/Cp (Tables 7-3 and 7-4). Here the index is calculated by dividing the reserve base by the average consumption of primary metal or nonmetal during 1974–1984. For asbestos and lead, however, averages for 1979–1984 are used, because the decline in consumption of the two commodities is likely to be permanent. The RB/Cp index has the same limitations as the RB/P index. It does not tell us how much of the reserve base can be recovered ultimately at acceptable costs. It does not tell us the "life" of reserves

TABLE 7-3 U.S. Reserve Base/Consumption Indices for Nonmetallic Minerals

	Reserve Base[a] 1984	Average Consumption[a] 1974–1984	RB/Cp
Salt	∞	44,561	∞
Cement	Very large	79,478	Very large
Clays[b]	Very large	48,212	Very large
Lime	Very large	18,792	Very large
Sand and gravel	Very large	807,556	Very large
Stone (crushed)	Very large	940,963	Very large
Sodium carbonate	36,600,000	7,043	5196
Diatomite	500,000	533	938
Perlite	200,000	555	360
Feldspar	200,000	700	288
Boron[c]	18,000	106	170
Phosphate rock	5,400,000	38,327	141
Titanium[d]	50,000	482	112
Potash	360,000	6,425	56
Gypsum	500,000	20,252	25
Barite	60,000	2,987	20
Asbestos	4,400	303	15
Sulfur	175,000	12,911	14
Fluorspar	8,000	1,112	7

[a]Thousands of tons.

[b]Clays of all types. Data for individual types are not available.

[c]Boron content of boron ores or concentrates.

[d]Titanium content of titanium minerals and slags used for nonmetallic purposes.

Source Data for reserve base and consumption from U.S. Bureau of Mines, *Commodity Data Summaries*.

TABLE 7-4 U.S. Reserve Base/Consumption Indices for Metals

	Reserve Base[a] 1984	Average Consumption Primary[a] Metal 1974–1984	RB/Cp 1984
Magnesium	∞	92.5	∞
Silicon	Very large	570	Very large
Molybdenum	5,900	28	211
Iron ore	24,800,000	120,000	207
Lithium	460	3.4	135
Copper	90,000	1,755	51
Zinc	53,000	1,036	51
Lead	27,000	661	41
Tungsten	320	9.6	33
Vanadium	240	7.6	32
Cobalt[b]	192	7	27
Nickel[c]	2,800	160	18
Antimony	100	14.9	7
Mercury[d]	200	35	6
Aluminum	12,000	4,881	2
Tin	55	44	1
Chromium	0	485	0
Manganese	0	1,115	0

[a]Thousands of tons, except mercury.

[b]The cobalt reserve base consists entirely of subeconomic resources.

[c]The nickel reserve base consists largely of subeconomic resources.

[d]Thousands of 76-pound flasks.

Source: Data for reserve base and consumption from U.S. Bureau of Mines, *Commodity Data Summaries.*

and resources, because that is a function of rates of production. However, it is useful if its limitations are kept in mind, and it is the only measure we have of the size of U.S. reserves and subeconomic resources relative to rates of consumption.

What of the future? This depends, of course, not only on the reserve base but also on whether U.S. consumption increases, decreases, or remains the same. In trying to decide which will be the case, one looks for trends in consumption in recent years, as we did in appraising the future of world mineral supply, but trends are difficult to discern in most of the graphs of U.S. mineral consumption presented in Figs. 7-9 to 7-13. The U.S. Bureau of Mines (Mineral Commodity Summaries, January 1985) predicts that there will be annual increases in consumption of most mineral commodities through 1990, but Figs. 7-2, 7-3, and 7-9 to 7-13 show that during 1974–1984 annual consumption of 28 of the 37 mineral commodities either did not increase or actually declined. For the present purpose, let us assume that average U.S. annual consumption during 1985–2005 remains the same as during 1974–1984. The question is, can present rates of U.S. primary production be maintained? Tables 7-5 and 7-6 are pertinent to this question. The tables show, for each commodity, average annual primary production during 1974–1984, RB/Pp indices as of 1984, total production during 1985–2005, the effect on the reserve base of each mineral, and RB/Pp indices calculated accordingly for 2005.

Table 7-5 suggests that in the year 2005 the mineral position of the United States with respect to nonmetallic minerals will still be very strong. There should be no difficulty in maintaining production of most of the nonmetals at 1974–1984 rates not only through 2005 but considerably beyond, although costs of some of them are likely to rise. There is a major deficiency in potash, but ample supply is available from the huge deposits of Saskatchewan. There may also be a problem with phosphate, owing to rising costs and environmental problems in the Florida phosphate field. However, there is some hope that very large resources identified off the coast of the Carolinas (see Chapter 14) can be brought into production by the end of the century. Maintaining production of gypsum, barite, sulfur, and fluorspar will require substantial additions to the reserve base through discovery. Chances of discovery are good for the first three minerals, and it is possible that new deposits of fluorspar will be found in Alaska. Asbestos is a special problem. As noted previously, the U.S. reserve base consists almost entirely of short-fiber asbestos. There is scant prospect of discovering deposits that will yield the higher grades; heavy dependence on imports will continue.

Table 7-6 indicates that unless there are substantial additions to the reserve base, dependence on imports of metals will increase. As noted in Table 7-6, only 150,000 tons of the reserve base for nickel is economically extractable; the rest is in low-grade, high-cost resources. If the latter are subtracted from the reserve base, the RB/Cp index for 1984 becomes 12.5, and the index for 2005 becomes zero, which means that the current rate of nickel production cannot be maintained for the rest of the century. The reserve base for copper calculated for 2005 is in the marginal range. Considering that the reserve base includes large amounts of material currently uneconomic, it is very doubtful that the 1974–1984 rate of production can be maintained. There are already predictions that U.S. primary copper production will drop to between 500,000 and 800,000 tons by the end of the 1980s. The reserve base figure for tungsten is like those of nickel and copper, including much low-grade, high-cost material. Only rarely in the past 20 years have domestic mines produced as much as 50 percent of U.S. requirements; in terms of the present reserve base, increased dependence on imports is inevitable. The 2005 RB/Pp indices for

TABLE 7-5 U.S. Nonmetals Reserve Base/Production Indices, 1984 and 2005

	Reserve Base 1984[a]	Average Annual Production 1974–1984[a,b]	Reserve Base/ Production Index 1984	Production 1985–2005[a,c]	Remaining Reserve Base 2005[a]	Reserve Base/ Production Index 2005
Salt	∞	38,380	∞	805,980	∞	∞
Cement	Very large	70,053	Very large	1,471,113	Very large	Very large
Clays[d]	Very large	45,396	Very large	953,316	Very large	Very large
Lime	Very large	17,202	Very large	361,242	Very large	Very large
Sand and gravel	Very large	747,202	Very large	15,691,242	Very large	Very large
Crushed stone	Very large	870,975	Very large	18,290,475	Very large	Very large
Sodium carbonate	36,600,000	7,411	4939	155,631	36,444,369	4918
Diatomite	500,000	648	772	13,608	486,392	751
Perlite	200,000	615	325	12,915	187,085	304
Feldspar	200,000	651	307	13,671	186,329	286
Potash	360,000	2,088	172	43,848	316,152	151
Titanium (ilmenite)	90,000	596	151	12,516	77,484	130
Talc and pyrophyllite	150,000	1,126	133	23,646	126,354	112
Phosphate rock	5,400,000[e]	47,840	113	1,004,640	4,395,360	92
Boron	18,000	206	87	4,326	13,674	66
Asbestos[f]	4,400	77	57	1,617	2,783	21
Fluorspar	8,000	173	46	3,633	4,367	25
Gypsum	500,000	11,657	43	244,797	255,403	22
Barite	60,000	1,501	40	31,521	28,479	19
Sulfur	193,000	11,592	17	243,432	0	0

[a]Thousands of tons.

[b]Except asbestos, for which the average for 1979–1984 is used.

[c]Assuming annual production at rates shown in column 3.

[d]All types.

[e]Revised figure, U.S. Bureau of Mines, 1985.

[f]Short fiber only.

Source: Data for production and 1984 reserve base from U.S. Bureau of Mines, *Mineral Commodity Summaries, 1984.*

TABLE 7-6 U.S. Metals Reserve Base/Production[a] Indices, 1984 and 2005

	Reserve Base[b]	Average Annual Production[a,b]	Reserve Base/ Production Index 1984	Production[a,b,c] 1985–2005[c]	Remaining Reserve Base[b] 2005	Reserve Base Production Index 2005
Magnesium	∞	132	∞	2,772	∞	∞
Vanadium	2,400	5	480	105	2,295	459
Silicon	Very large	481	Very large	10,101	Very large	Very large
Iron ore	24,800,000	74,684	332	1,568,364	23,231,636	311
Nickel[d]	2,800	12	233	252	2,548	212
Zinc	53,000	384	138	8,064	44,936	117
Tungsten	320	3	107	63	257	86
Molybdenum	5,900	57	104	1,197	4,703	83
Lithium	460	6	77	126	334	56
Copper[e]	90,000	1,433	63	30,093	59,907	42
Lead	27,000	559	48	11,739	15,261	27
Aluminum	12,000	356	34	7,476	4,524	13
Tin	55	small	~50	50?	5	5?
Mercury	200[f]	26[f]	8	546[f]	0	0
Antimony	100	13	8	273	0	0
Chromium	0	0	0	0	0	0
Manganese	0	0	0	0	0	0
Cobalt[g]	—	0	—	0	—	—

[a]Primary metal production.

[b]Thousands of tons, except mercury.

[c]Assuming average rate of production is equal to the 1974–1984 average. Exception mercury, for which the 1979–1984 average is used.

[d]Only 150,000 tons of the reserve base is currently ore.

[e]Reserve base includes a large amount of high-cost copper, currently uneconomic.

[f]Thousands of 76-pound flasks.

[g]The reserve base for cobalt is given by the U.S. Bureau of Mines as 950,000 tons. None is economic at present, and "most domestic resources are in subeconomic concentrations that will not be economically usable in the foreseeable future."

Source: Data for production and 1984 reserve base from U.S. Bureau of Mines, *Mineral Commodity Summaries.*

lead, mercury, and antimony clearly indicate increasing dependence on imports. Complete or virtually complete dependence on imports of tin, manganese, and chromium will continue.

The question of costs requires more than passing mention, because there are disquieting trends both in costs of mineral discovery and in costs of mineral production. There is much evidence that both dollar costs and energy costs of finding and producing minerals, especially the metals, are increasing substantially in the United States. We have already seen the potential impact of rising costs on phosphate production. Costs of mineral discovery are rising, because searching for concealed deposits at increasing depths becomes progressively more difficult. Costs of production are rising as lower-grade and deeper-lying deposits are tapped. For most metals, there is no breakdown of the reserve base into low-cost and high-cost ores. We do not, therefore, have the basis for calculating future costs of production and predicting their impact on availability of metals from domestic sources.

To sum up, the U.S. mineral position is weak with respect to a number of important metals; in terms of the present reserve base it will become progressively and significantly weaker through 2005. This is a serious problem; to a large extent the solution must lie in additions to reserves through discovery and development of economically viable new deposits. The western states including Alaska offer possibilities of large additions for lead, zinc, copper, molybdenum, and possibly tungsten, mercury, antimony, and even nickel. However, exploration in the United States for metals, except for silver and gold, is virtually at a standstill. This is an unhappy situation because of the long lead times required for discovering major mineral deposits and bringing them into production. The availability of new reserves by the end of the present century will depend on the level and success of exploration undertaken in the remaining years of the 1980s and the early 1990s. Time is running out.

The major deposits needed will not be found and developed quickly, easily, or cheaply. If the United States wishes to maintain its present levels of mineral supplies from its own resources or to increase them, it will have to provide, during the remainder of the 20th century, a political, social, and economic environment unusually favorable to mineral exploration and development. In succeeding chapters we will examine the elements of that environment and their impact on the availability of minerals from domestic sources.

REFERENCES AND ADDITIONAL READING

Brobst, D. A., and Pratt, W. P., eds., 1973, United States mineral resources. U.S. Geological Survey, Professional Paper 820, 722 pp.

Cameron, E. N., 1982, Non-fuel mineral problems of the United States. *Resources and Conservation*, Vol. 9, pp. 1–16.

DeYoung, J. H., Jr., Sutphin, D. M., and Cannon, W. F., 1984, International Strategic Minerals Inventory summary report—manganese. U.S. Geological Survey, Circular 930-A, 230 pp.

Hewett, D. F., 1929, Cycles in metal production. American Institute of Mining and Metallurgical Engineers, Yearbook, 1929, pp. 65–98.

McKelvey, V. E., 1973, Mineral potential of the United States. In *The Mineral Position of the United States, 1975–2000,* E. N. Cameron, ed., University of Wisconsin Press, Madison, pp. 67–82.

Morgan, J. D., Jr., 1975, The mineral position of the United States, 1974. In *Politics, Minerals, and Survival,* R. W. Marsden, ed., University of Wisconsin Press, Madison, pp. 3–26.

National Academy of Sciences, 1975, *Mineral Resources and the Environment.* National Academy of Sciences, Washington, D.C., pp. 77–183d.

President's Materials Policy Commission, 1952, *Resources for Freedom,* Vol. 1. U.S. Government Printing Office, Washington, D.C. 184 pp.

Rickard, T. A., 1932. *A History of American Mining.* McGraw-Hill, New York, 419 pp.

Spencer, V. E., 1972, *Raw Materials in the U.S. Economy: 1900–1969.* U.S. Bureau of the Census and U.S. Bureau of Mines (Bureau of the Census Working Paper No. 35), U.S. Government Printing Office, Washington, D.C., 66 pp.

Trainer, F. E., 1982, Potentially recoverable resources. *Resources Policy,* Vol. 8, pp. 46–48.

U.S. Bureau of Mines, 1965–1985, *Commodity Data Summaries and Mineral Commodity Summaries,* annual publication. U.S. Government Printing Office, Washington, D.C.

U.S. Bureau of Mines, *Minerals in the U.S. Economy, 1964–1973, 1965–1974, 1966–1975, 1968–1977.* U.S. Government Printing Office, Washington, D.C.

8 Mineral Conservation

INTRODUCTION

The previous chapter considered the present mineral position of the United States in terms of present rates of production and consumption and changes in the structure of American mineral-based industry. An attempt was made to forecast future demand for minerals and the future availability of supplies of minerals from domestic resources. However, both future demand and future supply will be heavily influenced by various elements of the economic, political, and social framework within which mining industry must operate in the United States, and the elements of that framework now require attention. We must consider the problems of conservation of mineral resources. We must examine the nature of mining law and land management policy, since these provide much of the legal framework of mineral exploration and mineral production. We must consider the regulations aimed at environmental protection and their impact on the present and future availability of minerals from domestic sources. We must examine taxation of mineral enterprise and its influence on mineral development. Finally, the trade policies of the United States require discussion, because they have a substantial impact on the nature and vigor of important segments of domestic mining industry. We begin with mineral conservation. Heavy demand for minerals and the depletion of known mineral deposits set a premium on conservation as the means of assuring adequate mineral supplies for the future. Not only is conservation of mineral raw materials essential but also conservation of the energy that is consumed in mining, mineral processing, and mineral use.

MINERAL CONSERVATION

A first necessity is a definition of the term "mineral conservation," for conservation of minerals is not the same as conservation of other resources. The dictionary definition of conservation is preservation, protection, or safeguarding. It is directly applicable to practice in our national parks, where every effort is directed toward preserving the park areas in their pristine condition. It is not applicable to conservation of wildlife or forests, where conservation means maintaining a balance between annual losses and the creation of new resources. Conservation of soils means minimizing losses of soil due to agricultural activities. Conservation of mineral resources is still different, because mineral deposits are "wasting assets." Once they are mined, most of them are gone forever.

There have been many definitions of mineral conservation, but one of the most satisfactory is that offered by the National Resources Board, Committee for Mineral Resources, in 1934 (National Resources Board, 1934, p. 32): "Conservation of minerals is orderly and efficient use in the interest of national welfare, both in war and in peace, without unnecessary waste either of the physical resources themselves or of the human resources involved in extraction." This definition is acceptable today, except that we would add "and with due regard for the impact of mining and mineral use on the environment."

During the colonial period and during the first hundred years of the Republic, there was little concern for conservation of resources, mineral or otherwise. The nation was concerned with opening up and settling vast tracts of land and developing whatever resources could be found. Measured against the needs of the times, resources of lands, forests, and minerals seemed inexhaustible. During the 19th century, one great mineral discovery after another was made in the midwestern and western states, and there seemed no end to the mineral treasures of the nation.

The early attitude toward resources began to change toward the end of the 19th century. By that time it was clear that forests could be rapidly depleted. As population grew, the areas favorable for settlement no longer seemed unlimited. The rate of discovery of new mineral deposits declined. The conservation movement in the United States was born. It arose out of a growing fear that the timber, mineral, and other natural resources of the United States were in danger of exhaustion, and out of dissatisfaction with the manner in which some of those resources were being exploited. It came to a head in 1908, in the Governors' Conference called by President Theodore Roosevelt. The conference appointed a National Conservation Commission. It was to consider the public lands, waterpower, forests, and mineral resources. The commission sponsored the first inventory of our natural resources and in 1909 reported this, with various recommendations, to the Governors' Conference. The conference approved the report and passed it to the President.

For minerals, the most important recommendations had to do with disposal of the public lands. The public domain at that time was available to all comers for only a few dollars an acre. Coal-bearing lands were sold for as little as $10 an acre, a very small price even at that time. In 1909, pursuant to the commission's report, a regulation was passed basing the selling price of coal lands on the tonnage and quality of underlying coal deposits. Further agitation led to the landmark legislation of the Alaska Coal Lands Leasing Act of 1914 and the Mineral Leasing Act of 1920. These are discussed in a later section on mining law.

The early pressure for conservation of mineral resources arose from fear that those resources were in danger of exhaustion. The fear proved unfounded, and interest in mineral conservation slackened in the 1920s. Late in the period, however, the real danger became apparent. The real danger is not exhaustion, but rather increase in cost due to depletion of the richer and more easily accessible deposits. As mines extend deeper and ore bodies grow leaner, costs of mineral production rise. As costs become economically unacceptable, rates of production must decline.

CONSERVATION IN PRACTICE

The definition of conservation quoted above, and accepted for purposes of this volume, is the expression of a philosophical concept. It contains no statement of how conservation is to be achieved, and we must now address this problem. In practice, conservation involves a series of activities—discovery and development of mineral deposits, efficiency in mineral extraction, efficiency in use of mineral materials after extraction, including recycling, and substituting more abundant minerals for the scarcer ones. All of these contribute to orderly and efficient use.

THE ROLE OF MINERAL EXPLORATION IN CONSERVATION

Exploration for mineral deposits is a critical aspect of mineral conservation. Unless mineral deposits can be discovered, there will be nothing to conserve. Mineral exploration clearly cannot be a one-time affair, because as known reserves are depleted by mining, new reserves must be discovered and developed to provide the mineral supplies of the future. Discoveries may be extensions of mineral deposits that are already being worked. In many mining districts exploration for such extensions and for new ore bodies close by is steadily pursued along with mining of ore bodies already found. The purpose is to maintain reserves at a constant level, as long as possible, in order that mining can be held at a desired rate over a period of years. In the old lead district of southeast Missouri, for example, this procedure, followed over many decades, was a major factor in prolonging the life of the district. In any event, mineral exploration must be a continuing process. Furthermore, since additional mineral deposits become more difficult to discover as time goes on, methods of exploration must be steadily improved. Research and development of new exploration techniques are critical. So is expansion of geological knowledge of the occurrence of mineral deposits in the earth, since this is the basis of modern mineral exploration.

The need for new mineral discoveries was brought out clearly by a study of the American copper industry under the auspices of the National Academy of Sciences (1975). Analysis of production rates, rates of discovery, and changes in reserves showed that to maintain U.S. copper production at the 1972 rate of 1.7 million tons per year, it would be necessary to discover each year a new copper deposit containing 250 million tons of ore averaging 0.8 percent copper. This means a vigorous and continuing exploration effort. Similar efforts are required to maintain reserves of a number of other mineral commodities required by industry today.

CONSERVATION THROUGH IMPROVEMENTS IN TECHNOLOGY OF EXTRACTION

Once deposits have been discovered and brought into production, conservation requires that mining and milling be as efficient as economically possible. The qualification is important. In theory it should be possible to extract all or very nearly all of whatever valuable mineral or minerals are present in an ore. In practice, as the recovery percentage is increased, for any ore a point is reached at which further improvement becomes unacceptably expensive, in dollars or energy or both. Recovery of oil by conventional methods is an example. As previously indicated, average recovery of oil is currently not more than 40 percent of the total oil in the ground. In some oil fields recovery could be greatly improved by injecting solvents into the wells. However, a point is quickly reached at which the energy cost of solvents offsets the gains in recovery percentage.

At any particular stage of the progress of technology, there are thus limits to efficiency in extraction, but the limits can change significantly with time. There is a tendency in some quarters today to denigrate the achievements of the U.S. mining industry. That is both unfortunate and unfair, because the industry led the world for decades in improving extractive methods; its achievements are one of the brightest pages in the history of American technology. Were it not for those achievements, for example, the United States would be almost completely dependent on imports for its supplies of copper, lead, zinc, molybdenum, iron, lithium, beryllium, tungsten, and vanadium. In the petroleum industry, improved technology has made it possible to tap resources of oil and natural gas from deeper and more difficultly discoverable reservoirs both onshore and beneath the continental shelves. The results are enormous contributions to the supply of minerals, not only from domestic deposits but also, since American technology has been widely shared, from the mineral resources of the world. Progress has not come easily; it is the result of research and development over many decades. Except in the petroleum industry, as described later on, conservation has been mostly the work of private industry, but it has been aided and stimulated by research carried on in government and in academic institutions.

As the cost of energy rises, it becomes a more and more important factor in the cost and availability of minerals, hence in mineral conservation. Energy requirements per ton of product are shown for various metals and nonmetals in Table 8-1, as of 1973. The table shows that production of nonmetallic minerals in general requires much less energy than production of metals. Of the metals listed, aluminum is by far the most "energy-intensive"—that is, the most consumptive of energy. Increases in the cost of energy since the Arab oil embargo of 1973 have had a heavy

TABLE 8-1 Energy Consumption per Ton of Selected Mineral Commodities in 1973

Commodity	Kilowatt Hours Equivalent per Ton
Bauxite (mining and processing)	687
Alumina	5,600
Aluminum metal	30,000
Cement	1,760
Clays	900
Coal (bituminous and lignite)	50
Copper	
Ore and concentrates	8,000
Smelters	10,000
Ferroalloys	12,800
Gypsum	750
Iron ore	313
Lime	2,200
Phosphate rock	190
Salt	
Solar, rock, and brine	14
Vacuum pan	1,230
Stone (crushed and broken)	14
Sand and gravel	16
Sulfur	1,920
Zinc	
Ore and concentrate	2,000
Smelters	10,000

Source: Mining Engineering (1975).

impact on the American aluminum industry. More efficient use of energy in mining and processing minerals, and in manufacturing consumer goods from them, must have a high priority in future efforts toward conservation of mineral resources.

CONSERVATION OF MINERALS IN USE

Conservation of mineral resources does not stop with extraction and processing. Conservation is possible at almost every stage of mineral use, from the elimination of waste in manufacturing operations to the recovery of materials from products that have reached the end of their useful lives.

The availability of minerals in large amounts and at relatively low cost in the past encouraged wasteful habits. Waste was further encouraged because harmful effects of mineral use were not taken into account. Oil and coal were burned, and industrial wastes were discarded without regard to the effects on the environment. Social costs that might have inhibited wasteful use were ignored and passed on to present and future generations.

The increased costs of energy that have followed the Arab oil embargo of 1973, together with rising concern with environmental protection, have set a premium on reducing waste in use of both energy and raw materials. Oil consumption in the industrial world, which had been increasing at an exponential rate, flattened out in 1978–1979 and declined during 1980–1982. This was due partly to a turn toward lighter-weight automobiles and partly to efforts to save energy in industrial processes, in transportation, in home heating and air conditioning, and in home construction. The results of increased attention to environmental protection have been mixed. On the one hand, introducing equipment and processes for control of pollution has increased use of energy and raw materials. On the other, attention has been focused on the recovery of useful materials from what have hitherto been wastes of mining, processing, and manufacturing industries. As one example consider materials produced in the processing and burning of coal. The annual production of coal in the United States in 1980 was about 800 million tons. About 400 million tons of the coal mined was processed in coal preparation plants (Rose, 1982), and about 25 percent of this, nearly 90 million tons, was rejected as waste. Studies by the U.S. Bureau of Mines have indicated, however, that at least some of this material can be converted into a sintered aggregate usable for making portland cement concrete, bituminous concrete paving mixtures, and concrete masonry blocks. Such usages could help relieve the problem of waste disposal and at the same time relieve shortages of natural aggregate materials that are appearing in various parts of the country. Beyond this, the burning of coal in power plants yields over 50 million metric tons of fly ash per year. Processes for using the ash for the production of concrete have already been developed, and processes for recovery of iron and aluminum from the ash are being studied.

In the fertilizer industry, a major achievement has been the recovery of fluorine from the rocks mined as sources of phosphate. Fluorine has many uses in metallurgical and chemical industries. Fluorine is present in all phosphate rocks but is not used in making commercial phosphate for fertilizers. It was formerly treated as a waste product, but it is now being recovered to such an extent that it contributes more than half the domestic production of fluorine raw materials. On the other hand, much of the artificial gypsum produced in processing phosphate rock is wasted because it cannot be marketed economically.

The potential for recovery of useful materials from present wastes is very great, and research of processes is being vigorously pursued in industry, in government agencies, and in universities. It offers the double reward of adding to the mineral reserves of the nation and eliminating or ameliorating serious problems of waste disposal.

RECYCLING

One of the most attractive means of conservation is recycling (reuse) of mineral-derived materials from discarded industrial products. It is doubly attractive, because it can mean partial recycling of the energy used in producing the materials in the first place. Aluminum is the most striking example. As indicated above, the energy cost of extracting the metal from bauxite is very high. When aluminum is recycled, 90 to 95 percent of the energy is recovered. Recycling is nothing new. It has been an important industrial activity for decades, but both the energy crunch and environmental considerations have brought efforts to improve and broaden it.

Whether mineral materials can be recycled depends on their patterns of use, which may be either dissipative or conservative. The burning of coal, oil, and natural gas for transport, heating, or power generation, for example, is dissipative. The principal products, carbon dioxide and water, are useless as sources of further energy. Phosphate, potash, and nitrogen compounds spread on fields as fertilizer are largely dissipated and cannot be recycled. The uses of some metals, however, lend themselves to recycling. Steel from automobiles and many other products, from milling machines to worn-out freighters, is recovered as scrap, and this is a valuable commodity. Manganese, however, is consumed mostly as a purifier and deoxidizer in steelmaking and passes off into furnace slags. Very little manganese can be recycled. The scope and importance of recycling of metals are indicated in Table 8-2. Comparison of the data for 1974 with data for 1981 indicates improvement for some metals but not for others.

TABLE 8-2 Old Scrap Recovered in the United States as a Percentage of Consumption

Commodities	Percent of U.S. Consumption 1974	1981
Major Metals		
Iron	24	~25
Lead	34	44
Copper	22	26
Aluminum	4	15
Zinc	6	5
Chromium	16	10
Nickel	19	23
Tin	19	13
Antimony	52	47
Magnesium	3	20
Minor Metals		
Mercury	17	11
Tungsten	3	15
Tantalum	13	5
Cobalt	<1	10
Precious Metals		
Silver	24	37
Gold	17	59
Platinum group	22	
No Significant Recovery		
Molybdenum		
Titanium		
Beryllium		
Vanadium		
Manganese		
Tellurium		
Scandium		
Bismuth		
Most nonmetallics		

Sources: U.S. Bureau of Mines, *Status of the Mineral Industries*, 1974, and *Mineral Commodity Summaries*, 1982.

The large increase in the percentage of lead recycled is due largely to reduction of the use of tetraethyl lead as an additive in gasoline. The major use of lead is now in storage batteries; most of it is recycled repeatedly. The figure for gold refers only to gold recycled from industrial products; it does not take into account the large amounts of gold stored in jewelry and coinage. Actually, as mentioned earlier, most of the gold mined in the world so far is still with us, because its high value and great durability have favored husbanding gold ever since it began to be used by man. In sharp contrast, very little of the nonmetallic minerals used can be recycled, although recycling of glass is becoming important, and some concrete and asphalt are being recycled in road making. The costs of recycling nonmetals are generally far beyond their value in reuse.

Some of the barriers to increased recycling are institutional or related to industrial procedures. Rail cargo rates, for example, may encourage primary production at the expense of recycling. The rate per unit of iron is higher for scrap iron than for iron ore. The multiplicity of specifications for metal alloys used in industry creates numerous difficulties. Recycling of alloys from automobile parts is difficult, because the variety of alloys used makes it hard to sort parts into individual batches of scrap having uniform composition. Simplification and, where possible, standardization could help resolve this problem.

SUBSTITUTION

Minerals can also be conserved in use through adaptations of technology. Substitution of one material for another is a simple form of adaptation. If the material replaced is scarcer than the substitute, there is a gain to society in terms of the availability of adequate supplies. There is a broad range of possible substitutions of materials. Nickel, zinc, titanium, aluminum, molybdenum, and vanadium, for example, can be substituted for chromium in certain applications. Substitution of synthetic materials for mineral materials can contribute greatly to conservation of the scarcer minerals. The development of synthetic diamonds and other very hard abrasive materials has been very important to industry. Supplies of mineral abrasives of this class, such as emery and corundum (alumina), are limited. Plastics have supplanted metals in many products of industry. Sometimes there are actually savings in costs. There are, however, limits to substitution. Use of a substitute may involve a cost in the form of lower performance. Almost every mineral or material is uniquely qualified by its properties for certain uses. Chromium is an essential constituent of certain stainless steels required for chemical processing equipment. Asbestos is another essential material. Its chemical inertness, low heat and electrical conductivity, high heat resistance, and capability of being spun into thread uniquely qualify it for use in brake linings and other friction-resistant and fire-resistant products. Despite much concern over the medical problems caused by asbestos fibers, substantial amounts of asbestos are still being used. Similarly, the largest single use of silver is in photochemicals; no satisfactory substitute has yet been found.

Substitution is only one form of adaptation of technology. Adaptation of design can achieve equal or sometimes greater savings of material. The outstanding modern examples are in the field of calculating machines and computers. Many older Americans will remember the Burroughs adding machine, constructed of many pounds of metal and capable only of addition. Today such machines have been replaced by calculators only a fraction of their weight and capable of performing many kinds of mathematical operations. The giant computers of the 1950s have given way to assemblages of tiny chips. The savings of raw materials are enormous, and here there has been an equally enormous improvement in performance.

As supplies of the scarcer mineral raw materials diminish, adaptation of technology must inevitably move in the direction of use of the more abundant materials of the earth. Actually, such movement has been going on, though slowly for the most part, throughout human history. Gold is a particularly striking example. Gold has a wonderful set of properties—softness, malleability, ductility, resistance to chemical reagents, high thermal and electrical conductivity—that qualify it for many uses in industry. Yet owing to its scarcity and high cost, its use in industry has been severely restricted. It could, for example, replace copper in many of the uses of that metal. We use copper because there are many deposits containing hundreds of millions of tons of ore with 8 to more than 50 pounds of copper per ton. In large gold deposits, the ores contain some fraction of an ounce of gold per ton. As a result copper sells for about 65 cents per pound, and gold sells (1985) for $3900 per pound. Technology has been adapted to the use of the cheaper and more abundant metal.

It should be understood that major adaptations of technology require very long periods of time, usually a generation or more. Solar energy is a case in point. Early enthusiasts envisioned rapid and large-scale substitution of solar for other forms of energy. That has simply not taken place, partly because of the inherent limitations of the use of solar energy, partly because the costs of rapid substitution would be unacceptable. Adaptation through successful research and development seldom involves strokes of genius and sudden miracles. It is the result of effort continued over a period of time. Adequate financing is necessary, but problems are not solved overnight simply by spending huge sums of money in crash programs. Implicit in this is the need for long-term commitment to research and development. It is trite but true that industry today is founded in large part on the research and development of preceding generations, and the industry of tomorrow will rely heavily on research and development being done today.

THE IMPACT OF ECONOMIC INSTABILITY ON MINERAL CONSERVATION

The record of the mineral industries is one of continuing progress in the technology of mineral extraction. Many mineral deposits that were totally uneconomic in 1900 or even in 1959 have been converted into reserves and are now important sources of mineral supplies. At the same time, physical losses of minerals have been reduced. The mine operator has control over processes involved in his operation, and the profit motive provides a powerful incentive toward reducing physical losses and making the fullest possible use of the materials present in a deposit. Unfortunately, there are losses over which he does not have control, because they result from instability of mineral markets. There is no quantitative measure of such losses, but they are known to be very serious. Continuity in the mining process over a period of years is essential to orderly and efficient extraction of mineral materials. Mining cannot be turned on and off at will like the water in a tap. Continuity is very difficult to maintain in an economically unstable environment. The problems brought on by instability are well illustrated in the histories of the petroleum and coal industries in the United States. The two histories are interesting to compare, because the petroleum industry is one in which instability has been countered with considerable success, whereas in the coal industry the problems of instability have never been resolved.

In the early days of the petroleum industry in America, wells were drilled as fast as money and equipment to drill them could be found, and without regard to efficiency in production. The climax of such activity was reached in the late 1920s and early 1930s, when tremendous oil pools were discovered in Texas, Oklahoma, and California. There were no restrictions on production; new wells were drilled at a fantastic rate. They flowed without control, and overproduction and cutthroat competition ensued. Efforts of the more responsible producers to control the situation failed; the industry was at the mercy of wildcat operators interested only in quick profits. By late 1932, oil was being sold in the East Texas field for as little as 4 cents a barrel, and price wars in other states were rampant. The oil industry was on the brink of chaos. At this point the states stepped in. In Oklahoma, the governor called out the National Guard and ordered it to close every oil well in the state until production was brought into line with demand. The governor of Texas ordered the East Texas field to be shut down. Control of the industry in the state was handed over to the Texas Railroad Commission.

Such measures helped, but it soon became apparent that there was a problem with bootleg oil—oil produced without conforming to state conservation regulations and shipped across state borders to be sold in neighboring states. The federal government now intervened. After an abortive attempt under the National Recovery Act, in 1935 Congress passed the Connally Act, which banned the movement of illegally produced oil in interstate and foreign commerce. The act became an important support of state conservation bodies. In the same year, the Interstate Oil Compact Commission was formed by Texas, Kansas, Oklahoma, New Mexico, Illinois, and Colorado. Those states were later joined by most of the other oil-producing states. The commission held no power, but served as an important forum for exchange of information among the oil-producing states and for discussion of problems of petroleum conservation. It supported proration, the system of establishing production quotas for each state. Many of the member states in consequence established quotas for the fields within their borders. Quotas were adjusted periodically to demand. The Texas Railroad Commission, for example, told the operators in each field each month the amounts of oil they would be allowed (the "allowables") to produce. Thus in the petroleum industry, regulation of production by government became established practice. It was accepted by the industry. Government did what industry could not do—namely, adjust supply to demand and prevent wastes due to economic fluctuations.

Stabilization made it practicable to adopt certain practices, either required or encouraged by state conservation commissions, that have contributed heavily to petroleum conservation. One is the practice of unit operation of oil reservoirs that are under multiple ownership. Maximum recovery of oil requires careful control of gas and water pressure in an oil reservoir. There was no means of doing this so long as individual owners of parts of a reservoir operated independently. Each operator drilled wells and pumped out oil as fast as he could to be sure of getting as large a share of the oil as possible. Operating a field as a unit permits close control and maximum recovery. Secondary recovery by water flooding (pumping water under pressure into wells at selected points to drive the oil to the producing wells) becomes possible and further increases recovery. Costs and yields of unit operation are prorated among the owners in proportion to their land holdings.

Secondary recovery by water flooding has been of great importance. In Pennsylvania, for example, the fields were considered exhausted after production of about 250 million bbl of oil. Water flooding brought forth another 250 million bbl. Tertiary recovery methods are now being actively pursued—steam injection, carbon dioxide injection, ignition, and others. These methods are applicable only under unit operation.

In the early days of the petroleum industry there was no market for the natural gas produced as a by-product of oil. The gas was burned at the wells simply to get rid of it. In the United States and also in Europe, the use of pipelines to transport gas to market has made possible widespread use of natural gas as a fuel. The result is a major contribution to conservation of energy resources. In countries that do not have pipelines, more and more gas is being liquefied and transported in special tankers or locally in the form of bottled gas.

Improvements in methods of cracking and refining petroleum have contributed greatly to conservation. Cracking is necessary to produce more of the lighter hydrocarbons such as gasoline and kerosene from heavy crude oils.

Conservation practices largely developed in the United States have been extended in varying degrees to oil fields in other parts of the world. Conservation in the oil industry has worked and has worked well. As Ridge (1964) has pointed out, it has worked because industry learned that it was more profitable to get the maximum amount of oil and gas from reservoirs through conservational practices. At the same time, the relative stability and profitability that were introduced into the industry through regulation encouraged conservation. Conservation is affordable only in a profitable industry.

In the 1930s, 1940s, and much of the 1950s, stability in the American petroleum industry meant guarding against overproduction from domestic fields and its consequences in cutthroat competition. Conservation and control of production were largely in the hands of the oil-producing states. The situation is now changed. The rise of production from the Middle East and other countries means that the United States no longer has a dominant position in world petroleum markets and that domestic demand can no longer be supplied from domestic fields. The federal government has assumed an important role in regulating the industry. Its actions, unfortunately, have been motivated more by politics than by conservational objectives. Prices of natural gas were for years held at artificially low levels that discouraged exploration, and thus far only partial deregulation has taken place. The windfall-profits tax and a total heavy burden of taxation have diminished the profitability of the industry. The net result is that the industry is now more vulnerable to changes in demand and price for petroleum than in the past. This is currently being demonstrated in graphic fashion. Exploration activity in 1981–1982 was at an all-time record level. The fall in world oil prices in early 1983, however, was followed by an abrupt decline in exploration activity.

The history summarized above indicates that on balance there have been substantial accomplishments in the conservation of petroleum and natural gas since 1930. Quite different is the story of the coal industry of the United States. Since 1918 the American coal industry has been in a state of chronic distress. It was not always so. From 1900 to 1918, as the industrial sector expanded, U.S. coal production grew, in general in an orderly fashion, at a rate of about 18.8 million tons per year, and by 1918 it had reached 568 million tons per year. Between 1918 and 1923, however, the number of coal mines in operation increased from 3245 to 9331. Productive capacity rose to 970 million tons per year, far more than the market could possibly absorb. Development of excess capacity had several causes, but the main one was strife between labor and management. Strikes in the unionized eastern and midwestern coal fields led to shortages of coal. This led to opening of new mines in coal fields, especially in the western states, which were not unionized. When a strike ended and the closed mines were reopened, production rapidly exceeded demand. Cutthroat competition ensued, coal prices fell, and reductions in wages were made. These led to new strikes in the unionized fields, and the cycle was repeated. The result was economic hardship both for the coal-mining companies and for the labor force involved.

The coal industry has never really recovered from the setback of the early 1920s and has been prosperous for only short periods of time. Management-labor strife has continued, and overcapacity has continued to plague the industry. Mechanization of coal mining has reduced employment; between 1943 and 1973, owing to mechanization, output per manshift increased from 8 tons to 19.2 tons, whereas annual production during the period was less than in the 1920s. Rising competition from oil and natural gas exacerbated the problems of the coal industry and was stimulated because repeated strikes in the industry made coal unreliable as a source of energy.

From the standpoint of conservation, the important thing is that instability has caused serious loss of coal resources, particularly in underground mines. Coal commonly occurs in groups of seams, separated by layers of sandstone and mudstone and limestone in various proportions. Seams within a group may differ in thickness and quality. In times of glut and low prices, only the best and thickest seams can be mined profitably. If an underground coal mine is shut down, costs of maintenance and capital continue and accumulate against the coal. Workings deteriorate, and cave-ins may occur. Costs of rehabilitation may become prohibitive. Mining of the poorer seams may never be resumed. The coal involved in such mines becomes a lost resource.

The effective action taken by the states to control instability in the petroleum industry and to promote conservation has no counterpart in the coal industry. There is no Interstate Coal Compact, and there is little cooperation among coal-mining states. Political maneuvering has been an obstacle to joint action. The eastern and midwestern states have been more concerned with protecting markets for their coal than with conservation. There has been no effective action at the federal level. Even in the 1930s, when conditions in the Appalachian coal fields became acute and the region was on the verge of anarchy, only very limited action was taken by government. The obvious need has been for measures aimed at stabilization, by adjusting production to demand. Since the Sherman Antitrust Act strictly limits action by private industry, only government can provide control.

Several studies of the coal industry appeared in the 1920s, as the troubles of the industry developed, but the only action taken was that embodied in the Bituminous Coal Conservation Act of 1935 (the Guffey Act) and its successor, the Bituminous Coal Act of 1937, which created a seven-man National Bituminous Coal Commission representing the government. The commission had power to set minimum and maximum prices for coal, and those operators subscribing to a code of regulations established by the commission could be empowered to form cooperative sales agencies. The commission was charged with investigating means of improving markets for coal, means of conservation, and means of lowering costs of coal.

The Guffey Act was renewed at intervals until 1943, when it lapsed in the face of the rise in demand for coal during World War II. The commission was largely ineffective. It had no power to adjust production to demand, and it did little to stimulate coal conservation. Since World War II periodic distress has been characteristic of the coal industry. Overcapacity has been a recurrent problem. Labor-management strife has continued, and there has been a struggle between union and nonunion operators. Vacillations in national energy policy have contributed to the difficulties of the industry. Congressional action reflects an attempt by midwestern and eastern coal-producing states, much of the coal from which is high in sulfur, to limit competition from low-sulfur western coals. It also reflects the lobbying of environmental groups.

This is not to say that there has been no progress in coal conservation during the past 60 years. Substantial progress has been made in coal preparation. The development of uses for the fine fractions of coal alone has made a very large contribution. In earlier times, the fine fractions were discarded as waste. Laboratory investigation of the complex materials of coal by the U.S. Bureau of Mines, the Illinois Geological Survey, and industry-funded research laboratories has led to great improvements in the utilization of coal and has alleviated concern about exhaustion of special grades of coal such as coking coal. Greater efficiency in the conversion of coal to energy has been achieved. The development of large-scale excavation machinery has made it possible to mine in open-cut seams that could not otherwise be utilized, and the increase in open-cut mining has improved the average recovery of coal.

The laboratory studies mentioned, and advances in safety and health of coal miners have been noteworthy contributions by government to conservation of coal and of the human resources involved, but government has not addressed the basic conservational problem—the stabilization of the coal industry by adjusting production to demand. Nor has it addressed the problems of strife between labor and management.

CONSERVATION IN THE NONFUEL MINING INDUSTRIES

In the nonfuel-mining industries, government has played only a minor role, or none at all, in promoting stability. Yet certain of these industries, the metal-mining industries in particular, have been plagued by instability of markets and prices. The copper-mining industry is a prime example. Since 1900 there has been a trend toward development of large-scale, mass mining of copper deposits, mostly by open-pit mining. Such operations require very large capital investments, generally in the hundreds of millions of dollars. Their financial viability is heavily dependent on maintaining high rates of production. Capital and overhead costs are fixed costs that must be spread over the units of production. The larger the number of pounds of copper produced per day, the smaller the fixed costs per pound. Mine operators have a stake in stabilizing production at an optimum level. Stabilization, however, has been impossible to achieve. Since World War II, world capacity for production of copper has always exceeded maximum demand, and demand itself has fluctuated with the vagaries of the U.S. and world economies. In the United States, the problems of industry have been aggravated by a series of prolonged strikes. Fixed costs continue during strikes and must be added to the costs of production after a strike is over. Material that was ore before a strike may not be minable afterward. The result is loss of reserves. Strife between labor and management contributed significantly to the demise of the great copper-mining operation at Butte, Montana. It contributed to rising costs of copper production in other mines and in consequence to added difficulties in competing in international markets. The situation is not confined to the copper-mining industry. In 1980, after a prolonged struggle between labor and management, the operations of the Bunker Hill and Sullivan Company, in the Coeur d'Alene district of Idaho, closed down. A major source of lead and zinc was lost, perhaps forever.

PROSPECTS FOR DISCOVERY OF MORE MINERAL DEPOSITS IN THE UNITED STATES

When we have taken into account all the possibilities of substitution, adaptation of technology, elimination of waste in mining and processing, reduction of the waste due to economic causes, and greater attention to economy of use, we are left with the harsh fact that the mere maintenance of modern civilization, *without further growth or progress*, will require large, newly mined supplies of a broad array of fuel and nonfuel minerals. There is simply no escape from this conclusion. What we cannot get from domestic deposits we will have to get from abroad, but imports can be only a partial answer to the problem. For many minerals, transportation costs involved in importation would cause large increases in prices of mineral raw materials in the United States. As pointed out earlier in this volume, rising prices may stimulate supply, but they impose restrictions on mineral use. In short, for many minerals there is no good substitute for supply from domestic resources. Since our present productive deposits are being steadily depleted, this means that there must be a continuing effort to find new mineral deposits within our borders and to convert them into reserves.

What are the prospects for new mineral discoveries in the United States? There is no simple or single answer to this question. Prospects for discovery of significant deposits of tin, bauxite, tantalum and high-grade manganese and chromium, for example, are poor. For many other minerals and metals, prospects for discovery are bright. The period since World War II has seen many major discoveries of lead, zinc, copper, molybdenum, gold, sodium carbonate, and other minerals in the United States. There are no geological reasons to suppose that additional discoveries will not be made. Quite the reverse is true; our knowledge of the parts of the crust of the earth included within the 50 states is incomplete, and so is exploration of our lands for minerals.

The portions of the earth crust underlying the United States display a remarkable diversity of geological formations, because in different areas of the country varied geological processes have operated in the past. Different processes produce different mineral deposits, and that is why the United States possesses such a broad range of mineral resources. Some of these deposits formed at or close to the surface; others were formed far below the surface. Of the latter, some have been exposed by erosion of the overlying rocks, but others lie buried beneath hundreds or thousands of feet of rocks.

Much of the exploration thus far done has been relatively shallow, confined to the upper few thousand feet of the crust. Exploration even to those depths is slow and very expensive. It cannot be done everywhere. Exploration must be selective. Targets of exploration (i.e., parts of the crust that are likely to contain particular mineral deposits) must be selected on the basis of geological and geophysical information. Much information is available. Geological mapping and related investigations have been carried on in various parts of the country for about 150 years. The work has been done mainly by the U.S. Geological Survey and various state geological organizations, but special agencies such as the Tennessee Valley Authority and the Department of Energy have been involved at times. Substantial contributions have been made by geologists and geophysicists from academic institutions, and mining organizations have actively investigated many areas of the country. By projection from surface relations, by geophysical studies, and by drilling, we have learned something of relations in depth. An enormous amount of information is available, and this provides the background and the guides for mineral exploration. However, the information is both incomplete and uneven in quality. V. E. McKelvey, former director of the U.S. Geological Survey, has pointed out that only about half of the United States is geologically mapped in enough detail to provide an adequate background for mineral exploration. Since our knowledge of the crust is incomplete, so is our knowledge of its mineral deposits.

As the search for mineral deposits goes to greater depths, reliance on geophysical methods increases. Great advances in geophysical methods have been made since World War II, but there is general agreement that new and improved methods are needed.

The very fact that our information is incomplete and our methods of mineral exploration are imperfect is grounds for optimism about the prospects for discovery. There is an additional reason: we are still learning to recognize the kinds of mineral deposits present in the crust. Each new advance in that knowledge enlarges the scope of mineral exploration and leads to new discoveries. There have been dramatic advances since World War II. Discovery of two new types of gold deposits, in California and Nevada, has added significantly to U.S. gold reserves. It is now recognized that certain deposits of gold, copper, lead, zinc, and silver are associated with rocks produced by past submarine volcanic activity. Recognition has led to a rash of new discoveries. A new type of deposit of the then rare metal niobium was recognized in the early 1950s. The metal is no longer rare. The list of new discoveries based on new knowledge is an impressive one. The discoveries are a fascinating chapter in the long history of man's effort to reveal the mineral treasures of the earth.

Among those who are unfamiliar with the status of geological knowledge and mineral exploration, there is a widespread misconception about the prospects for discovery of additional mineral deposits. During the debate over the proposed creation of MX missile sites in a large area of northwestern Utah and northeastern Nevada, statements were made that the area has been so thoroughly prospected that chances of further discoveries are very slight. Nothing could be farther from the truth. Until the onset of recession in 1982, the area was a hotbed of exploration activity, and new deposits of gold and barite were discovered there. Old mining districts were being investigated in the light of modern knowledge of mineral deposits. Future mineral production from the area may well exceed the mineral production of the past. New knowledge has also led to intense interest in the mineral deposits of Alaska. Despite some statements to the contrary, the mineral potential of Alaska is enormous.

In short, prospects for discovery of additional deposits of many minerals in the United States are very bright so long as society provides a political, economic, and social environment that encourages mineral exploration. Our task in succeeding chapters is to examine the elements of that environment in the United States.

REFERENCES AND ADDITIONAL READING

Alter, H., ed., 1982, Resource recovery and environmental issues of industrial solid wastes. *Resources and Conservation*, Vol. 9, 365 pp.

Allsman, P. T., 1962, Conservation in metal mining. In *Minerals and Energy*, Part I, *Colorado School of Mines Quarterly*, Vol. 57, No. 4, pp. 87–95.

Brooks, D. B., 1976, Conservation of minerals and of the environment. In *World Mineral Supplies*, G. J. S. Govett, ed., Elsevier, Amsterdam, pp. 287–314.

Burton, I., and Kates, R. W., 1965, *Readings in Resource Management and Conservation*. University of Chicago Press, Chicago, 609 pp.

Ellis, P. L., 1970, Conservation—a geological rationale. *Queensland Government Mining Journal,* May 1970, pp. 1–22.

Flawn, P. T., 1966, *Mineral Resources.* Rand McNally, New York, pp. 183–194.

Goeller, H. E., and Weinberg, A. M., 1976, The age of substitutability. *Science,* Vol. 191, No. 4227, pp. 683–689.

Interstate Oil Compact Commission, 1964, *A Study of Conservation of Oil and Gas in the United States.* Interstate Oil Compact Commission, Oklahoma City, 230 pp.

Kellogg, H. H., 1977, Sizing up the energy requirements for producing primary materials. *Engineering and Mining Journal,* Vol. 178, No. 4, pp. 61–65.

Landsberg, H. H., 1976, Materials: Some recent trends and issues. *Science,* Vol. 191, No. 4227, pp. 637–640.

Leith, C. K., 1935, Conservation of minerals. *Science,* Vol. 82, No. 2119, pp. 109–117.

Lovejoy, W. F., 1976, Conservation. In *Economics of the Mineral Industries,* 3rd Ed., W. F. Vogely, ed., American Institute of Mining, Metallurgical, and Petroleum Engineers, New York, pp. 684–692.

McKelvey, V. E., 1962, National goal for mineral resources: Efficient development and full use. In *Minerals and Energy,* Part I, *Colorado School of Mines Quarterly,* Vol. 57, No. 4, pp. 143–152.

Mining Engineering, 1975, Minerals industry news. Vol. 27, No. 7, p. 30.

National Academy of Sciences, 1975, *Report of the Committee on Mineral Resources and the Environment.* Natural Academy of Sciences, Washington, D.C., pp. 17–76 and 143–152.

National Resources Board, 1934, *A Report on National Planning and Public Works in Relation to Natural Resources and Including Land Use and Water Resources.* U.S. Government Printing Office, Washington, D.C., 455 pp.

Ridge, J. D., 1964, Conservation and Stabilization. In *Economics of the Mineral Industries,* 2nd Ed., E. H. Robie, ed., American Institute of Mining, Metallurgical, and Petroleum Engineers, New York, pp. 593–638.

Rose, J. G., 1982, Processing and material properties of energy-efficient coal refuse lightweight aggregate. *Resources and Conservation,* Vol. 9, pp. 119–129.

Smith, G. H., ed., 1958, *Conservation of Natural Resources,* 2nd Ed. Wiley, New York, 474 pp.

U.S. Bureau of Mines, 1982, *Mineral Commodity Summaries.* U.S. Government Printing Office, Washington, D.C., 178 pp.

U.S. Bureau of Mines, 1974, *Status of the Mineral Industries.* U.S. Dept. of Interior, Washington, D.C., 20 pp.

9 Mining Law and Land Policy

MINING LAW

Mining law (Dempsey, 1973) sets the rules governing the ownership of mineral deposits and the relation between miners and other users of lands, and it provides for government enforcement of those rules. Mining law parallels other property law; it reflects the fact that each mineral deposit has a specific location. The miner always has the choice as to developing a particular deposit, but when the choice is made, the location of the mine becomes an irrevocable fact. Furthermore, the choice is not a broad one, because only a few mineral deposits have the economic prerequisites for development. The rules set by mining law are therefore of vital importance. On the one hand, no individual or corporation can afford to invest in mineral discovery and development unless he has security of tenure of the land during the period of time necessary for production of minerals from whatever deposits may be found. On the other hand, the rules must be such that the interests of the general public and the state are adequately protected.

There are three systems of mining law in the United States. The oldest, largely inherited from English mining law, applies to private lands and land owned by state and local governments. State law governs the use of such lands for mining. In general, mineral rights go with the land, along with surface rights, water rights, and timber rights. The owner of the land may lease it for mining purposes in return for a royalty on production or some other form of compensation, he may sell the land, or he may sell the mineral rights separately from the other rights. This system of mining law was that of the original 13 colonies. It applies also to the state of Texas, which on joining the Union in 1845, retained ownership of its public lands. It further applies to grants of land made to individuals under the Spanish territorial government in New Mexico, Arizona, and California. Private ownership of this land was confirmed by a Supreme Court decision in 1845. In Alaska, the laws of private ownership apply to the approximately 44 million acres assigned to Native Corporations by act of Congress in 1971.

The remainder of the lands of the United States, consisting largely of lands acquired by purchase or conquest, constitute the public domain; originally this amounted to about 1.8 billion acres (about 2.81 million square miles). Under a series of acts beginning in 1796 and continuing until the 1870s, about 1.1 billion acres was alienated from the public domain, meaning that ownership passed into private hands or was transferred to state or local governments. Public lands owned by the states other than Alaska now amount to about 103 million acres. In general, on the alienated lands, mineral rights go with the land. In 1958, the Alaska Statehood Act provided for transfer of up to 103.5 million acres of the public domain to ownership by the State of Alaska. In 1971, the Alaska Native Claims Settlement Act provided for transfer of ownership of about 44 million acres of the public domain in Alaska to native regional corporations. Meanwhile lands purchased by the Federal Government (the Acquired Lands) have been added to the public domain. In 1982, the public domain was reported to amount to about 737 million acres, more than 90 percent in the western states including Alaska. The percentages of lands owned by the Federal Government in the various western states and Alaska are indicated in Fig. 9-1. Jurisdiction is distributed among various federal agencies as follows:

Bureau of Land Management	320 million acres
U.S. Forest Service	191
National Park Service	74
Wildlife refuges	87
Department of Defense	24
Other	41
Total	737 million acres

There are two systems of mining law applicable to the public domain, the discovery-location system (claims system) and the leasing system. The discovery-location system was established under several acts, especially the Mining Law of 1872. It applies generally to the public domain in the western or public-land states. The law of 1872 provides that any U.S. citizen or corporation or an individual who expresses his intention of becoming an American citizen may enter the public domain and, upon making a discovery of a valuable mineral deposit, mark a claim. There are two main classes of claims: lode claims for minerals in firm bedrock, and placer claims for minerals in unconsolidated surficial materials. A

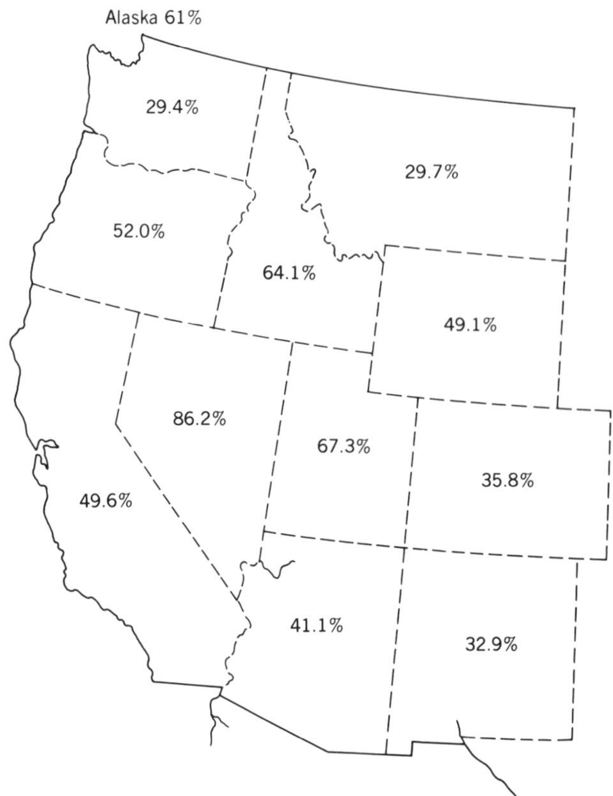

Figure 9-1 Percentages of the areas of the western states including Alaska that are federal public lands. The total area of the western states including Alaska is about 1200 million acres, of which about 565 million acres consists of federal public lands. Modified from Bailly (1966).

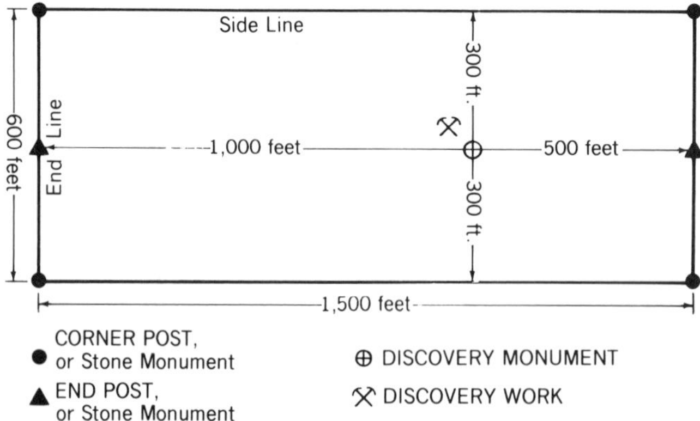

Figure 9-2 Diagram showing features of a lode mining claim as required in California. From Bailly (1966), by permission.

placer claim may not exceed 20 acres; a lode claim (Fig. 9-2) may be slightly larger than 20 acres. The sides and ends must be parallel. The corners must generally be marked by monuments or stakes (hence the term "staking a claim"), but the exact requirements for marking, for size of discovery pits, and for recording of a claim are set by the state in which the claim is located. In general, any number of claims may be staked, but there must be a "discovery" on each, and some states have limited precious-metal locations. The requirement of a discovery is a federal one, and a discovery can be challenged by the appropriate federal agency—the Bureau of Land Management, the U.S. Forest Service, or the National Park Service. The claimant must perform $5 per acre of beneficial assessment work each year; this work can include drilling, trenching, and underground development work. Geologic, geochemical, and geophysical surveys can be done on a limited basis if they are not duplicative. Assessment work can be aggregated on contiguous claims, and in modern practice thousands or tens of thousands of dollars is spent annually on significant blocks of mining claims on federal lands.

Having fulfilled the various requirements, the claimant in effect becomes the owner of the land, holding mineral and mining-related surface rights. Special provisions of the federal mining laws also allow the acquisition of mill sites and tunnel sites. The laws also provide that with completion of at least $500 worth of improvements to a claim, an approved survey, and proven discovery of a mineral deposit, the claimant may apply for a patent on the claim. If granted, the claim can be purchased and completely alienated from the public domain.

As previously noted, the law of 1872 distinguished between placer deposits and lode deposits. True placer deposits are deposits of valuable minerals in sand or gravel accumulations along stream channels or along beaches. Neither in nature nor in the mining law, however, is the distinction between placer deposits and lode deposits clearly drawn. Borderline cases have had to be settled in the courts.

The Mining Law of 1872 was designed by the individual prospector and was intended to assure him of the rewards of his discoveries. In general, that purpose was achieved, but experience has shown that the law has certain defects. In the first place, it rested on a very simple concept of mineral deposits. Most of the lode deposits discovered in the early days of mining in the West were vein deposits or other tabular bodies dipping into the ground (Fig. 9-3). Since the early days, many nontabular deposits have been discovered and developed; likewise some tabular deposits, such as bedded sedimentary deposits, that are essentially parallel to the surface of the ground. Another difficulty of the law is the apex provision. If a claim includes the apex of a deposit, the locator has the right to follow the deposit underground beyond the vertical downward extensions of the sidelines of his claim, although only within the vertically projected end lines. In many mining districts, owing to faults and other

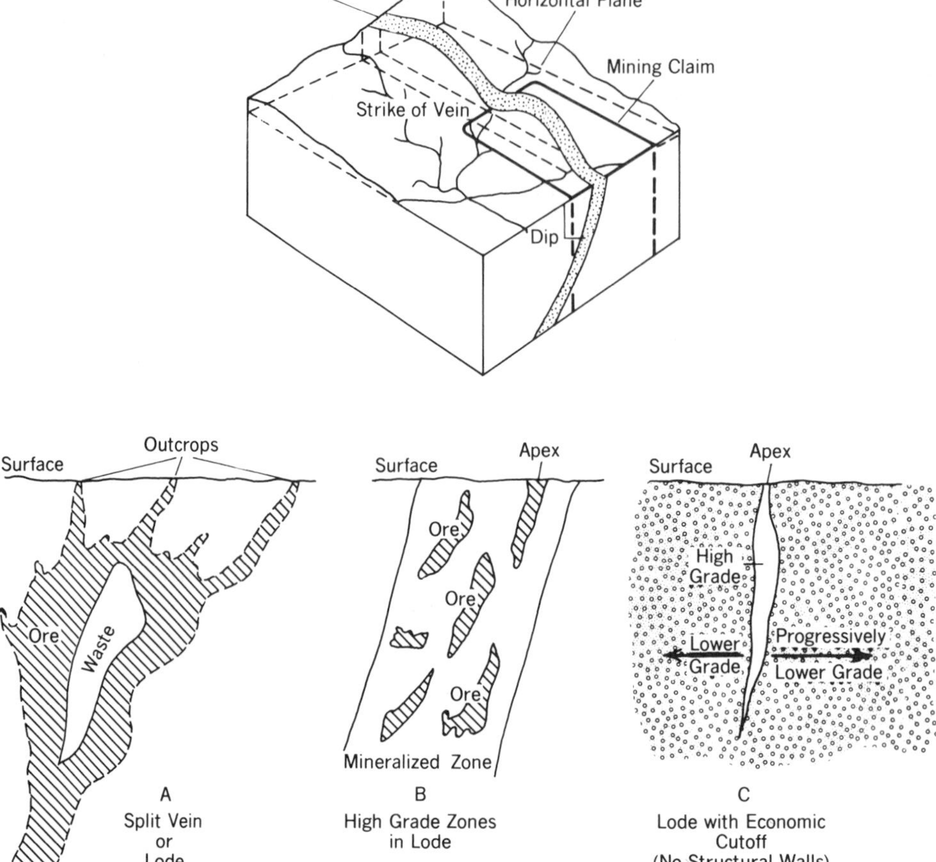

Figure 9-3 (A) Sketch showing strike, dip, and outcrop of a vein, and a mining claim. (B) Sketch showing cross sections of veins and other lodes of the kinds sought by prospectors in the nineteenth century. Diagrams from Bailly (1966), by permission.

structures (Fig. 9-4) the apices of lodes have proved difficult to determine. In such districts the provision has led to protracted litigation that often could be settled only by consolidation of competing claims under a single ownership or by agreement among claim owners to accept the vertical extensions of their claims as boundaries.

Another difficulty has been the requirement of a discovery. It seemed logical enough when most discoveries were made in outcrop, but it inevitably caused difficulty in securing tenure of deposits that did not crop out. The provision was ill suited to the search for hidden deposits that soon became necessary and now constitutes the great bulk of mineral exploration. Bailly (1966) points out that the 1872 law was conceived as postdiscovery law and was adequate for situations where deposits were discovered prior to the staking of claims. Modern exploration, however, may require tenure of large areas before a discovery is made. The dis-

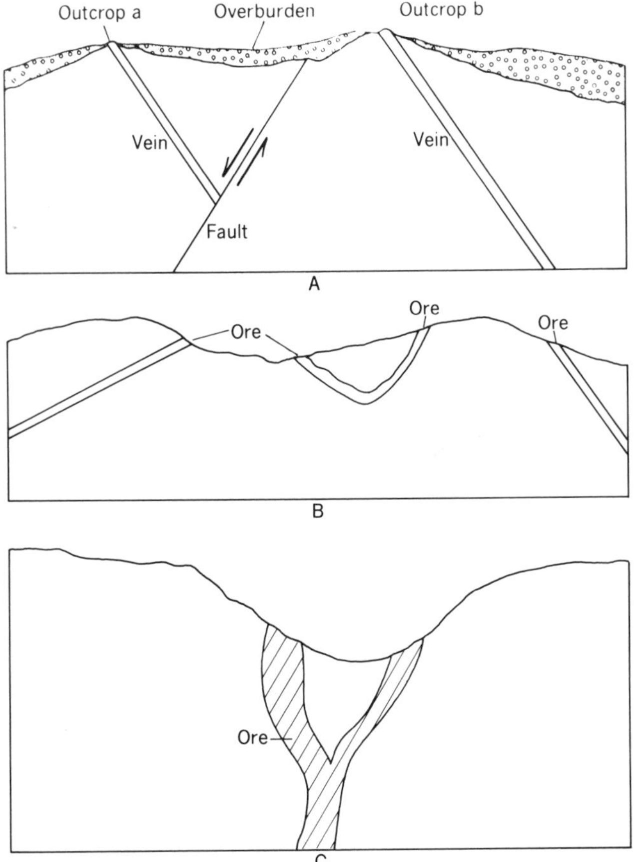

Figure 9-4 Three examples of situations in which application of the apex provision causes difficulties. In (A), outcrops *a* and *b* of the vein are discovered and staked by two different prospectors, each believing that his claim covers the apex of the vein. Later on, the fault is discovered, and it becomes apparent that the true apex is at *a*, that the claim staked over *b* is invalid, and that both segments of the vein belong to the prospector who staked outcrop *a*. In (B), a bedded deposit has been dismembered by erosion. Which outcrop is the apex? In (C), the ore body has two apices. If the two apices have been staked by different prospectors, who owns the deposit?

covery requirement in such cases may be a source of serious difficulties and an obstacle to mineral exploration (Fig. 9-5).

There are still other defects. Unless patented, the boundaries of many claims were not accurately surveyed, and the exact locations of such claims with respect to official land surveys were not established. Claim notices were often vague in describing the claims, and corner stakes could be lost or moved by "claim jumpers." Boundaries of some claims were overlapping. Litigation was the result in many cases. The claim system lent itself to certain abuses. Enforcement of requirements was often lax. Claims were staked and sometimes patented as a means of acquiring land for nonmining purposes, such as grazing. On many claims assessment work

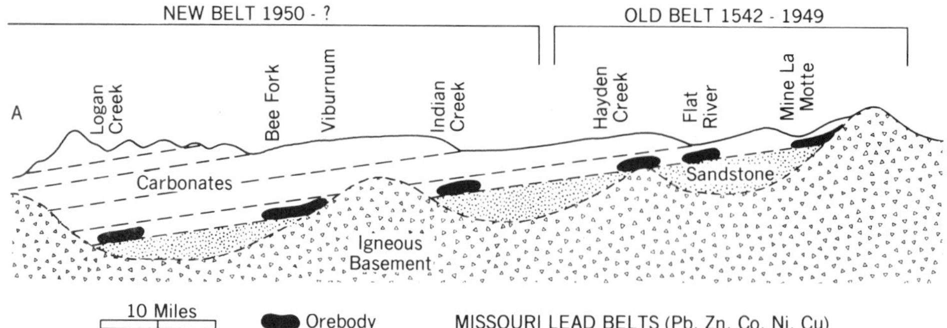

Figure 9-5 The lead belts of Missouri. Outcropping ore bodies such as the one shown at Mine La Motte could be covered immediately by leases, but buried ore deposits in the old and new belts were discovered only after extensive exploration. Tenure of land in such cases must be obtained long before any discovery is made. From Bailly (1966), by permission.

was never done. Since under the 1872 act the claims did not have to be recorded with any federal agency, the Federal Government for many years had no record of the number or status of various claims on the public domain. However, recording requirements under the Federal Land Management Policy Act of 1976 have largely corrected the problem.

Most but not all problems of the federal mining laws have been resolved by legislation or by application of the court-recognized concept of *pedis possessio*. In effect, *pedis possessio* allows a senior locator to maintain claims against a third party in the absence of a discovery, as long as discovery is pursued diligently. For the last two decades, almost all exploration on public lands has been based on this concept. In practice, a discovery is made on one or a few claims, and claims are located to the outer limit of suspected mineralization. The discovery is pursued first by geologic, geochemical, and geophysical means, then by drilling or excavation. The system is not perfect, but it does work. Its most fundamental defect emerges when public lands are withdrawn from location before a discovery is made. *Pedis possessio* cannot be asserted against the Federal Government. Even in this difficult kind of case, Congress has given some flexibility. For example, in the Alaska National Interests Land Act, locators in two separate withdrawals in southeastern Alaska were given 5 years to pursue discovery in peripheral claims.

The Mining Law of 1872 is still the basic law governing most lode and placer deposits, but there have been modifications over the years. The Coal Act of 1873 covered the acquisition of patents on coal lands, and rules for the acquisition of deposits of building stone were set forth in the Building Stone Act of 1892. In 1914, passage of the Alaska Coal Lands Leasing Act marked the beginning of a second system of mining law applicable to the public domain. The act provided for coal leases not exceeding 2560 acres at a minimum royalty of 2 cents per ton and an initial annual rental of 25 cents per acre. Title to the leased coal lands was reserved to the Federal Government. The act marked the introduction of the doctrine of regalian or sovereign rights into federal mining law. The doctrine had its origin in the monarchies of western Europe, where rights to mineral resources were reserved to the sovereign. The 1914 act was the forerunner (Ely, 1964) of the much more important Mineral Leasing Act of 1920, under which the Federal Government reserved title to all deposits of coal, oil shale, oil, natural gas, sodium minerals, and phosphate rock on the public domain, and prospecting and mining of such deposits were thenceforth permitted only under lease. In 1926 sulfur in New Mexico and Louisiana was added to the list of minerals covered by the act, and in 1927 potassium minerals were added. The leasing principle was extended by the Acquired Lands Leasing Act of 1947, applicable to lands acquired by the Federal Government and added to the public domain. The Outer Continental Shelves Act of 1953 and the Submerged Lands Leasing Act of 1953 added to the leasing system the continental shelves beyond 3 leagues off Texas and Louisiana and beyond 3 miles off the rest of the coast. In effect the various leasing acts have established a third system of mining law in the United States.

There are other important acts. The Surface Resources Act of 1947 (amended in 1955) authorized the Secretary of the Interior to sell certain materials on the public domain and expressly excluded from the mineral location (claim) laws such materials as sand, gravel, cinders, stone, and pumice. The Multiple Use Act of 1954 and the Common Varieties Act of 1955 authorized mineral location on the same tract under both the Mining Law of 1872 and the leasing acts. The Federal Land Management Policy Act of 1976 has led to regulations requiring the recording of unpatented claims with the Bureau of Land Management and to the establishment of procedures by which inactive claims can be invalidated. Affidavits of assessment work must be furnished annually to the bureau.

Dissatisfaction with the Mining Law of 1872 has been repeatedly expressed, and almost every Congress since World War II has entertained proposals for modifying the law. Some of the defects of the original law have been remedied by legislation and court actions going back more than 100 years. The new requirements for recording claims and the procedures for invalidating inactive claims, for example, are important changes, but certain defects remain. The distinction between lode claims and placer claims ignores the fact that some deposits fit in neither category. The apex provision of the law should be eliminated; no other nation that maintains a mining claim system has such a provision. The requirement of a surface discovery is outmoded. Geological and geophysical evidence and evidence from drilling should be accepted as evidence of a discovery. Actual discoveries are now made almost entirely by drilling or by underground workings. Yet under present law the prospector has no firm protection during the period when these expensive investigations must be done. Under existing law, acquiring and maintaining control over sufficiently large blocks of ground can be very difficult and are complicated in some districts by the confusion of overlapping and conflicting claims. Changes in the law are needed to remedy these defects. Tenure of claims for prospecting should be limited to a reasonable period. If no discovery is made, a claim should be terminated. There must, however, be some room for exercise of judgment. In principle, allowing claims to be held indefinitely, without mining activity, is not justifiable, but in practice there may be valid reasons, such as adverse market conditions, for inactivity. In such cases some provision is needed for holding claims for longer periods at relatively low cost, or for deferring assessment work. The 20-acre limitation on the area of a claim, particularly during the exploration period, flies in the face of the realities of modern mineral exploration. However, apart from the leasing acts, attempts to replace the Mining Law of 1872 have failed. It has not been possible to devise a more suitable alternative.

The fundamental mining law described above is the result of congressional legislation. However, the growth of administrative law since the 1920s, through rules and regulations laid down by various federal, state, and local agencies, has had a profound effect on the land law framework: "The intricate web of federal, state, and local laws and restrictions dealing with mining is further complicated by inconsistencies in policy goals . . . uneven enforcement and overlapping jurisdiction of the enforcement agencies" (Dempsey, 1973, p. 119). The difficulties and costs of exploration and mining development are thereby greatly increased.

LAND POLICY

Land policy determines the availability of public lands for mineral exploration and mining. Mining law is a basic expression of land policy, but other expressions, in various legislative and administrative acts, may have the effect of modifying mining law. There is a sharp contrast between the evolution of mining law in the United States and the evolution of land policy as a whole. The basic premise of American mining law, ever since the Republic was founded, has been that the public lands should be open to mineral exploration and mining. In other areas of land policy, however, evolution has been in the direction of restricting access to the public lands.

Restrictions have taken the form of "withdrawals from mineral entry"—that is, closure of lands to location of claims or to location under the leasing acts. One category consists of withdrawals of the areas of the national parks, beginning with the creation of Yellowstone National Park in 1872. In 1902 withdrawals of land were authorized for irrigation works and associated irrigable lands. In 1910 the Pickett Act gave the President authority to withdraw lands for a variety of public purposes. Petroleum reserves were established under the act, and oil shale lands were withdrawn, although they remained open for the location of metalliferous claims and uranium-bearing claims. Substantial acreages of land were set aside for federal installations, military and naval reservations, and other purposes, but mineral development is not necessarily excluded from these. Classification of public lands according to potential use was also authorized under the Pickett Act and further under the Taylor Grazing Act of 1934. Additional acts excluded the sites of towns, villages, and cities from mineral entry. Many new national parks and monuments were created. Until 1976, mining was permitted in six national parks but excluded in all others. Public Law 94-429 of 1976, however, excluded mining in the six parks, except for existing claims and mining operations, which are to be strictly regulated. By 1975, the acreage withdrawn in national parks, historical parks, national monuments, wildlife refuges, and wild and scenic rivers had increased to about 47 million acres, approximately 6 percent of the public domain. The pace of withdrawal accelerated after 1975. By 1982 the National Park system included 74 million acres, wildlife refuges amounted to 87 million acres, and wild and scenic rivers totaled 4 million acres. Accessibility of wildlife refuges is dependent on the terms of the individual acts establishing them. Exploration for oil and gas or nonfuel minerals, or both, is permitted in some refuges, but most are closed to mineral exploration and development. National forests are legally open to mineral exploration, but obtaining permits for this has not been easy.

The Wilderness Act of 1964 initiated a new series of withdrawals, one that is still in progress. The act set aside 9.1 million acres, mostly in Idaho, as a permanent National Wilderness Preservation System. The lands were legally open to mineral entry through December 31, 1983, after which lands not claimed or leased for minerals were to be closed to entry. The act stipulated, however, that activities (including prospecting) for purposes of gathering information about mineral or other resources were to be permitted provided they were compatible with preservation of the wilderness environment. It was further provided that wilderness areas should be surveyed for their mineral potential on a planned, recurring basis "consistent with the concept of wilderness preservation." Surveying was to be done by the U.S. Geological Survey and the U.S. Bureau of Mines.

The first withdrawals under the Wilderness Act have been followed by withdrawals of additional lands. Section 3(c) of the act required the Secretaries of the Interior and Agriculture to review every roadless area of 5000 or more contiguous acres within units of the national park system, national wildlife refuges, and national forests and to make recommendations for inclusion of such areas in the wilderness system. Since access to national parks is already prohibited, the chief impact on availability of land for mineral exploration is through withdrawals from national wildlife refuges and wilderness areas in national forests. As of the end of 1984, 88.6 million acres were included in the wilderness system, 56 million of these in Alaska. An additional 78 million acres, mostly lands under jurisdiction of the National Forest Service and the Bureau of Land Management, were under review for possible inclusion in the wilderness system.

The Wilderness Act of 1964 provided for a recurring review of the mineral potential of wilderness lands and also posed economic tests for wilderness areas. The Federal Land Management Act of 1976 and the Alaska National Interests Land Act of 1980 also authorized wilderness areas but essentially dropped economic tests. In theory, wilderness lands are not permanently closed to mineral exploration and mining but can be reopened by Congress. Actually there are problems in implementing provisions of the act with regard to mining. Appraising the mineral potential of 88.6 million acres of wilderness lands is an enormously difficult and expensive task. As of 1983, nineteen years after passage of the Wilderness Act, the Geological Survey and Bureau of Mines had assessed the mineral potential of 45 million acres of National Forest lands included in wilderness areas or being considered for inclusion. Assessment required approximately 1000 man-years of work. The results are summarized in Geological Survey Professional Paper 1300 (Marsh et al., 1983). "Mineral-resource potential" is assessed for 800 areas, the majority in the western states including Alaska. The report is illustrated by 332 maps, each covering one or more individual areas. Areas of "substantial" or "probable" mineral resource potential appear on 221 of the maps.

The report is an impressive and valuable work, but it is important to recognize its limitations. The assessments are based on compilation of all available data on mines, prospects, and mineral deposits, plus geological mapping, surface geochemical sampling, and geophysical surveys in some of the areas. However, the authors state (p. 7):

> *The credibility of all assessments of mineral-resources potential is a matter of concern. The data available for virtually all wilderness lands are incomplete; subsurface data are lacking. Assessment of mineral-resource potential is by its nature speculative and involves considerable uncertainty. Construction of uniform quantified assessments is currently impossible.*

This comment stems from the basic difficulty confronting all assessments of "mineral-resource potential" that are based primarily on surface work. Assessing mineral resources is different from assessing most other natural resources. The floral and faunal resources of an area can be quantified, because trees, other plants, and animals are accessible and can be counted. Soils are also easily accessible and can be quantitatively appraised by simple methods. The volumes of surface bodies of water can be measured. Hidden mineral deposits, however, cannot be "counted" until they have been found and until sufficient subsurface exploration has been done to define them in terms of position, shape, attitude, tonnage, and contents of valuable minerals. Yet under the terms of the Wilderness Act the necessary exploration appears for all practical purposes to be foreclosed. Unless an area is taken out of wilderness status by act of Congress, mineral entry, leasing, and exploration are forbidden. The horns of a dilemma now emerge. On the one hand, an area is not likely to be removed from wilderness status unless positive evidence of the existence of a significant mineral deposit can be produced. On the other hand, the necessary information cannot be obtained until the area is released.

To sum up, assessment of resource potential does not define the mineral resources of an area. Except in rare cases, such as coal seams that have already been explored to some extent, resource assèssment by surface methods cannot give quantitative or even semiquantitative estimates of mineral resources and reserves on wilderness lands. So long as surface assessment is not followed by subsurface exploration, the true importance of mineral-bearing wilderness lands as potential sources of mineral supply will not be known.

It will be evident from the above review that access to the public domain for purposes of mineral exploration and development has become increasingly difficult in the past 20 years. It is not possible, however, to state how much of the public domain is currently closed to mineral entry. The status of mineral entry on some federal lands is confused and uncertain. In the Final Report of the Task Force on the Availability of Federally Owned Mineral Lands (Office of Technology Assessment, 1976), it is stated that in 1974 location of minerals on federal lands under the Mining Law of 1872 was formally prohibited on 305.5 million acres (41.9 percent of the total public domain), severely restricted on 118.5 million acres (16.2 percent), and moderately restricted on 76.3 million acres (11.4 percent). Exploration and development of minerals under the mineral leasing laws were formally prohibited on 312.5 million acres (36 percent of the public domain), severely restricted on 183 million acres (22.7 percent), and moderately restricted on 53.8 million acres (6.6 percent). The figures were considered to be approximations. In view of subsequent withdrawals, access must be even more restricted at the present time.

It is important to understand what is at stake in the controversy over land policy, in terms of the future availability of minerals from domestic sources. Most of the public domain (Fig. 9-1) is in the western states including Alaska. The western states have yielded the major U.S. production of copper, molybdenum, tungsten, mercury, silver, gold, uranium, beryllium, arsenic, antimony, and bismuth. The same states have also been richly productive of lead, zinc, sodium carbonate, lithium, boron minerals, clays, and many other nonmetallic minerals. Prospecting is not a thing of the past. It has been vigorously pursued in the western states since 1848, and many notable discoveries have been made in recent years. More are to be expected as exploration continues in the future.

What about Alaska? Since the discovery of oil at Prudhoe Bay in 1968, the status of public lands in Alaska has been a focus of conflict between environmental groups and those concerned with development of mineral resources. For environmental groups, Alaska is a last frontier, a region in which there are vast areas of unspoiled wilderness that should be preserved in their pristine state for future generations. For those concerned with mineral resources, Alaska is also a last frontier, a last frontier of mineral exploration in the United States. The mineral potential of Alaska is only beginning to be known. The Klondike gold rush in northwest Canada in the final years of the 19th century soon led to gold discoveries at Nome and Fairbanks in Alaska. The national attention was suddenly directed to a territory that had been largely neglected since its purchase from Russia in 1867. However, owing to its harsh climate, the inhospitable nature of its terrain, and a meager system of roads and railways, investigation of the geological and mineral resources of Alaska has been slow. Only about 5 percent of Alaska has been mapped geologically on scales of 1 mile to the inch or larger. A mile to the inch is considered the minimum scale of geological mapping necessary as a basis for mineral exploration. After the gold rush, large coal resources were discovered. In subsequent years, deposits of a number of metals and nonmetallic minerals were discovered, and some were worked on a small scale. Coal mining was begun near Healy, south of Fairbanks, along the Alaska Railroad. Oil fields were discovered in 1954 around Cook Inlet, and in 1968 the great discovery at Prudhoe Bay was made. Several new coal-mining projects are now under way. A porphyry deposit at Quartz Hill, in southeastern Alaska, now under development, may be the largest molybdenum deposit in the world. The Red Dog zinc-lead-silver deposit, in extreme northwestern Alaska, is said to be the largest undeveloped zinc ore body in the western world. Drilling suggests (*Engineering and Mining Journal,* 1985) that the Lik deposit, in the same mineral belt, is second only to the Red Dog deposit. Some 244 major mineral deposit areas (Dayton, 1979) have been identified in various parts of the state, and, given the great geological diversity of Alaska, more major discoveries are to be expected. Exploration for oil and gas continues.

It is clear that Alaska, despite some statements to the contrary, has the potential for yielding large amounts of a variety of minerals important to the economy of the United States, including certain strategic minerals, but this potential can be realized only to the extent that Alaskan lands are open to mineral exploration. The ultimate allocation of Alaskan lands as stipulated by the Alaska Statehood Act of 1958, the Alaska Native Claims Settlement Act of 1971, and the Alaskan National Interests Land Act of 1980 is shown in Table 9-1. The allocation process (as of January 1985) is not complete; the 12 Native Regional Corporations established under the 1971 act have received only 30.6 million acres of their allotment, and the State of Alaska has received only 78.1 million acres. About 80 percent of the land allotted to the state will probably be available for mineral entry under a location system leading to a leasehold, but ownership remains with the state. About 60 to 75 percent of Forest Service land will probably be available for entry under federal mining law, together with about 80 percent of the land assigned to the Bureau of Land Management. Some of the fish and wildlife lands are open to oil and gas leasing, but none are open to entry for nonfuel minerals. Lands allotted to native corporations are available for mineral entry under leases that must be negotiated with regional or village corporations or both. All National Park Service lands (national parks and monuments), U.S. Bureau of Land Management Conservation System lands, and forest service wilderness lands are closed—a total of 141.3 million acres. Including state lands and forest service nonwilderness lands that are closed, total lands

TABLE 9-1 Allocation of Alaskan Lands[a]

	Millions of Acres
State of Alaska	103.5
Native lands	44.0
Other private lands	0.3
U.S. National Parks Service	54.7
U.S. Fish and Wildlife Service	77.3
U.S. Forest Service	
Wilderness	(7.1)
Nonwilderness	(15.7)
Total Forest Service	22.8
U.S. Bureau of Land Management	
Conservation system lands	(2.2)
U.S. public lands	(75.0–77.0)
Total U.S. BLM lands	≈ 80.0
Total	382.6

[a]The author is indebted to Mr. C. C. Hawley for compiling (January 1985) data obtained from the various agencies in Anchorage. The figures are not exact, particularly those for lands under the U.S. Bureau of Land Management. The total area of the state is somewhat uncertain. The total given above compares with the 375–378 million acres generally given as the total area.

closed to mineral entry amount to at least 150 million acres, roughly equal to the combined areas of the states of California and Washington.

In selecting the areas to be set aside for national parks and wildlife refuges, there was only limited consideration given to mineral potential. Large portions of some of the major mineral belts identified in Alaska lie within the areas withdrawn. For example, much of the belt in which the Red Dog and Lik deposits were discovered now lies within the Noatak National Preserve and the Gates of the Arctic National Park.

In considering the impact of land policy on access to public lands for mineral exploration, it is essential to recognize that whereas legislation sets the basic ground rules for use of the public domain, the laws give wide latitude to the agencies that administer them. The Federal Land Management Policy Act of 1976, for example, grants very broad powers to the Bureau of Land Management of the Department of Interior and to the Forest Service of the Department of Agriculture. Administrative regulations and rulings can place formidable obstacles in the path of the would-be prospector and miner. Law may authorize access, but administrative regulations can prevent access simply by making it too time-consuming and expensive. The coal leasing acts call for granting leases on federal coal-bearing lands, but during 1968–1977 coal leases were severely limited by the Department of the Interior. Attempts to liberalize leasing during James Watt's tenure as Secretary of the Interior were strongly opposed. The Forest Service has an elaborate set of regulations governing the large areas of land under its jurisdiction in Alaska. Dayton (1979) has charted the steps involved in complying with the surface regulations of the Forest Service. His analysis shows that the time from reconnaissance prospecting to mining may range from 7 to 24½ years, the most likely time being 14 years. Until the whole process is complete, there can be no return on the investment of funds in exploration, development, and compliance. Such regulations are major barriers to discovery and development of mineral resources.

If the United States is to maintain a strong mineral base, as much as possible of the public domain must be open to discovery of mineral deposits that will take the place of those that are now known and are being steadily depleted. Exploration must be a continuing process. It has sometimes been argued that wilderness areas should be left untouched against a time of national need. The argument is very appealing, but it rests on a misunderstanding of how mineral resources are discovered and developed, and of the time that those processes require. In the first place, when a crisis arrives, it is too late to begin exploration. In the second place, and just as important, is the fact that a major by-product of continuing, active exploration is a better understanding of the nature and occurrence of mineral deposits in the crust. This knowledge cannot be generated by abstract studies. It is generated when scientific studies are related directly to the realities of mineral deposits that are faced when mineral deposits are explored and mined. Active exploration and development generate new ideas that may lead to discovery of new kinds of mineral deposits; they also provide the acid tests of those ideas. Much of what we know today about our mineral resources, and much of our ability to discover and develop them, is our heritage from the exploration and mining of the past. The time factor is critical. As we have seen in earlier chapters, years are required for the discovery and development of a significant mineral deposit.

Access to lands for mineral exploration is not just a problem of public domain. The Chase Manhattan Bank has estimated that 30 percent of the lower 48 states has been effectively withdrawn from exploration by the need to comply with environmental laws and regulations. There is little consideration of mineral resources in urban and regional planning. For example, the rapid increase in the population of central Florida and the consequent spread of housing developments threaten the future of the phosphate-mining industry of the area, the nation's most important source of phosphate.

Much of the evolution of federal land policy in recent years has been a reflection of environmental concerns. Besides land policy acts, there have been other acts that affect the development of mineral resources in greater or less degree. Under the Occupational Safety and Health Act of 1969, many regulations of the procedures of the mining industry have been put in place. The Mine Safety Act of 1969, the National Environmental Protection Act of 1970, the Clean Air Act of 1973, the Endangered Species Act of 1973, and the Surface Mining and Reclamation Act of 1977 all have consequences for mineral development. These matters are discussed in a later chapter.

REFERENCES AND ADDITIONAL READING

Bailly, P., 1966, Mineral exploration and mine developing problems related to use and management of other resources and to U.S. public land laws, especially the Mining Law of 1872. Statement to the Public Lands Law Conference, University of Idaho, Oct. 10, 1966, 43 pp.

Bundtzen, T. K., 1983, Overview of Alaska's strategic minerals. In *International Minerals,* A. F. Agnew, ed., Westview Press, Boulder, Colo. pp. 37–70.

Dayton, S., 1979, Alaska: A land and people in search of a future. *Engineering and Mining Journal,* Vol. 180, No. 5, pp. 72–87.

Dempsey, S., 1973, Land management and mining law: The framework of mineral development. In *The Mineral Position of the United States, 1975–2000,* E. N. Cameron, ed., University of Wisconsin Press, Madison, pp. 109–125.

Department of Agriculture, 1979, National Forest System, Roadless Area Review and Evaluation Program (RARE II). U.S. 96th Congress, 1st Session, House Document No. 96-119, 99 pp.

Department of the Interior, 1979, National Wilderness Preservation System, 14th Annual Report, U.S. 96th Congress, 1st Session, House Document No. 96-51, 19 pp.

Ely, N., 1964, Mineral titles and concessions. In *Economics of the Mineral Industries,* 2nd Ed., E. H. Robie, ed., American Institute of Mining, Metallurgical, and Petroleum Engineers, New York, pp. 81–130.

Engineering and Mining Journal, 1985, Red Dog has pups. Vol. 186, No. 1, p. 17.

Flawn, P. T., 1966, *Mineral Resources,* Rand McNally, New York, pp. 148–181.

Maley, T. S., 1983, *Handbook of Mineral Law,* 3rd Ed. Mineral Land Publications, Boise, Idaho, 711 pp.

Marsh, S. P., Kropschot, S. J., and Dickinson, G. G., 1983, Wilderness mineral potential. U.S. Geological Survey, Professional Paper 1300, 2 Vols., 1300 pp.

Office of Technology Assessment, Technology Assessment Board of the U.S. Congress, 1976, *Mineral Accessibility on Federal Land,* U.S. Government Printing Office, Washington, D.C.

10 Environmental Regulation

One of the most striking developments of the years since World War II has been the rise of concern about damage to the environment and to public health and safety that results from industrial activities in the United States. Many factors are responsible—the growth in population to 220 million persons, the expansion of urban areas, the greater accessibility of all areas of the United States in the age of the airplane and the automobile, the enormous growth of industrial activities of all kinds, the attendant greater impact of industrial activities on the environment, and a change in social attitudes toward the development of natural resources. Although the Environmental Decade did not begin until 1970, in retrospect the Wilderness Act of 1964 was the forerunner of many things that would strongly affect the activities of the mining industry in the United States.

The environmental movement developed out of the conservation movement of the earlier part of the twentieth century and soon turned its attention to the environmental consequences of mining and mineral use. During the 1960s and 1970s the mining industry became identified as a major source of damage to the environment. Mining is essentially a destructive process, and few mining operations can be conducted without some damage to the land. The effects of past mining operations in certain areas, particularly in certain coal-mining regions, are both conspicuous and consequential. The result has been one of the major social controversies in the history of the United States. The issues are complex, and debate has been confused by much angry rhetoric on both sides. There are strong differences in the priorities assigned by different groups to preservation versus development of natural resources, and there are disputes over the facts of environmental damage.

In approaching the environmental problem in relation to the mining industry, we must first examine specific environmental consequences of mining activities. There are potential impacts at almost every stage, but depending on the type of mineral deposit, the type of mining operation involved, and the climatic and topographic conditions under which each mining operation is being performed, they range from negligible to serious in consequences.

Different types of deposits differ greatly in the nature and composition of the ores they contain. Some, like sand and gravel, crushed stone, barite, most iron ores, and cement materials, consist of inert, nontoxic minerals. Processing such materials so as to prevent environmental damage consists mostly in controlling dusts produced during crushing and grinding. Waste rock and tailings must be disposed of in such a manner that they do not enter into local drainage or underground water supplies. Revegetation of areas of waste rock and tailings with grass or trees is now standard practice at these and other mining operations.

The mining and processing of some ores, particularly sulfide ores, produces potentially toxic substances. Iron sulfides are waste products of some sulfide ores and pass into the tailings. In contact with air and water, iron sulfides yield sulfuric acid, and special measures are required to keep the acid from contaminating streams and underground water supplies. Mine and mill wastes may also contain toxic trace elements; wastes from the processing of uranium ores present special problems of disposal. The smelting of sulfide ores produces sulfur dioxide, which forms sulfuric acid if allowed to escape into the atmosphere. Fumes from smelters at some localities, such as Sudbury, Ontario, and Ducktown, Tennessee, have had devastating effects on vegetation of surrounding areas and have been contributors to the acid rains that are currently of concern in the northern hemisphere.

The burning of coal presents some of the most serious problems of environmental damage and pollution. All coal contains sulfur—partly bound up in the organic matter, partly in the form of sulfides. Coal "cleaning" procedures can remove much of the sulfide sulfur but cannot remove the organically bound sulfur. Illinois coals, for example, contain up to 7 percent sulfur. Cleaning removes the roughly 3 percent present as iron sulfides, but the organically bound sulfur remains in the cleaned coal. Unless measures are taken to remove it from stack gases, sulfur dioxide is released to the atmosphere when coal is burned and may be returned to the earth in acid rain. As noted previously, coals differ in sulfur contents. Some Appalachian coals and much western coal contain less than 1 percent sulfur. For western coals the advantage is partly offset by the fact that much of it is lignite or subbituminous coal. As much as 2 tons of lignite may have to be burned to yield as much energy as that from a single ton of eastern bituminous coal. The problems of atmospheric pollution become more serious with the increase in world consumption of coal. Sulfur in coal is also the source of acid mine waters, which have caused serious pollution of streams in certain coal-mining areas. Coal dust and fines from coal operations have also caused pollution of streams. Finally, when coal is burned, fine ash (known as fly ash) goes up the stacks. Unless it is removed from the stack emissions, it is a significant source of air pollution.

Let us turn now to the consequences of surface mining operations. Let us distinguish two types of surface mining—open-pit and stripping. Open-pit mining is exemplified by the surface iron mines of the Lake Superior district, most operations for sand and gravel, stone quarries, clay pits, and the phosphate mines of Florida. Their scale ranges from small for many sand and gravel operations to very large for iron ore mines and open-pit copper mines. The general results are excavations of various sizes. Waste rock must generally be piled in dumps at sites close to the mines. Associated with the excavations, sites are required for office buildings, machine and equipment shops and sheds, mills, and dumps for disposal of wastes from the mills. When mining ends, buildings at the mine site can be razed, and the land around the mine can be partially or wholly reclaimed. Mine dumps can be contoured to acceptable shapes, and grass and trees can generally be made to grow over them. Current practice commonly calls for reclamation concurrent with mining, but the open pits remain. In regions of adequate rainfall, pits generally become ponds, and if the products of mining are nontoxic, the ponds can generally be converted to recreational use. Otherwise they remain as scars on the landscape that will only be obliterated by natural processes over considerable periods of time.

Strip mining for coal as practiced in the past resulted in extensive damage to lands in Illinois, Indiana, Kansas, and other areas, where it left alternating ridges and linear depressions (Fig. 10-1). The landscapes in such areas may remain scarred for hundreds of years or more and, except where the depressions have filled with nontoxic waters, unfit for further use.

In central Tennessee, former practice in phosphate mining consisted of stripping overburden and mining the underlying phosphate-rich material down to a highly irregular limestone bedrock surface (Fig. 10-2). No reclamation was attempted, and many square miles of land were left unfit for agricultural use. In 1967, the Tennessee legislature passed an act requiring reclamation of land mined for phosphate, and reclamation is now standard practice. In mining an area, the first step is stripping the topsoil, which is stored beside the mine area. Subsoil is then stripped, and phosphate ore is then mined. The subsoil is next restored to the mined area, which is then graded to roughly the original contour. The topsoil is then spread over the surface, fertilized, and seeded. The procedure is based on studies by the University of Tennessee Experiment

Figure 10-1 In the foreground, strip mining of coal is in progress. In the background, a mined-out area has been left unreclaimed. Department of the Interior (1967).

Station, which showed that an area restored in this manner will yield a good crop beginning with the first year after restoration. In 1971 the cost of reclamation was about $300 per acre for land yielding up to 10,000 tons of phosphate per acre, worth roughly $100,000. The cost was more than offset by the resale value of the land for agricultural use. In Indiana, essentially the same procedure has meant that land used for farming before coal mining is equally valuable for farming afterward (Fig. 10-3, top). In Florida, after years of unsuccessful experiments in reclamation, a similar procedure has been developed and is being used successfully to restore pine lands after the dredging operations for ilmenite and associated heavy minerals described in Chapter 2 (Fig. 10-3, bottom). In all these operations, reclamation is part and parcel of the total mining plan. Pits made in extracting sand and gravel can be converted into ponds (Fig. 10-4). This makes it possible to minimize environmental damage at reasonable cost.

Reclamation of mined land is most successful in areas of good rainfall, warm or temperate climate, and flat or rolling topography. In arid or semiarid climates revegetation is not so easy, because rates of growth may be outstripped by rates of erosion. There is still much to be learned about reclamation under such conditions. Even in moist climates, revegetation of steep slopes may be very difficult. An expensive seeding may be washed away during a single heavy rain. There is much concern over the problem of reclamation posed by the large strip-mining operations in the coal fields of North Dakota, Montana, and Wyoming. Only experience can indicate the possible degree of success that can be achieved.

Figure 10-2 Photographs taken in the phosphate-mining area near Columbia, Tennessee. Top: Land unreclaimed after mining. Bottom: Three years after the end of mining, the land has been reclaimed to pasture and a pine planting.

Reclamation in arctic climates, not only in Alaska but on high mountains in the lower 48 states, is difficult because rates of growth of vegetation are very low. There are no easy answers to the problems of reclamation in such areas.

One of the most troublesome areas for reclamation of mined land is the coal-mining region of the Appalachian Plateau in Pennsylvania, West Virginia, and the eastern parts of Ohio, Kentucky, and Tennessee. The region is dissected by deep valleys. The coal seams lie flat and crop out on the valley slopes. As shown in Fig. 10-5, a seam too thin to be mined underground is mined in an open cut that runs along the seam outcrop.

Figure 10-3 Top: Land mined for coal has been regraded and is ready for restoration of soil stripped off before mining and put aside. Bottom: Land dredged for beach sands at Trail Ridge, Florida, has been regraded, recovered with stripped-off topsoil, and planted in grass and pines. It will revert to its original state, a pine forest.

The overburden is removed, leaving a bench capped by the coal seam. The coal is then stripped from the bench. Much spoil is dumped downhill from the excavation. Given the heavy rains of the region and the tendency toward rapid erosion, revegetation is very difficult.

In general, surface damage due to underground mining is far less than that due to surface mining. The chief problem is disposal of wastes that cannot be put back underground. At some mines as much as 70 to 80 percent of the waste can be backfilled into the workings. In the Park City district of Utah, for example, most of the waste of extensive past operations has now been put back underground. Backfill can prevent subsidence of ground over underground workings. However, backfilling is

Figure 10-4 A sand-and-gravel operation. At the right is an active pit. Above to the left is a pit in process of conversion to a pond. Below it are completed ponds, and plantings around the margins. Department of the Interior (1967).

often not economical and is not always possible, particularly if the percentage of waste in the ore is relatively small, as it is in most underground coal mines. Subsidence has been a very serious problem in the Scranton–Wilkes Barre anthracite coal basin of eastern Pennsylvania and in other parts of the state. Another hazard of underground coal mines is fire. Once started, underground coal fires may be impossible to control. Fires in coal seams underlying the town of Centralia, Pennsylvania, have been burning since 1962 and have recently forced evacuation of the town. Withdrawal of fluids has caused subsidence over many oil fields. Generally it is only a few feet, but serious subsidence has taken place over the Inglewood and Wilmington oil fields of the Los Angeles coastal plain area.

How much damage has been done? Obviously, most of it is done by surface mining, which in 1975 accounted for about 93 percent of the total tonnage of solid minerals (including coal) produced in the United States. Under the Appalachian Regional Development Act of 1965, the Secretary of the Interior was required to investigate the effects of surface mining in the United States. The results were reported to Congress (Department of Interior, 1967). It was found that approximately 4985 square miles, not quite 0.14 percent of the total area of the United States, had been disturbed by surface mining. Of this, approximately 1780 square miles had already been reclaimed, leaving 3205 square miles or 0.09 percent of the total area of the United States still disturbed and unreclaimed. Of the total area disturbed, coal mining accounted for 41 percent, sand and gravel mining 26 percent, stone quarrying 8 percent, phosphate mining 6 percent, and

Figure 10-5 Top: A coal seam is exposed above a bench from which another seam has already been mined. Near Middlesboro, Kentucky. Bottom: The result (contour mining) of the mining process illustrated above. The lighter-colored materials below the open cuts are waste rock. Department of the Interior (1967).

iron mining 6 percent, together amounting to 87 percent of the total area.

In two later studies by the U.S. Bureau of Mines (Johnson and Paone, 1982), a more comprehensive assessment of land utilized for mining was made. The period covered was 1930 to 1980. Underground mining operations were included in the assessment. The survey included land occupied by mining and milling installations, land actually excavated, land occupied by waste dumps, land occupied by tailings dumps, and land affected by subsidence after underground mining. The total area of land involved was 5.7 million acres, or 0.25 percent of the total land area of the United States. Land reclaimed by industry during the same period was 2.7 million acres, or 47 percent of the land disturbed. Land used for disposal of waste from surface mining of coal during 1971–1980 was not included, because under federal and state environmental regulations, almost all the land so used was reclaimed. Land disturbed by subsidence during 1971–1980 was also not included; the area involved is very small. For nonfuel-mining operations during 1971–1980 there are no data for land reclaimed, but most if not all nonfuel-mining operations during the period involved reclamation.

Several things are evident from the studies cited. One is that only a very small fraction of the country has been affected by mining, and a substantial part of the disturbed ground has already been restored to agricultural, forest, or other uses. The second is that mining is not the principal disturber of the environment. Table 10-1 shows the allocation of land to various uses in 1980. The principal disturber of the lands is agriculture, followed by urban development and highway construction. Mining is at the bottom of the list, along with railroads and airports. If Americans are truly concerned with the impact of civilization on the environment, attention should be directed primarily to soil conservation, the consequences of urban and suburban sprawl, and the extravagant use of land for interstate highway systems. The third thing is that since

TABLE 10-1 Land Use in the United States in 1980

	Millions of Acres
Agriculture	
Cropland	413.0
Grassland pasture, and range	985.7
Forest land grazed	179.4
Farmsteads, farm roads	10.9
Total, agriculture	1589.0
Wildlife refuge system	88.7
National park system	77.0
Urban and built-up areas	68.7
Forest Service wilderness	25.1
Highways (1978)	21.5
Mining	5.7
Airports (1978)	4.0
Railroads (1978)	3.0
Other	388.1
Total, all uses	2270.8

Source: U.S. Bureau of Mines, 1982.

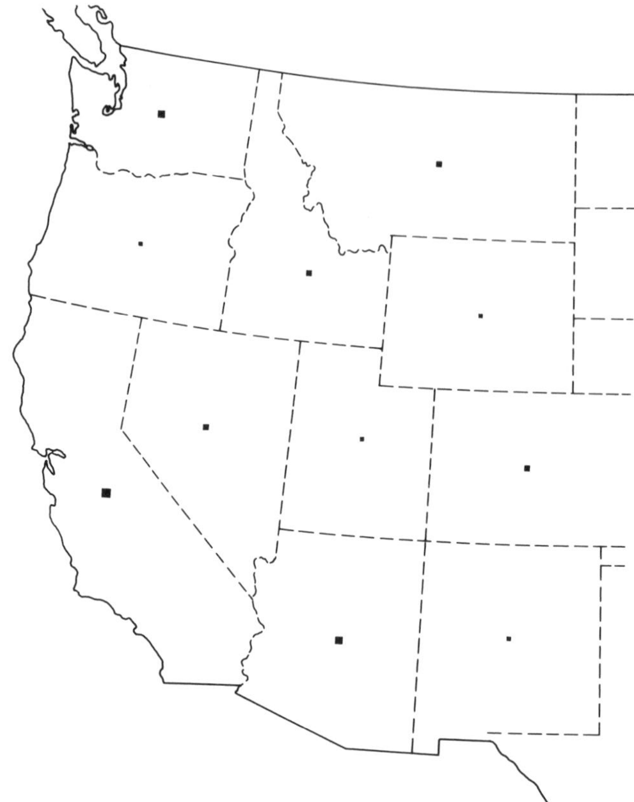

Figure 10-6 The total area occupied by mining operations in each western state is indicated, to scale, by a black rectangle. From E. L. Ohle (*Economic Geology*, 1975), by permission.

1970 great progress has been made in minimizing environmental damage due to mining and mineral processing.

In the 1960s and 1970s, much attention was focused on the impact of mining on the environment in the western states. Metal-mining operations, mainly those for base metals and gold and silver, were particularly singled out. They were frequently accused of "rape" of the western states. The interests of a nation that is heavily dependent on supplies of metals are not served by such irresponsible and unfounded statements. E. L. Ohle (1975) determined the areas disturbed at each of the copper, lead, zinc, gold, and silver deposits mined in the western states. Included were waste rock piles, tailings dumps, and other surface features. From the data for individual operations, a total area disturbed was determined for each state. This area was plotted to scale, as a square, on a map of the states, with results shown in Figure 10-6. In no state is the area as much as 0.2 percent. About 40 percent of the disturbed areas, on average, has been reclaimed.

As indicated above, mining does not necessarily mean permanent loss of land for other purposes. Substantial amounts of mined land can be restored for agricultural use or for use for recreational, forest, or wildlife purposes. Some, however, cannot be restored. In such cases, the value of mineral production must be weighed against the value of the land for other purposes and against the costs in terms of damage to the environment. Mining is generally a high-value use of land. Ohle (1975) points out that 1 acre of a 0.6 percent porphyry copper deposit 300 feet thick contains metal worth, at 80 cents per pound, $10,460,000, and that an acre of virtually any economic metalliferous deposit is sure to contain metal worth at least $200,000. The copper deposit at Ladysmith, Wisconsin, is a specific example. Kennecott Copper Company discovered the deposit in 1965 and proposed to acquire 2603 acres of land for a mining operation. Of this, 378 acres was actually to be used for an open-pit mine (55 acres), waste disposal, and surface installations. The remaining acreage was to serve as a shelter belt around the mine area and would be continued in normal use. In the 11-year first stage of operation, production was planned at a rate of 365,000 tons per year of ore containing 80 pounds of copper per ton. At 80 percent recovery of copper from the ore, the 11-year yield would be 256,960,000 pounds of copper. At the March 1985 price of copper of 70 cents per pound, this would be worth $179,872,000. The yield per acre of land actually used (378 acres) would be $475,850. This would be approximately equal to the value of the corn produced from an average acre of Wisconsin farmland over a period of 1070 years. If a second, underground stage of mining were undertaken, there would be an additional yield per acre.

To catalog in detail the environmental impacts of mining and mineral processing is beyond the scope of this book; they are graphically portrayed in the report of the Department of Interior (1967). The aim here is only to indicate the kinds of damage that may result from mining. Enormous progress has indeed been made in the past 20 years in developing successful reclamation practices. Environmental protection is now part of the plan of every mining operation. Under environmental laws that have been enacted by the states and by the Federal Government, the nature of environmental impacts must be determined in advance of mining, and plans for minimizing damage must be approved before mining is begun. So great is the progress that much of the criticism leveled at the mining industry in the 1960s and early 1970s is now irrelevant. This is not to say that damage to the environment has been eliminated. It cannot be eliminated; it can only be minimized within limits that are essentially defined by the nature of each deposit, the type of mining operation that is required, and local topographic and climatic conditions. A certain amount of damage to the environment has to be accepted if society is to be provided with adequate supplies of minerals.

It is essential that environmental regulations be carefully framed and realistically implemented. This has not always been the case. A ruling that led to closing of a lead-zinc mine in southwest Wisconsin is an example (Barton, 1980). The Environmental Protection Agency set 0.5 parts per million zinc as the maximum allowable content of effluent water from the mine, despite the fact that the Public Health Service recommends a maximum of 5 parts per million. Effluent from the mine averaged about 2.5 parts per million, approximately the same as that of normal groundwater that was draining into the mine and had to be pumped out during mining operations. In effect, the mining company was required to reduce, to an unnecessarily low level, a pollution that originated outside the mine and that was beyond the control of the company. In this connection it was pointed out that the zinc content of the drinking water used in the EPA offices in Washington was about 20 parts per million. The ruling held, and the mine was forced to close. Actions of this kind are patently unrealistic. They are not trivial matters, since they result not only in loss of investment on the part of mining organizations but also in loss of resources to society.

It is encouraging to note that the costly and sometimes bitter confrontations that characterized the early years of the environmental movement have been giving way to collaboration of industry, environmental groups, and agencies of state governments and the Federal Government. The phosphate-mining operation of Chevron Resources north of Vernal, Utah, is one of many examples (Wells, 1985). The ore body there is the largest phosphate deposit in the western United States and will contribute importantly to the agriculture of the region. It is large enough to be mined for at least a century, a 700-million-ton deposit beneath about 90 feet of overburden. Reclamation of mined land is an integral part of the mining operation. All mined land will be restored to a condition suitable for grazing by wildlife and cattle. There is close collaboration with the Utah Division of Wildlife; the area will ultimately become a state wildlife refuge. Conservation of water resources, a critical matter in the area, is also part of the mining plan. Hydrogen sulfide, a contaminant of natural gas being produced in the Overthrust fields of southwestern Wyoming, will be recovered and employed in processing the phosphate rock for use in commercial fertilizer. Constructive collaboration of this kind is clearly in the interests of both society and the mining industry.

Environmental protection does not come free. There are cases in which environmental protection has yielded a profit, but they are not common. Billions of dollars have been spent by mineral industries on the preparation of environmental-impact statements, on proceedings involved in obtaining permits, on reclamation, and on pollution control. Such costs are tangible, and data for them have become available. From 1973 to 1975 (Dulaney, 1976), the steel industry spent $850 million on air and water pollution abatement, and the copper industry spent $1.5 billion. By 1982, costs to the steel industry had risen to $8 billion, and expenditure of $10 billion more was anticipated. In these industries and others that compete in international markets, such costs are significant, particularly because competition in considerable part is with mining industries in other countries, in which there is little or no environmental protection. All industries in the United States have shared in the costs of environmental protection, but costs to some segments of industry have been particularly heavy. DeYoung and Yasnowsky (1980) have computed average pollution-control expenditures during 1973–1978 as a percentage of total new plant and equipment expenditures (1) for all industries, (2) for nonferrous metal industries, and (3) for blast furnaces and steelworks. The averages are 5.2 percent, 20.9 percent, and 15.6 percent, respectively.

The impact of environmental regulations appears to be particularly significant for the nonferrous metal industries. From a study of the copper industry, Arthur D. Little, Inc. (1978), concluded that owing to environmental regulations domestic smelter capacity and domestic copper supply will be constrained in the next decades, that costs of domestic copper will rise, and that imports of copper may increase. The prophecy is already being fulfilled. In 1984 (Strauss, 1985), the Tacoma smelter of ASARCO closed; $100 million had been spent in an effort to comply with environmental regulations. In late 1984, Phelps Dodge announced that owing to the low price of copper and difficulty in meeting environmental standards, the Morenci, Arizona, smelter would close. Investments made in an effort to meet the standards amount to $246 million. In late March 1985, the same company announced that its smelter at Ajo, Arizona, would be closed down in early April. Cost of installing environmental-control equipment was cited as one factor in the decision (*Wall Street Journal*, 1985). Costs of environmental protection have thus been a factor in the decline of the American copper industry since 1980.

Part of the cost of environmental protection consists of changes in productivity. In the coal industry the enactment of the Federal Coal Mine Health and Safety Act of 1969 was followed by a welcome decline of 55 percent in the frequency rate of fatal accidents (U.S. Bureau of Mines, 1975), but the decline was accompanied by a decrease in output per man-day in underground mines from 15.6 tons in 1969 to 9.5 tons in 1975. The decline has been only partly reversed by subsequent improvements in mining methods.

Environmental protection is now demanded, and rightly, by the American people. One might expect that in return the American consumer would be willing to accept higher costs for minerals produced from domestic mines. There are few indications that this is the case. The American consumer's purchases are controlled primarily by price. If minerals can be obtained more cheaply from abroad, the consumer buys the imports. Domestic mining industries are expected somehow to absorb environmental costs and still remain competitive. Consumer attitude is reflected in government indifference to environmental costs.

Besides tangible costs of environmental regulation, there are costs that cannot be measured fully. These are costs in terms of exploration and development programs that have not been carried out and in terms of deposits that have been found but cannot be brought into production. The time costs of conforming to environmental regulations have undoubtedly become a significant obstacle to the development of the mineral resources of the United States. The almost endless procedures of application for permits, public hearings, and preparation of environmental impact statements add to the time gap between discovery of a new deposit and the start-up of production. Expenditures bring no return during the period. The Duluth gabbro of northern Minnesota contains the nation's largest resource of nickel, and several mining companies have invested large amounts in exploration and in efforts to meet environmental requirements. In 1975, one of them, International Nickel Company, after expenditures of $9 million in exploration since 1952, assessed the status of its program. The company determined that 31 permits would have to be obtained from various state and federal agencies, each application requiring reports and public hearings. It was estimated that if results of all hearings were favorable and all the permits were granted, a minimum of 5 years would elapse before construction of a mine could begin. In the face of this and proposed new restrictions on mining in the area, the project was abandoned.

Part of the problem is that there can be no assurance in advance that permission to mine will be granted, even if all the required procedures are followed. The previously mentioned copper deposit at Ladysmith, Wisconsin, was discovered in 1968. Control of the necessary property was acquired, exploration was undertaken, engineering planning was done, environmental studies were conducted, and plans for reclamation were made. In 1978, after protracted hearings, a permit to mine was refused by the state. In 1974, a much larger zinc-copper deposit was discovered at Crandon, Wisconsin. Again exploration was completed, engineering studies were made, and environmental investigations were carried out. An environmental-impact statement was submitted to the Department of Natural Resources, in December 1982, after protracted public hearings. In December 1983, the Department announced that more information on effects of mining on water conditions in the area was needed. It was estimated that the company could provide the necessary information by May 1985 and that after further public hearings and review of permit applications, a final decision on the mining permit could be made by May 1987. There is thus added to the normal risks of mineral exploration a new and significant element of uncertainty that is bound to discourage exploration and the creation of new reserves. In the future, marginal deposits are less likely to be considered for development. There is evidence, also, that the role of the smaller mining organizations in mineral development in the United States is being diminished, since such organizations are less able to bear the costs of environmental studies and the protracted permitting procedures.

In summary, the expansion of U.S. and world production and consumption of minerals has created environmental problems that demand serious and continuing attention. Every effort must be made to improve procedures for minimizing damage to the environment and for reclaiming mined land. Research and development of methods for controlling pollution and for utilizing or disposing of mineral wastes should be given strong support. At the same time, costs of environmental protection must be kept to the lowest possible levels. Such costs are costs to society in terms of prices of minerals to consumers and in terms of the availability of mineral supplies. It is particularly necessary that the administration of environmental regulations be simplified to eliminate costly delays in the permitting process.

ENVIRONMENTAL REGULATION AND CONSERVATION

We have given a good deal of attention to the problem of mineral conservation, but that is only one aspect of the broader problem of conservation of natural resources. Environmental regulation is another part of the broader problem. The treatment of the broader problem has led to serious conflict in the United States. The conflict arises because individuals, and groups, differ in their perceptions of what should be conserved. Flawn (1966, p. 184) remarks that some conservationists are really preservationists, that they are opposed to any change:

> *They resent the march of new houses onto the rural scene, they fight subdivisions that rout wildlife from its woodland homes, and they bitterly oppose the extractive industries which consume natural resources and convert them to material and energy products . . . they are strongly motivated to preserve the past as the heritage of the future. . . . But however worthwhile the preservationist movement might be in some instances . . . in others it is unrealistic in asking present society to pay too high a price for the past. The preservationist lives in our modern industrial society and enjoys its benefits. These are not without their price.*

The extreme preservationist ignores or derogates the role that minerals play in modern civilization. He does not address the dismal consequences of failure of the world's mineral supplies, the wholesale misery and death that would ensue. He ignores the fact that mineral industry performs a vital social and economic function and that society has a heavy stake in the vigor of the industry. He prefers to paint the extractive industries not as the means of providing society with the materials it needs, but as plunderers of the nation's resources. He seems to overlook the fact that preservation, carried to extremes, can be self-defeating. Only an economically healthy society can support preservation on a significant scale.

There are limits to the amount of land that we can set aside and still be capable of supplying the minerals we need. We must be capable of finding new mineral deposits, converting them to reserves, and extracting them efficiently. Part of the heritage that future generations must have is that same capability. If we lock up too much of the mineral resources of the United States and continue to deplete the resources thus far discovered, the heritage of future generations will be poor indeed. Today we use the mineral reserves that were created by preceding generations. The present generation has the responsibility for creating reserves for its successors. It cannot escape that responsibility. The desire for complete preservation is a nostalgic longing based on regret that the world must change. Yet change it will.

A realistic and constructive approach is necessary in dealing with mineral and other natural resources. It involves a chain of difficult decisions as to the use of those resources and the lands on which or under which they occur. All of us are interested in preserving as much of pristine North America as possible, but civilization cannot survive without the use of land for production. Somehow, enough land must be tilled to provide the nation's food supply. Enough forest land must be maintained in production to supply our needs for timber and related products. The water needs of the nation must somehow be met. Mineral supplies must be provided, and this means that huge areas of potential mineral-producing land cannot be forever closed to mineral exploration and mineral extraction.

REFERENCES AND ADDITIONAL READING

Arthur D. Little, Inc., 1978, Economic impact of environmental regulations on the United States copper industry. Report to U.S. Environmental Protection Agency, xviii + 429 pp., with appendices.

Atwood, G., 1975, The strip mining of western coal. *Scientific American,* Vol. 233, No. 12, pp. 25–29.

Barton, P. B., 1980, Public perspective of resources. *Economic Geology,* Vol. 75, pp. 801–805.

Brock, S. M., 1976, Preservation of the environment. In *Economics of the Mineral Industries,* 3rd Ed., W. A. Vogely, ed., American Institute of Mining, Metallurgical, and Petroleum Engineers, New York, pp. 693–711.

Carpenter, R. A., 1976, Tensions between materials and environmental quality. *Science,* Vol. 191, No. 4227, p. 665–668.

Department of the Interior, 1967, *Surface Mining and Our Environment.* U.S. Government Printing Office, Washington, D.C. 124 pp.

DeYoung, J. H., Jr., and Yasnowsky, P. N., 1980, Capital formation in the nonfuel mineral industries—a literature survey. NTIS #PB 80-147 549, 56 pp.

Dulaney, T., 1976, Industry is shifting to pollution spending. *Iron Age,* Vol. 218, Sept. 13, 1976, pp. 59–60.

Energy Information Administration, 1985, Labor productivity in coal mining, 1949–1983. In *Annual Energy Review 1984*, p. 161.

Flawn, P. T., 1966, *Mineral Resources*. Rand McNally, Chicago, 406 pp.

Flawn, P. T., 1973, Impact of environmental concerns on the mineral industry, 1975–2000. In *Mineral Position of the United States, 1975–2000*, E. N. Cameron, ed., University of Wisconsin Press, Madison, pp. 109–126.

Johnson, W., and Paone, J., 1982, Land utilization and reclamation in the mining industry, 1930–1980. U.S. Bureau of Mines, Information Circular 8862, 22 pp.

Lutjen, G. P., 1971, The curious case of the Puerto Rican copper mines. *Engineering and Mining Journal*, vol. 172, No. 2, pp. 74–84.

Molotch, H., 1970, Santa Barbara: Oil in the velvet playground. In *Eco-Catastrophe*, by the editors of *Ramparts*, Harper and Row, New York, pp. 84–105.

Ohle, E. L., 1975, Economic geologists, SEG, and the future. *Economic Geology*, Vol. 70, pp. 612–622.

Oil and Gas Journal, 1971, Oil-spill control: A hard fight but industry is winning. Vol. 69, No. 34, p. 69–87.

Poland, J. F., 1969, Land subsidence in western United States. In *Geologic Hazards and Public Problems*, May 27–28, 1969, Conference Proceedings, Olson, R. A., and Wallace, M. W., eds., U.S. Government Printing Office, Washington, D.C. pp. 77–96.

Risser, H. S., 1971, Environmental quality control and minerals. Illinois Geological Survey Environmental Notes No. 49, 10 pp.

Strauss, S. D., 1985, Copper: A year of paradox. *Engineering and Mining Journal*, Vol. 186, No. 3, pp. 40–43.

U.S. Bureau of Mines, 1975, Bituminous coal and lignite. In *Minerals Facts and Problems*, Bulletin 667, pp. 166.

Wall Street Journal, 1985, Phelps Dodge Corporation plans to shut down Ajo, Arizona, smelter. March 28, p. 22.

Wells, H., 1985, Expansion at Vernal. *Chevron Focus*, January–February, pp. 1–5.

11 Taxation of Mineral Enterprise

The fruits of industry are divided between capital, labor, and government. Capital takes its redemption and remuneration through profits; labor takes its share through wages; governments take their share through taxes. Each must have a just share, and the question of what constitutes a just share, and the question of who shall make the allocation are vital current political and economic issues. (Borden, 1959, p. 463)

There is no better statement of the problem of taxation; the statement is as timely now as it was in 1959. We might add that the influence of taxation on investment in mineral exploration and mining is now far greater than in 1959, because the burden of taxation at federal, state, and local levels has substantially increased. The prime purpose of taxation of mineral industry is to raise revenue for support of government, but it can be used for other purposes—to encourage or discourage mineral enterprise, or to achieve particular political or social goals. It is so used both in the United States and in other countries of the world.

The subject of taxation has become a life study in itself. Each country of the world and each of the 50 United States has its own system of taxation. In the present volume only a brief survey of taxation of mineral enterprise is possible. We will examine the important types of taxes and consider their impact on mineral conservation and on the availability of minerals to society.

TAXATION BY LOCAL AND STATE GOVERNMENTS

Taxes imposed by local governments are mostly property taxes. The value of the property belonging to a mineral enterprise is assessed, and a rate of so many dollars per thousand dollars of assessed valuation is applied to calculate the tax. The assessed valuation may be equal to the estimated fair market value or to some fraction of the fair market value of the property. The property assessed may be only the surface plant and equipment or it may include the ore reserves, the assessed value of which is discounted to allow for the fact that income from reserves will only be realized over a period of time.

State taxation of mineral industry shows a bewildering diversity. In a fairly recent compilation (*Engineering and Mining Journal*, 1975) taxes in 13 western states—Arizona, Colorado, Idaho, Kansas, Minnesota, Missouri, Montana, Nebraska, Nevada, New Mexico, South Dakota, Utah, and Wyoming—are tabulated. The taxes are listed under four categories—corporate income, severance, sales, and property taxes— but no two state taxes under any of the categories are the same in their provisions. All the states except Nevada, South Dakota, and Wyoming have income taxes. The rates range from 2.5 to 12 percent. Federal income tax is deductible in some states. All the states have property taxes, and in some the tax is based on the value (or a fraction of the value) of all property of a mining enterprise, including reserves. In other states, a mining operation is assessed on the value of either the net proceeds or the gross output, plus the value of mine and mill property and equipment. Property taxes are criticized as anticonservational, because the tax burden is equally distributed over low-grade and high-grade ores. If the property tax makes low-grade ore unprofitable to mine, it will be left in the ground. Rapid depletion of reserves is encouraged, since the shorter the period of mining the less the taxes that will be paid. Taxation of reserves discourages orderly exploration, since reserves begin to be taxed as soon as they are outlined by exploration. From the standpoint of a state or local government, property taxes are advantageous because they yield a stable income during the main period of operation of a mine, but they are difficult to administer, chiefly because of the difficulties and complexities of calculating assessed valuation. They are regressive, because they are levied without regard to the profitability of an enterprise. In periods of depressed prices the tax burden continues, and if depressed prices persist over an extended period, a mine may be closed prematurely, with loss of resources to society.

Severance taxes are defined by the Bureau of the Census (Graham, 1983) as "taxes imposed distinctively on removal of natural products—e.g., oil, natural gas, other minerals, timber, fish, etc.—from land or water and measured by value or quantity of products removed or sold." A pure severance tax is levied under the authority of a state to license activities within its borders, but the term is often loosely applied to other taxes that are not legally severance taxes but have their characteristics. In the simplest case, a severance tax is a tax of so much per pound or per ton of mineral, per barrel of oil, or per thousand cubic feet of natural gas, but the tax may be levied as some percentage of the net or gross value of the mineral produced. There is no uniformity. In Kansas, the severance tax on gas is 0.006 cent per thousand cubic feet; in Louisiana the tax is 1.3 cents to 7 cents. In Kentucky, the severance tax on oil production is 4.5 percent of market value; in Louisiana it is 12.5 percent. Rates vary with the commodity. In Louisiana rates range from 3 cents per ton for sand and gravel to $1.03 per ton for sulfur.

True severance taxes, like property taxes, are regressive, being unaffected by profitability. Copper produced from low-grade ore at high cost is taxed at the same rate as copper produced from high-grade ore at low cost. In times of low prices for minerals severance taxes will encourage high-grading (mining only the richer portions of a mineral deposit). In principle, therefore, severance taxes are anticonservational. The cost to society will be loss of reserves. Sales taxes, based on the sale value of mineral products, are levied by some states and can have the same effect as severance taxes. In theory, so can royalty taxes, which are levied on royalties received from the producers by the owners of mineral properties. In many cases royalty taxes are simply passed on to the mining companies.

Whether taxes are actually conservational or anticonservational depends, of course, on how high they are. If the contribution of a tax to the total cost of an operation is insignificant, its effect on operations will also be insignificant. Nor will the effect be significant if the tax can be passed on to the consumer in the form of increased prices. This is often the case with bulk nonmetallic minerals such as sand and gravel. Sand and gravel producers serve local markets that are protected from outside competition by costs of transportation. Occasionally taxes can be passed on to consumers in wider markets. Montana currently has a 30 percent severance tax on surface-mined coal rated at more than 7000 Btu per pound; much of the proceeds is to be set aside in trust for the future. Wyoming now has a 17 percent severance tax. Such unusually high taxes represent a judgment as to what the traffic will bear. In the case of Montana and Wyoming coal, the traffic will bear quite a lot, because efforts to control sulfur dioxide emissions from coal have created a large demand for low-sulfur coals produced in the two states. The taxes have brought indignant protests and threats of retaliation from the governors of the midwestern states, where the principal consumers of Montana and Wyoming coals are located. The impact of high taxes is especially severe for the producers of mineral commodities that must compete in international markets. High taxes can force selective mining of deposits or can make certain deposits totally uneconomic.

Corporate income taxes or the closely related net-proceeds taxes are currently levied by 37 states. They are based on ability to pay. In calculating taxable income, income tax laws generally permit deductions from gross income for expenses of exploration and development, construction of mines and mills, and operating costs, together with deductions for depreciation of equipment and depletion of reserves. They are considered conservational; they impose no penalty on the operator for mining low-grade ores and therefore do not encourage high-grading of deposits.

TAXATION BY FEDERAL GOVERNMENT

The heaviest taxes on mineral industry, as on other industries, are federal taxes, and of these the most important are income and Social Security taxes. The burden of these taxes has grown enormously over the years. The first federal income tax law was enacted in 1913 subsequent to the passage of the 16th amendment to the Constitution. The corporate rate was 1 percent of taxable income. At the present time the corporate tax is 22 percent on taxable income, with a surtax of 26 percent on income above $25,000. In calculating taxable income, a mining company is allowed to deduct operating expenses, depreciation of plant and equipment, and certain other costs that are allowed manufacturing operations. There is also an allowance for depletion of reserves, which for any given year is calculated as a percentage of the gross income from a mining property. Since the 1930s, depletion allowances for nonfuel minerals have ranged from 5 percent for low-cost commodities such as sand, gravel, common clay, and chlorides produced from brine to 23 percent for a group of minerals including asbestos, bauxite, chromite, fluorspar, and titanium minerals. Until 1969, the depletion allowance for petroleum and natural gas was 27½ percent, the highest allowance of all.

The depletion allowance is an important deduction from taxable income and has been the subject of debate since the passage of the original income tax in 1913. The concept on which the allowance is based is a simple one. A mineral deposit, once discovered and developed, is a capital asset. In mining it, the mine operator is considered to be disposing of it in annual installments. The capital value of the deposit is therefore depreciating over time. To compensate, a depletion allowance from taxable income is allowed. There is nothing wrong with this in principle. A mineral deposit represents an investment of capital. From the proceeds of the minerals produced, the mine owner must recover his invested capital during the life of the mine. A portion of the annual income represents this recovery, as distinguished from operating profits. The same principle applies in federal taxation of industrial operations. The physical assets of an industrial concern consist of its plant and equipment. These are depreciating capital assets, and the law grants the concern the right to recover, out of each year's income, a portion of his capital investment.

Despite the above, the depletion allowance has been controversial for many years. In the 1960s and 1970s the controversy became focused on the high allowance for oil and gas, which was felt to contribute to excessive profits. The matter was finally settled by near abolition of the oil and gas allowance in 1975, but depletion allowances for other mineral commodities are still permitted. They constitute an important incentive toward discovery and development of mineral resources, simply because these activities must be financed largely out of capital recovered from operating mines.

The Social Security tax is a tax equal to a percentage of the wages of each employee, levied on a prescribed amount of the employee's wages. Employer and employee pay equal amounts. Increases in the tax and the amounts of wages taxable in recent years have made it a significant element in the tax burden of all industry, including the mining industry. It is another regressive tax, since it is unrelated to profitability.

EFFECTS OF TAXATION ON THE AVAILABILITY OF MINERALS

We are concerned here primarily with the effects of taxation on the availability of minerals to society. There are two major aspects of this problem. One is the effect of taxation on existing mining operations. The other is the effect of taxation on mineral exploration and the creation of new reserves. The two effects are most significant for minerals that must compete in international markets. As noted earlier, industries that command internal markets, such as the sand-and-gravel industry, can often pass the burden of taxation on to the consumer. The metal-mining industries are heavily involved in international markets, but so are some of the higher-value nonmetallic mineral industries. Heavy taxation may force selective mining and the loss of reserves, particularly in times of depressed markets and prices. Taxation may also destroy the incentives toward creation of new reserves through exploration. If sufficiently high, taxes levied without reference to ability to pay may have the same effect. As profitability diminishes, the ability of industry to generate capital internally and from outside sources declines. Funds available for research and development, essential to the maintenance of a vigorous mining industry over any long period, may be the first to shrink when profit margins narrow.

Unfortunately, however, although it is easy to recognize the likely consequences of heavy taxation, what constitutes heavy taxation and what are its actual effects are difficult to quantify (DeYoung, 1977, p. 107). One problem is simply the lack of adequate data, particularly for the United States. A second problem is that of isolating the effects of changes in tax laws from the many other elements (land policies, changes in demand and price, labor conditions, environmental problems, etc.) that enter into decisions to invest or not to invest in mineral enterprise. A third problem is that full evaluation requires knowledge of what has not been done owing to unacceptable tax burdens. How many exploration programs have not been carried out and how many discoveries have not been developed because of unfavorable taxation? The data on which an answer to this question might be based are mostly buried in the files of mining companies of the present and the past.

There are specific instances, of course, in which tax legislation has had an identifiable effect. Prior to World War II, Minnesota taxes, state and local, on the iron ore industry were much higher than those on other industries in the state. This was possible because the iron ore deposits of the Mesabi and other ranges in Minnesota had a commanding position as sources of iron ore. By the end of the war, however, the high-grade deposits on which the industry was based had been seriously depleted. After the war alternate sources of ore were developed in Canada, Liberia, and Venezuela. The Minnesota iron-mining industry turned toward use of lower-grade taconite ores, which were taxed at a lower rate. However, the companies were not willing to make the huge investments required for development of taconites without assurance of stable and reasonable taxes over a period of time. Under the threat of loss of the iron ore industry, in 1964 the Constitution of the State of Minnesota was amended to guarantee that taxes on taconite production during the next 25 years would not be increased over the general level of corporation taxes. That action triggered the subsequent major expansion of the taconite industry in the Mesabi Range. Over $500 million was invested in plant construction.

The recent history of Canadian mining offers a much broader picture of the effects of changes in tax policy. Prior to 1970 Canadian tax laws offered strong incentives to new mineral enterprise (Ely, 1964), but from 1971 to 1975, changes in federal and provincial tax laws diminished or removed those incentives and increased the tax burden on the mining industry, especially in British Columbia and Ontario. DeYoung (1977, 1978) concludes that the changes in taxation caused a decline in exploration and capital expenditures by the mining industry in those provinces imposing the heaviest tax burdens, a shift of exploration to the most tax-favorable provinces, and some shift of capital into nonmining ventures.

Another study of Canadian mineral taxation (Mackenzie and Bilodeau, 1979) is based on data for 124 Canadian base-metal mines. In their calculations the net value of mining to society was defined as the increase in society's real wealth from investing in minerals. Assuming a no-tax situation, it was calculated that 86 of the 124 deposits mined, with a net present value to society of $2.751 billion, would be economically viable. Applying the Canadian national mineral taxation structure of 1969 (generally regarded as favorable to investment in mining), 82 deposits with a net present value of $2.646 billion remained economically viable. Government would collect 36 percent of the net present value, leaving 64 percent for the private investor. Under the early-1976 taxation system, 74 deposits with a net present value to society of $2.350 billion would be economic: the government share would be 62 percent, the share for the private investor 38 percent, the reverse of their shares in 1969. The late-1976 taxation system would produce 77 economic deposits, with a net present value of $2.484 billion, shared 61 percent for government and 39 percent for private investors.

The findings summarized above refer to the overall mining investment and taxation environment in the Canadian base-metal sector, but because of differences in provincial tax systems, that environment varies considerably from province to province. The authors conclude that investment incentive is low in British Columbia, Quebec, and Manitoba. The damage to society is incomplete redemption of the potential value of its mineral resources. It is also clear from the studies that since 1969 the incentive to invest in mineral development in Canada has been greatly diminished.

In state taxation of minerals there periodically arises a tendency to tax the mining industry more heavily than other industries, because mineral resources are considered the heritage of the people and, once mined, cannot be restored. This philosophy was expressed, for example, in the Wisconsin 1977 law governing taxation of metal-mining enterprise in the state. The standard corporation income tax rate in Wisconsin is 7.9 percent. The law of 1977 established a net-proceeds occupation tax, applicable solely to individuals or corporations occupied in mining metallic minerals. Tax rates ranged from 3 percent of net proceeds averaging, over a 3-year period, $250,000 to $5 million, to 15 percent of average net proceeds exceeding $25 million. The higher rates were later recognized as unreasonable, and in 1981 they were substantially reduced.

The petroleum industry in the United States has been particularly singled out for heavy taxation. In 1979, in return for partial deregulation of domestic prices of oil, a special "windfall profits" tax was levied. This and other special taxes such as severance taxes are said to be responsible for taxation of the oil industry at much higher rates than those applied to other industries. In 1981, federal and state taxation of oil companies is reported to have taken 54 percent of net pretax income compared to only 26 percent for other major industrial companies. Given the fact that discovery of oil and gas in the United States is becoming more difficult and more costly, and the fact that funds for exploration must be generated out of profits, the wisdom of the present tax structure is questionable.

There are other recent tendencies in state taxation that could have a substantial effect on mineral enterprise. One is embodied in proposals that only expenses of operation within a state be deductible from state taxable income. This is unreasonable because a portion of a company's overhead must be charged to the local operation. The other development consists of proposals that a state be allowed to tax a mining company on its earnings both outside and inside the state. If enacted by all or even a number of the states, such laws could greatly increase the burden of taxation.

From the above it will be clear that there is no easy answer to the problem of mineral taxation. There is no simple definition of unduly heavy taxation. A great deal of wisdom is required in framing tax measures. When taxes become so high that they destroy incentives toward the creation of new reserves through mineral exploration and mining development, the availability of minerals to society declines. Exploration will not be pursued unless those who assume the risks of exploration are rewarded by commensurate returns on invested capital.

Taxation of mineral industry in the United States is a monstrous jumble of 50 state systems and a federal system, a great boon to the legal and accounting professions. Tax legislation at the federal level shows some recognition of national problems with respect to mineral supplies, but national interests and national needs have little or no influence on taxation of mineral industry by the states. In a time when the mineral position of the United States is steadily deteriorating, one wonders how long the nation can afford such haphazard treatment of taxation of its mineral industries.

REFERENCES AND ADDITIONAL READING

Borden, G. S., 1959, Taxation of mineral properties. In *Economics of the Mineral Industries*, E. H. Robie, ed., American Institute of Mining, Metallurgical, and Petroleum Engineers, New York, pp. 451–496.

Conrad, R. F., and Hool, R. B., 1980, *Taxation of Mineral Resources.* D. C. Heath, Lexington, Mass., 109 pp.

DeYoung, J. H., Jr., 1977, Effect of tax laws on mineral exploration in Canada. *Resources Policy*, Vol. 3, June 1977, pp. 96–107.

DeYoung, J. H., Jr., 1978, Measuring economic effects of tax laws on mineral exploration. Council of Economics of AIME, 107th Annual Meeting, Denver, 1978, Proceedings, pp. 29–40.

Ely, N., 1964, Trends in natural petroleum and mining laws. In *The Role of National Governments in Exploration for Mineral Resources*, W. E. Bonini, H. D. Hedberg, and J. Kalliokoski, eds., Ocean City, N.J., Littoral Press, pp. 17–24.

Engineering and Mining Journal, 1975, Colorado's proposed tax sparks interest in other states' mining taxes. Vol. 176, No. 6, pp. 32–33.

General Accounting Office, Energy and Minerals Division, 1981, *Assessing the Impact of Federal and State Taxes on the Domestic Minerals Industry.* NTIS PB82, 117953, U.S. Government Printing Office, Washington, D.C.

Graham, A. P., 1983, Summary of state severance taxes on mineral production as of January 31, 1983. *Minerals and Materials*, February–March, 1983, U.S. Government Printing Office, Washington, D.C., pp. 41–53.

Mackenzie, B. W., and Bilodeau, M. L., 1979, *Effects of Taxation on Base Metal Mining in Canada.* Centre for Resource Studies, Queen's University, Kingston, Ontario, 190 pp.

Roberts, W. A., 1944, State taxation of metallic deposits. *Harvard Economic Studies*, Vol. 77, 400 pp.

Van Rensburg, W. C. J., and Bambrick, S., 1978, *Economics of the World's Mineral Industries.* McGraw-Hill, New York, pp. 81–123.

12 Trade Policy

The trade policy of the United States has an important influence on the mineral industries of the United States, most directly on those that compete in international markets, but indirectly on all others, since demand and price for any mineral are functions of the general industrial structure. Three aspects of trade policy are examined here: tariffs, sanctions, and stockpiling.

TARIFFS ON IMPORTS

Tariffs have been levied on imports since early times, and the right of a nation to levy duties on imports as a source of revenue has rarely been questioned. Controversy over tariffs relates to those that are set high enough to discourage imports of goods that would be in competition with domestic products. These are the protective tariffs. They act as restraints on international trade and as sources of friction in international relations.

There are two general stages of the development of a mining industry at which protective tariffs may be levied. One is in the early stage, when an "infant" industry may be struggling to compete with producers already established in the international marketplace. A tariff at this stage may give substantial aid. The other is a later stage, when, owing to declining grade of ore and rising costs of production from domestic mines, competition with imported minerals becomes difficult or impossible. A protective tariff levied at this stage stems from a decision that a particular industry, for reasons of employment or national security, must be maintained despite the fact that it may never be competitive in world markets. The tariff on copper in 1923 was levied at the urging of certain domestic mining companies, particularly a group in the Lake Superior district, that were faced with rising costs and could no longer compete with copper from South America. It served to delay the demise of mining in the older part of the Lake Superior district.

The actual effects of a protective tariff on a mineral commodity depend fundamentally on the size and quality of resources of the mineral. The case of manganese is an example. During World War I the United States was cut off from Russia, a major source of manganese. With encouragement from the Federal Government, intensive exploitation of domestic manganese deposits was undertaken in Arkansas, Montana, and elsewhere. Production rose to 300,000 tons in 1918, 35 percent of U.S. requirements. After the war, manganese again became available from abroad. Prices fell; the domestic manganese mining industry was hurt by the competition and applied to the Congress for a protective tariff. The Congress was impressed by the wartime production and in 1922 complied by levying a tariff of 1 cent per pound on imported manganese ore. Nevertheless production fell to 25 percent of requirements in 1925 and met only 5 percent of requirements in 1930. The reason was that the wartime effort had virtually exhausted U.S. deposits of higher-grade manganese ore. The United States has substantial resources of manganese in deposits in Maine, South Dakota, and Arizona, but the ores of these deposits are low in grade. Manganese could be produced from them only at far higher cost than permitted by the tariff. Requests for protective tariffs on manganese and tin after World War II were rejected. The United States has only small deposits of tin ore. A search during World War I and another during World War II failed to discover significant domestic tin deposits. Even a very high tariff on tin would have no significant effect on domestic production.

In the United States, tariffs on imports of minerals and other goods have been controversial ever since the founding of the republic. We can divide their history into two periods—the period from 1783 to 1933, and the period that began in 1934 and still continues. During the earlier period, tariffs on minerals were changed from time to time, but during most of it protective tariffs were part of U.S. trade policy. The period of protective tariffs ended in 1934. In 1930, with the onset of the Great Depression, Congress passed the Smoot–Hawley Act, which imposed the highest tariffs on imports, including mineral imports, in the history of the United States. In 1934, however, the policy of protectionism was abruptly reversed by passage of the Reciprocal Trade Agreements Act. That act has been a cornerstone of American trade policy ever since. The act empowered the President of the United States to make agreements with foreign countries for mutual reductions in tariffs, without submitting the agreements to the Senate for ratification. Reduction was at first limited to 50 percent of existing tariffs, but later the limit was raised to 75 percent. The act led to successive International Trade Conferences, beginning in 1947, and out of those have emerged the General Agreements on Tariffs and Trade. U.S. tariffs on minerals are currently 50 to 75 percent lower than in 1934.

An important feature of the Reciprocal Trade Agreements Act became known as the "most-favored-nation" clause. Suppose that the United States agrees with a particular nation to levy a tariff on manganese ores that is lower than the tariff imposed on manganese ores from other nations. That nation now becomes the "most favored nation." However, under the Reciprocal Trade Agreements Act, the tariff is automatically lowered to the same level for manganese ores from other nations with which the United States has reciprocal trade agreements. The "most-favored-nation" clause thus eliminates discriminatory tariffs, a cause of many trade wars in the past.

The reciprocal trade policy has had a stormy career, marked by heated debates each time the act has come up for renewal. It has been hailed on the one hand as an indispensable lubricant of the channels of international trade, on the other as a scheme to sell American industry down the river. It has been supported, however, by successive administrations, Republican and Democratic alike. It received strong support from the President's Materials Policy Commission in 1952 and, by implication, from the President's Cabinet Committee on Mineral Policy in 1954. In 1973, the National Commission on Materials Policy recommended only short-term protective tariffs, to be used to ease adjustment of a domestic industry to rising imports.

The General Agreements on Tariff and Trade (GATT) have aroused strong reactions in the Congress, reflected over the years in the bitter debates over the reciprocal trade policy. The agreements have never been ratified by the Congress and function only under executive authority. In 1954, the ninth GATT conference recommended establishment of an International Trade Organization. The idea found little favor in the Congress, and no administration has seen fit to submit the proposal to the Congress. The trade organization was intended to administer the General Agreements.

In 1962, the Reciprocal Trade Agreements Act was once again up for renewal. Renewal was made part of a Trade Expansion Act proposed by President Kennedy and passed by the Congress in September of that year. It gave the President broad powers to negotiate trade arrangements with other countries. Two provisions of the bill were especially important. One, the "orderly marketing" provision, gave the President the power to negotiate with other countries to form international marketing cartels in minerals and other commodities. Such cartels would set limits on annual exports of producing countries. Lead, zinc, and aluminum were suggested as ripe for such agreements. However, the provision has been implemented only once, as described in a later section. The second provision empowered the President to raise tariffs, set import quotas, or make other restrictions on imports that might be in the national interest.

The provisions of the act of 1962 were continued under the Trade Act of 1974.

The Reciprocal Trade Agreements Act and its extensions contain an "escape" clause allowing the President to raise tariffs when industries are severely injured by cuts made under that law. The President now acts on recommendations of the U.S. International Trade Commission. From time to time the commission or its predecessor, the U.S. Tariff Commission, has actually recommended tariffs on mineral commodities (e.g., on lead, zinc, copper, ferromanganese, and ferrochrome), but the only increase ever approved took the form of a tariff of 4 cents per pound levied in November 1978 on ferrochrome entering the country at a price of less than 38 cents per pound of contained chromium. The tariff was terminated in November 1982.

The American steel industry has been hard hit by competition from foreign steel manufacturers, some of which have been subsidized by their governments. In 1976, quotas were imposed on specialty steels, and these have been continued. In 1978, a trigger price mechanism was instituted by the Federal Government in lieu of a protective tariff. Trigger prices are those below which steel cannot be imported without automatically triggering an antidumping investigation. The purpose of the legislation is to ensure that prices of imported steel will reflect the actual costs of production and that dumping of foreign steel on U.S. markets at artificially low prices will be curtailed.

The deep recession of 1981–1982 exacerbated the plight of the steel industries of the world. Foreign governments subsidized exports of steel and other products to maintain production, employment, and export income. Late in 1982 (Aus, 1983), only a few of some 70 primary steel exporting nations were selling their steel products abroad at a margin above variable cost. Tax and other subsidies played a major role in most steel-exporting countries. In 1982–1983, negotiations between the United States and the European Economic Community (EEC) settled many of the complaints against European steel producers. Limits were set on shares of the U.S. market for certain steel products that can be filled by imports from the EEC. Despite these actions, the competitive position of the American steel industry has steadily worsened. New agreements on quotas were negotiated in 1984.

The copper-mining industry has suffered periodically from the instability of the world market for the metal, and with the onset of the recession in 1980 the industry entered a period of acute distress. The glut of copper in world markets caused the price to plummet from $1.41 per pound in February 1980 to 58 cents a pound in December 1984. The effects on domestic industry were severe. After nearly 120 years of operation, the great copper mines of Butte, Montana, were shut down. Other mines closed or curtailed production. Part of the problem is that some of the foreign copper mines with which domestic mines must compete are subsidized by their governments or by low-interest loans from the International Monetary Fund, to which the United States is a principal contributor. In January 1984, eleven U.S. copper producers, representing about 87 percent of the domestic copper industry, filed a petition with the International Trade Commission requesting imposition of temporary import quotas. Relief was subsequently recommended by the commission, but the recommendation was rejected by the President in September 1984. Subsequently (*Engineering and Mining Journal,* 1984) it was predicted that U.S. copper production would fall from 1.1 million tons per year to 585,000 tons per year within the next 5 years. This compares with an average production of about 1.5 million tons during 1977–1981.

As indicated above, other measures have occasionally been tried as substitutes for tariff protection. In 1954, for example, President Eisenhower rejected the Tariff Commission's recommendation of an increase in tariffs on lead and zinc imports. Instead, lead and zinc were purchased and added to the stockpiles of the metals, in an effort to relieve the pressure of world surpluses on metal prices. The action was unsuccessful. Import quotas on lead and zinc ores and metal were imposed in 1958 but were terminated in 1965. Quotas on oil imports in the 1960s limited imports to 12 percent of domestic production, but the quotas had to be abandoned when domestic oil production peaked in 1970, and surplus production capacity was exhausted in 1972. Other devices (barter deals, exploration assistance,international negotiations, special subsidies, guaranteed prices, tax relief) have been used from time to time, but protective tariffs have not been a significant part of U.S. trade policy since 1934. In his review of U.S. trade policies Pehrson (1964) remarked that in 1963 U.S. tariff policy was strongly oriented toward free trade, and this was true through the 1960s and much of the 1970s. The 1973 report of the National Commission on Materials Policy recommended that "traditional U.S. economic policy be maintained by relying on market forces as the prime determinant of the mix of imports and domestic production subject to considerations of public policy involving the health and viability of domestic materials industry, national security, and fair international competition." The commission recommended only temporary protection for distressed industries, to ease them through short periods of adjustment.

In a separate opinion attached to the commission's report, commissioner J. H. Liedtke sounded a note of caution:

> *We must realize that the United States stands almost alone as a true believer in and practitioner of free competitive market forces. The state-owned and [state-] controlled economies of communist, socialist and other authoritarian countries dictate prices and all matters of internal and external trade. Even in West European countries there exists a long tradition of marketing through the means of cartels and associations. In Japan a unique form of state and*

business protectionism is practiced with great success. While we may think our system superior to these other traditions, we must realize that in world trade we enjoy no extraterritorial sovereignty or jurisdiction to enforce compliance with free competitive market factors. Therefore, when we talk about reliance upon market factors we must think in terms of market factors as they in fact exist and not as they theoretically exist in traditional theories of economics.

The cogency of these remarks is underlined by developments since 1973—the proliferation of trade barriers in other countries of the world and the adoption of other measures (such as subsidies) by various nations to give their nationals advantages in international trade. The 1981–1982 round of GATT negotiations in Geneva ended in failure. Meanwhile there has been a rising tide of imports into the United States; the international trade deficit rose from $42.7 billion in 1982 to $62.4 billion in 1983 to $123.3 billion in 1984. In 1983, the value of net nonfuel mineral imports was $12 billion, and the value of fossil fuel imports was $56 billion—heavy contributions to the trade deficit. These developments and the depressed state of certain domestic industries have led the United States to modify its free trade position. The negotiations over textiles and steel imports are examples. However, apart from the brief tariff on ferrochrome, no protection has been given to other mineral industries. Title III of the Defense Production Act being still in force, in 1981–1982 there were requests for financial aid, in lieu of tariff protection, to the nickel-chromium-cobalt mining project at Gasquet Mountain in northern California, to nickel-cobalt mining in Missouri, and to cobalt mining in Idaho. After long debate in Congress, the requests were disapproved. The Reagan Administration has made it clear that mineral industry must compete on the same basis as other industries.

A free trade policy was advantageous to the United States during the first 15 years after World War II. The United States held a dominant position in the industrial world and was, as we have seen, the most important single customer for the mineral products of the underdeveloped countries. Much of the Free World was open to investment of American capital and technology, and these were used to discover and develop new mineral resources in many countries. For a time the newly discovered deposits served as sources of mineral raw materials for American smelters, refineries, and manufacturing industries, as well as those of western Europe. This condition could not last. The history of our own country tells us this. In the 1960s, the minerals once exported to the United States began to feed the development of native industries directly in competition with those of the United States and the older industrialized nations. The United States finds it more and more difficult to maintain a free trade policy.

MINERAL SANCTIONS

Embargoes and boycotts involving international trade have been used at times by various nations, generally for political or military purposes. The most conspicuous example of recent years is the Arab oil embargo of 1973. Embargoes have been used by the United States mainly in times of war, as part of schemes to cut off trade that would be favorable to its enemies. Boycotts and embargoes are unfriendly acts, and they long ago acquired an unsavory reputation in international affairs. When such measures came up for consideration during the formation of the League of Nations after World War I, a new term, "economic sanction," was coined. Article XVI of the Covenant of the League of Nations provided that if a nation were convicted, by unanimous vote of the Assembly of the League, as an aggressor against another nation, economic sanctions were to be applied. Application of sanctions would simply mean cutting off trade with the aggressor nation—that is, applying an embargo under another name.

Sanctions received their first real test when Japan invaded Manchukuo in 1931. That act aroused worldwide indignation and brought calls for the application of economic sanctions. It shortly became evident (1) that a unanimous vote of the Assembly of the League of Nations could not be obtained, and (2) that application of sanctions would result in severe loss in trade to the sanctioning nations as well as the aggressor nation. Nothing came of the attempt, even though the United States, not a party to the League but one of the signatories of the Nine-Power Treaty respecting China, took the lead in suggesting sanctions. Japan was still fighting in China when World War II began.

In the meantime, Sir Thomas Holland, Director of the Geological Survey of India, had been developing the idea of a mineral sanction. He first proposed this in 1929 and in 1935 developed the idea in more elaborate form (Holland, 1935). He accurately predicted failure of any attempt to apply general sanctions against an aggressor. He argued, however, that since no industrial nation is self-sufficient in mineral raw materials, aggression by such a nation could be stopped by shutting off its supplies of minerals. The idea was widely discussed, but the only attempt ever made to apply it was a failure. In the same year Holland's book was published, Italy invaded Abyssinia, and a great clamor for the application of sanctions arose both in the United States and abroad. Under overwhelming pressure from its public, the British government voted to apply sanctions to Italian marble, sulfur, and similar items and placed sanctions on still other items not essential to Italy's conduct of the war. Italy made it clear that sanctions on oil would be regarded as an act of war, and the British government, which owned stock in the Anglo-Iranian Oil Company and the Shell Oil Company, allowed those companies to supply Italy with oil for the Abyssinian campaign. The simple fact was that in consequence of the disarmament program, Britain was in no position to fight a war, and neither was the United States. Italy's conquest of Ethiopia continued without hindrance.

Holland claimed that mineral sanctions are a means of stopping a war without military action. The argument has two fatal flaws. One is that mineral or any other sanctions against a powerful aggressor, unless backed up by a willingness to undertake military action, are a hollow bluff that is readily recognized as such. The second flaw was disclosed upon the advent of World War II. Germany had long been dependent on other countries for vital mineral supplies. When Hitler began preparing for World War II, he stockpiled minerals as fast as possible. He misjudged the length of the war, and the stockpiles were adequate for only a year. However, the point is that by the time an aggressor declares war, he is already prepared or thinks he is prepared against cutoff of supplies. Sanctions applied after aggression come too late to prevent aggression.

The charter of the United Nations also provides for sanctions against an aggressor, and a test of the provision came when North Korea attacked South Korea in June of 1950. Hastily assembled, the United Nations voted to apply sanctions. The vote was unanimous, but only because the Soviet delegation had walked out of the General Assembly before the vote was taken. Since China continued to supply North Korea, the sanctions were not particularly effective.

The next attempt to apply sanctions involved Southern Rhodesia. Rhodesia's dispute with Great Britain over the future government of the colony culminated in 1965, when the de facto government of Rhodesia issued its Unilateral Declaration of Independence. On January 20, 1966, the United Kingdom made it a criminal offense to export to or import from Rhodesia certain specified products. On December 16, 1966, the United Nations Security Council adopted a resolution that member states cease imports from Rhodesia, and on January 5, 1967, the resolution was implemented by the United States by executive order. The case of Rhodesia was a curious one. No aggression of the kind considered in the framing of the United Nations charter was involved. Sanctions were applied in an attempt to force a particular solution of an internal political problem.

Strategically the situation seemed ideal for application of sanctions. Here was a militarily weak nation almost completely surrounded by hostile nations. It was dependent on international trade both as a source of manufactured goods and as an outlet for its agricultural and mineral products. Yet sanctions failed. In 1971, the American Universities Field Staff issued a report on the economy of Rhodesia under sanctions (Hooker, 1971). It was found that the economy suffered a temporary setback, but by 1969 problems were overcome and the economy was booming. Trade with other nations continued. Trade with African neighbors was brisk, and exports and imports moved over rail lines through South Africa and Mozambique.

The chief sufferer was the United States. The most important deposits of metallurgical chromite in the Free World were in Rhodesia. In 1965 they yielded 625,000 tons of metallurgical chrome ore, approximately 40 percent of Free World production. About 78 percent of the ore was produced by subsidiaries of Union Carbide Corporation. Union Carbide attempted to comply with the American executive order, but the Rhodesian government responded by ordering local management to produce and export chromite or be taken over. Chromite continued to flow into world markets through the Soviet Union and Japan. The United States was forced to import 80 percent of its requirements for metallurgical ore from the Soviet Union, some of inferior grade, at prices far above the normal range. A smelter was built in Rhodesia to produce ferrochrome, a first step in the subsequent demise of the American ferrochrome industry. In 1967, the United States imported 155,000 tons of ferrochrome, at least 105,000 tons of which came directly or indirectly from Rhodesia. The Japanese profited from the export of stainless-steel sheet made from Rhodesian chromite that was imported in covert defiance of the sanctions. In 1972 the U.S. embargo on Rhodesian chrome ore was repealed, but it was reimposed in March 1977. This was done despite clear evidence that sanctions were being flouted by other nations, including African nations that were demanding sanctions on the part of the United States.

The history of mineral sanctions is a sorry one indeed. For the most part they have been futile gestures that have flown in the face of the realities of international relations. U.S. policy as expressed in the Rhodesia sanctions can only be characterized as bizarre. It was totally ineffective and was pursued with complete disregard of consequences. Actions of this kind evoke only ridicule on the part of other nations. In view of the history of mineral sanctions in general and the American experience with them in particular, the present (1985) call for sanctions against South Africa is astonishing. The principal effect is likely to be a loss of American influence on South African affairs.

STOCKPILING

Closely related to trade policy, and at times an integral part of it, is U.S. policy governing stockpiling of strategic minerals and other materials. The problems posed by deficiencies in domestic mineral production were brought sharply into focus during World War I, when the United States was cut off from foreign sources of manganese, potash, tin, tungsten, and other essential minerals. Severe shortages of these minerals prompted desperate and costly efforts, mostly unsuccessful, to discover and develop domestic resources. Minerals that were essential to the war effort but could not be produced in adequate amounts from domestic sources became known as the strategic minerals. The need for stockpiles of these and other strategic materials was discussed in the aftermath of war, but nothing was done until 1939, when the clouds of war were already gathering over Europe. The Strategic and Critical Materials Stockpiling Act of 1939 was passed on June 7. It was too late. On September 1 the German army crossed the Polish border, World War II began, and U.S. access to sources of strategic minerals was progressively curtailed. After the United States entered the war, the mineral supply situation threatened the entire war effort. At that time most of the supply of aluminum ore (bauxite) came from the Guianas and West Africa. The shipping lanes along which the ore carriers moved were exposed to submarine attack, and in 1942 and early 1943 most of the ore-carrying fleet was lost. The supply of manganese ore from Brazil, West Africa, and India was similarly imperiled, as were supplies of chromite from southern Africa. Sheet mica, essential to the war effort, had to be flown from India and Brazil. There were problems with other minerals. Again there was a costly effort to develop supplies of strategic minerals from domestic sources. Like the World War I effort, it met with only limited success. In 1942 and the first half of 1943, before the German submarine offensive was brought under control, the situation was critical. As one example, in 1943 the supply of tin was just 100.1 percent of the most essential military and civilian needs. Only the most stringent allocation of mineral materials to the most essential needs enabled the United States to meet the requirements of the war effort.

The wartime experience led to passage of the Strategic and Critical Materials Stockpiling Act of 1946. The act defined strategic and critical materials as "materials that (a) would be needed to supply the military, industrial, and essential civilian needs of the United States during an emergency, and (b) are not found or produced in the United States in sufficient quantities to meet such needs." Unfortunately, only limited funding of the act was provided, and only $150 million was expended for stockpiling through fiscal 1948. In 1950 the Korean War began, U.S. consumption of minerals escalated, and shortages again developed. This time no disruption of trade routes took place, but the war came at a time when Free World capacity for mineral production was already straining to meet increased peacetime world demand. The United States therefore paid high prices for what minerals it could obtain from Free World sources and was once again forced to mount a very costly crash program to stimulate supplies from domestic deposits. The program was authorized by the Defense Production Act of 1950 (DPA), the belated response of the Federal Government to the onset of the Korean War in June of that year. Title III of the new act gave authority for expanding supplies of minerals through support of exploration, development, and mining of strategic and critical minerals, and through expansion of processing industries. The consequences have been summarized by Morgan (1984). Through June 30, 1956, DPA supply expansion programs reached $8.4 billion in gross transactions involving aluminum, titanium, nickel, copper, manganese, cobalt, tungsten, tin, molybdenum, magnesium, columbium-tantalum, chromium, and other minerals from both domestic sources and sources abroad. U.S. production of aluminum was doubled, copper mining capacity was increased by 25 percent, U.S. nickel mining was initiated, and a titanium metal industry was established. Tax incentives, mainly accelerated depreciation of plant and equipment, were used to stimulate mineral production and industrial expansion.

The lessons of three wars had finally sunk home. By 1957 $5 billion had been spent on stockpiles of strategic materials. Stockpile targets (amounts to be stockpiled) were generally set at 3 years' supply but have undergone numerous revisions since 1946. After 1957 expenditures for stockpiles declined, but they have been resumed under the Reagan Administration. The Strategic and Critical Materials Stockpiling Revision Act of 1979 reaffirmed the need for stockpiling, conservation, and development of domestic resources. A list of nonfuel minerals included in the national stockpiles is indicated in Table 12-1. Quantities range from tens of tons to millions of tons for various commodities. The list of commodities may be matched against the mineral deficiencies shown in Figure 6-4. Stockpiles are managed by the Federal Emergency Management Agency. Purchases, sales, and storage of stockpile materials are handled by the General Services Administration.

Stockpiles were initially conceived as protection against wartime disruptions in trade, but it was early recognized (and feared by industry) that they could be used as a means of influencing markets for minerals. The fears were justified. The purchases of lead and zinc under the Eisenhower Administration were one such use. Sales of tin from the stockpile have been used from time to time to influence world tin prices. In 1965–1966, President Johnson used a threat of releases from the stockpile to force a rollback of copper and aluminum prices. In the 1970s the use of stockpiling primarily for economic purposes was widely discussed. A stockpile of oil, the Strategic Petroleum Reserve, was provided for in 1978. The ostensible purpose is to protect against another embargo on oil from the Middle East, but the stockpile could readily be used to influence prices in world markets.

As indicated in Table 12-1, minerals stockpiled may be in the form of ores, concentrates, metals, or processed nonmetallics. They could even be in the form of finished products suitable for immediate use. From raw materials to finished goods there is an increase in costs and a decrease in time of availability for use, but there is also a decrease in flexibility of use. The selection of materials for stockpiles therefore involves difficult choices. The form of the material, the time necessary to process it for military or essential civilian needs, the energy needed for processing, the manpower required, and the plant and equipment capacity necessary

TABLE 12-1 Mineral Commodities in the U.S. Stockpile, 1982

Aluminum (metal, alumina, and bauxite)
Antimony (metal)
Asbestos (amosite, chrysotile)
Bauxite (refractory grade)
Beryllium (ore, beryllium-copper alloy, and metal)
Bismuth (metal)
Cadmium (metal)
Chromium materials (chemical, metallurgical, and refractory-grade ores, ferrochromium)
Cobalt
Columbium (concentrates, columbium carbide, ferrocolumbium, and metal)
Copper
Diamond (industrial)
Fluorspar (acid grade and metallurgical grade)
Graphite (various grades)
Iodine
Jewels (bearings)
Lead (metal)
Manganese (battery-grade oxide, metal, chemical and metallurgical ores, ferromanganese, ferrosilicomanganese)
Mercury (metal)
Mica (various grades of muscovite and phlogopite sheet, block, and splittings)
Molybdenum (sulfide and ferromolybdenum)
Nickel (nickel plus cobalt)
Platinum-group metals (palladium, platinum, iridium)
Quartz (crystals)
Rutile
Sapphire and ruby
Silicon carbide
Silver (metal)
Talc (block and lump)
Tantalum (metal, carbide, and tantalum mineral concentrates)
Thorium nitrate
Tin
Titanium (sponge)
Tungsten (metal, metal powder, ferrotungsten, tungsten carbide powder, ores, and concentrates)
Vanadium (metal, ferrovanadium, vanadium pentoxide)
Zinc (metal)

must all be considered. Periodic review of stockpiles is essential to ensure maximum usefulness in a time of national emergency.

Stockpiles were initiated as protection against shortages of the kinds that occurred during the protracted struggles of World War I and World War II. In a time when the nature of future wars is highly uncertain, their utility is more difficult to appraise. They can, however, provide some protection against disruptions of supply from politically unstable areas of the world. At the least they provide a period during which American industry may be able to adjust to changes in the availability of minerals.

The Arab oil embargo of 1973 focused attention on the problem of reliance on foreign sources of fuel supply, but more generally on the whole problem of mineral dependence. A series of studies was undertaken. The Nonfuel Mineral Policy Review led in 1979 to a draft report to the Senate Foreign Relations Committee in which issues were identified but no recommendations were made. A 1980 report by the Congressional Research Service to the Senate Committee on Foreign Relations (1980), Subcommittee on African Affairs, focused on chromium, vanadium, manganese, platinum-group metals, and gold, for all of which South Africa is the major Free World source. The conclusions were that South Africa's minerals are of significant but not critical importance to the United States and that disruptions of supply could be minimized if preparations for a possible cutoff were made in advance. In a 1980 report by the Subcommittee on Mines and Mining of the House Committee on Interior and Insular Affairs (1980), a broader survey of the strategic mineral problem was made. It was concluded that reliance on imports is a serious problem, placing in jeopardy the nation's economy, defense, and world stature. The recommendations of the House subcommittee led to passage of the National Materials and Minerals Policy Act of 1980. The act states that "it is the continuing policy of the United States to promote an adequate and stable supply of materials necessary to maintain national security, economic well-being, and industrial production with appropriate attention to a long-term balance between resource production, energy use, a healthy environment, and social needs." The act calls for a coordinated effort on the part of the Executive agencies to implement the policy, taking into account future materials demand, supplies, needs for materials research and development, and possible cooperation with other nations in conservational use of materials. Private enterprise is to be encouraged in the development of economically sound and stable domestic industries, and federal agencies are to be encouraged to facilitate availability and development of domestic resources. A review of federal policies affecting materials availability is mandated, as is an assessment of opportunities for cooperative agreements for materials development in foreign nations, with the aim of increasing the reliability of materials supplies from abroad.

A report by President Reagan, pursuant to the act of 1980, was made in April 1982 (Reagan, 1982). The statement of policy contained in the report calls for increased availability of public land for mineral exploration, elimination of barriers to deep-sea mining, collection of minerals data, government support of long-term high-risk materials research and development, review and reform of regulations adversely affecting domestic minerals industry, and revisions of stockpiles. National materials policy is to be coordinated through the Cabinet Council on National Resources and Environment. Despite criticisms of specific aspects of the report, the statement as a whole represents progress toward the development of a national minerals policy. The ouster of Secretary James Watt, however, may be followed by a retreat from some of the objectives stated in the report.

The problem of strategic minerals is far from resolved. There are several difficulties. One is the tendency of the American public, and the Congress, to take a very short-term view of problems of mineral supply. Again and again, and most recently in the case of the Arab oil embargo, periods of shortage have been quickly forgotten. American policy with respect to strategic minerals has too often been one of hasty response to crises. The crises have arisen directly out of failure to treat the problem as one that will periodically recur and therefore must be addressed on a long-term basis. We see this now in public reaction to the temporary glut of petroleum on world markets. For the public in general, the energy problem is a thing of the past. The desirability of achieving the target set for the Strategic Petroleum Reserve is now being questioned. Support for programs aimed at developing alternative energy sources has weakened. It is in this way that the stage is being set for the mineral crises of the future. Temporary responses bring temporary solutions; long-term problems go unresolved.

There are two significant developments in the strategic minerals situation since the Korean War—one unfavorable, one favorable. The unfavorable development is the change in American industrial structure. During both world wars and the Korean War, the strategic mineral problem was largely one of supplying mineral raw materials to American industries that were large enough and diversified enough to convert the raw materials into necessary goods. This was what was implied in the description of the United States as "the arsenal of democracy." However, the arsenal has now been partly dismantled, and the nation is increasingly dependent on industries widely dispersed over the world. The problem of materials supply is no longer merely a problem of supply of *raw* materials.

The second and favorable development is the rise and flowering of materials science, through which the quality of materials is being improved and, more relevant here, whole new groups of materials are being created. As we noted, plastics have already supplanted metals in many uses. A new generation of ceramics promises to replace metals in still other uses, including uses that consume strategic metals such as cobalt and tungsten. Other materials being developed enable strategic metals to be used in much smaller amounts without loss of performance. Along with these advances, studies of the uses of strategic minerals in industries such as the aerospace industry (National Bureau of Standards, 1981) are identifying truly critical uses as opposed to uses for reasons of present availability, low cost, and convenience. Out of all these studies could come reduction in use and change in technology toward the use of nonstrategic mineral materials. The implications of the new materials science are very broad; they are discussed in a later chapter.

Past attempts to deal with the strategic minerals problem have been beset with some thorny political and economic problems. Steps taken to reduce dependence on mineral imports have come into conflict with other objectives of foreign policy—promoting free international trade, containing the spread of Communism, and assistance to developing nations. Added to this are conflicts of interest within the United States, including conflicts within mineral industry, and uncertainties as to future political developments in various mineral-producing countries of the world. These compound the problem and hamper efforts to appraise the seriousness of the strategic minerals situation and to develop measures that would deal with it effectively. Though the National Materials and Minerals Policy Act of 1980 called for development of a coherent national minerals policy, the machinery needed for development of such a policy has not been created. There is no single policy-making body; responsibility is currently divided among 14 different federal departments and agencies. The result of such a hydra-headed approach to the strategic minerals problem is readily predictable. Over the years an almost endless series of studies of strategic minerals has been substituted for effective action. Those who have participated, like the author, come to recognize that such studies may sometimes be evasions of government responsibility for a serious national problem.

REFERENCES AND ADDITIONAL READING

Aus, R. M., 1983, Iron and Steel. *Engineering and Mining Journal*, Vol. 184, No. 3, pp. 67–68.

Committee on Foreign Relations, United States Senate, Subcommittee on African Affairs, 1980, *Imports of Minerals from South Africa by the United States and the OECD Countries*, a report prepared by the Congressional Research Service. U.S. Government Printing Office, Washington, D.C., 46 pp.

Committee on Interior and Insular Affairs, Subcommittee on Mines and Mining, U.S. House of Representatives, 1980, *U.S. Minerals Vulnerability*. U.S. Government Printing Office, Washington, D.C., 83 pp.

Curlin, J. W., 1984, The political dimensions of strategic minerals. In *American Strategic Minerals*, G. J. Mangone, ed., Crane Russak, New York, pp. 135–147.

Eckes, A. E., Jr., 1983, The global struggle for minerals: A historian's perspective. In *International Minerals*, A. F. Agnew, ed., Westview Press, Boulder, Colo., pp. 143–155.

Engineering and Mining Journal, 1984, No protection for U.S. copper, but market may improve. Vol. 185, No. 10, pp. 23, 83, 84.

General Services Administration, 1979, *Stockpile Report to the Congress, October 1978–March 1979.* Federal Preparedness Agency, General Services Administration, Washington, D.C., pp. 1–35.

Gordon, R. L., 1976, Government policies for mineral development and trade. In *Economics of the Mineral Industries,* 3rd Ed., W. A. Vogley, ed., American Institute of Mining, Metallurgical, and Petroleum Engineers, New York, pp. 735–738.

Holland, T. H., 1935, *The Mineral Sanction as an Aid to International Security.* Oliver and Boyd, London, 95 pp.

Hooker, J. R., 1971, Rhodesia revisited: Are United Nations sanctions Mr. Ian Smith's secret weapon? American Universities Field Staff, Report No. XV, April 1971.

Huddle, F. P., 1976, The evolving national policy for materials. *Science,* Vol. 191, Feb. 20, 1976.

Morgan, J. D., 1984, Future demands of the United States for strategic minerals. In *America's Strategic Minerals,* G. J. Mangone, ed., Crane Russak, New York, pp. 59–84.

National Bureau of Standards, 1981, Proceedings, *U.S. Department of Commerce Public Workshop on Critical Materials Needs in the Aerospace Industry, February 9–10, 1981.* U.S. Department of Commerce, Washington, D.C.

National Commission on Materials Policy, 1973, Final Report, *Material Needs and the Environment Today and Tomorrow.* U.S. Government Printing Office, Washington, D.C., 294 pp.

National Materials Advisory Board, 1972, *Elements of a National Materials Policy.* National Academy of Sciences, NMAB-294, Washington, D.C., 66 pp.

Pehrson, E. W., 1964, Minerals in national and international affairs. In *Economics of the Mineral Industries,* 2nd Ed., E. H. Robie, ed., American Institute of Mining, Metallurgical, and Petroleum Engineers, New York, pp. 511–560.

President's Materials Policy Commission, 1952, *Resources for Freedom,* Vol. 1. U.S. Government Printing Office, Washington, D.C., 184 pp.

Reagan, R., 1982, National materials and minerals program plan and report to Congress, Washington, D.C., 21 pp.

Van Rensburg, W. C. J., and Pretorius, D. A., 1977, *South Africa's Strategic Minerals.* Valiant, Johannesburg, 156 pp.

Van Rensburg, W. C. J., 1986, *Strategic Minerals,* Prentice-Hall, Englewood Cliffs, N.J., 2 vols, 552 + 362 pp.

13 Stabilization on the International Scene

CONTROL SCHEMES IN MINERALS

In examining conservation of mineral resources we have seen that fluctuations in markets for minerals have been a perennial problem for domestic mineral industries. They have also been a problem for the mineral industries of the rest of the world, and they are one of the principal problems facing those industries today. Instability has serious consequences. For the mineral industries it means difficulties in maintaining operations at profitable levels, in securing adequate returns on investments, and in carrying out orderly development of mineral deposits. For the countries involved, it means periods of vigorous activity in mineral industry alternating with periods of depression, with all that this means in terms of reduced employment, lower tax revenues, inhibited development of secondary and tertiary industries, and other symptoms of economic distress. The effects are especially severe for those countries that depend heavily on mineral industry for employment and national income, but they have also been severe for certain mining districts of the United States.

The response to instability has been the periodic organization of control schemes in minerals, with the purpose of stabilizing prices and markets. At various times during the past 150 years, there have been control schemes in aluminum, bauxite, coal, copper, diamonds, lead, mercury, nickel, nitrate, petroleum, potash, silver, steel, sulfur, tin, tungsten, uranium, and zinc. Control schemes are of two kinds—monopolies and cartels (Leith et al., 1943). A monopoly is control by a single individual or organization. A cartel is a control scheme involving an association of producers, with or without participation by governments of producing countries. One cartel, in tin, is an association of governments of producing countries and governments of consuming countries. Cartels exist at present in diamonds, tin, bauxite, iron ore, copper, petroleum, mercury and uranium.

Some control schemes have been successful for long periods of time, but many have been unsuccessful. A full presentation of their history is beyond the scope of this volume. It is instructive, however, to examine the histories of a few, because they give us insight into the motivations of cartels, the reasons for success or failure, and what can be expected from present cartels.

Deposits of native sulfur in Sicily were for a long time the principal source of sulfur for western Europe. The deposits were therefore of international importance. In 1838 the Sicilian government granted a private concern the exclusive right to extract and export sulfur; in other words, it created a monopoly. The monopoly abruptly increased the price of sulfur nearly 200 percent. This action was particularly resented by Great Britain, the principal importer of the mineral, and was held to be in violation of treaty rights and injurious to British consumers. The Sicilian government argued that its aims were to stabilize prices, prevent wasteful methods of production, and increase revenues. Two years of negotiations, reinforced on the British side by gunboats, was required to settle the debate. Meanwhile, Great Britain turned to pyrites (natural iron sulfides) as alternative sources of sulfur. The net result of the monopoly was that Sicily lost a significant portion of its market.

In 1887–1888, a notorious, almost ludicrous attempt was made to gain a monopoly of world copper supply. After its main agent, a French citizen named Sécrétan, it became known as the Sécrétan corner. The stage for the attempt was set in the 1880s by a large increase in output of copper in the western United States. In 1885–1887 this caused a decline in prices, and production of copper decreased at all mines except those in Spain and the United States. At the same time consumption was growing, and late in 1887 improvement in the demand for copper set in.

At this point Sécrétan, with the backing of the French Rothschilds, Baring Bros., and the Comptoir Escompte, began buying both copper metal and copper mining stocks on a large scale. By the end of 1887, Sécrétan controlled the Spanish mines and had contracted for three fourths of world output for a period of 3 years. The price of copper rose 100 percent, values of mining shares skyrocketed, and Sécrétan appeared to have the copper market well in hand. However, he had miscalculated. Copper production in the western United States rose 20 percent, and a flood of scrap copper entered the market. In an attempt to control the situation, Sécrétan had to increase his purchases, and by June of 1888 he had 75,000 tons of copper (an enormous amount for the time) on hand and owed $27 million. Copper still poured into the market; Sécrétan's credit was finally exhausted. He had to sell his holdings, and the market for copper crashed.

The Sécrétan monopoly failed because it violated three basic principles of a successful control scheme:

1. It did not have a publicly acceptable justification.
2. It had no real control over sources of production.
3. It adopted a predatory price policy; consumer reaction was to cut use of copper and send scrap into the market.

Whereas the Sécrétan corner was a monopoly, other control schemes in copper have been cartels. Two of the cartels were organized during the interval between the two world wars. One was a great success, the other a disastrous failure. Their histories are particularly instructive (Elliott et al., 1937). The first cartel was a direct result of World War I. Requirements for copper during the war were very high. Copper production capacity was increased in the United States, Chile, the Belgian Congo, and Mexico. In the United States, the increase in capacity was 65 percent. When the war ended, productive capacity was greatly in excess of peacetime demand. The major problem, however, was in Europe. An estimated million tons of recoverable copper lay on the battlefields of western Europe. Stocks in the hands of European governments contained 240,000 tons of virgin copper and 400,000 tons of copper in brass, the total being about equivalent to the annual prewar European consumption of copper during a period of heavy use for armaments.

When the armistice was signed on November 11, 1918, stocks and battlefield scrap overhung a market unable to sustain just the capacity for production of new copper. To make matters worse in the United States, in December 1918 the War Industries Board stepped in and ordered the American copper industry to freeze prices, production levels, and wages at wartime levels. It was an incredible blunder. Realizing this, the board shortly withdrew the order, but it was too late. The market for copper lapsed into a coma; no quotations for copper could be had in the New York market in late December 1918 and January 1919.

To meet this situation, a cartel, the Copper Export Association, was formed in late December of 1918. The association could not control production and markets in the United States, since such action would be a violation of the Sherman Antitrust Act. However, under the Webb-Pomerene Act of April 1918, it could allocate orders in the European market and fix prices in those markets. Orders could be prorated among the American producers based on their rates of production during the previous 12 months. Members of the cartel controlled 95 percent of U.S. production and most of the production outside the United States. In theory, the cartel could control only the export markets, but in fact all American producers were heavily dependent on export markets, and the cartel's allocations therefore strongly influenced American production.

Production was actually curtailed, apparently by common consent of the copper producers. Such action was contrary to the Sherman Antitrust Act but was wisely overlooked.

The cartel was successful in maintaining reasonable prices for copper from 1919 to 1923. By that time the European stocks had been disposed of, consumption of copper was rising, and the outlook for the copper industry in general was very good. At this point, however, the cartel broke up owing to a dispute between high-cost domestic producers, who wanted a protective tariff, and producers who had large foreign mines, were importing copper from them, and therefore opposed a tariff.

During its rather brief life the cartel was a notable success. It had adequate control over the sources of new copper supply. It adopted and maintained a reasonable price policy. There were sound reasons for the cartel; it was recognized as an essential step by industry to meet a crisis that might have had severe repercussions throughout the American economy. Public reaction to the cartel was very favorable.

Quite different was the history of the second cartel. From 1923 to 1926 there was no cartel, but there were developments in the copper industry that set the stage for formation of a new one. In Chile and the Belgian Congo, increases in production capacity amounted to 500,000 tons, nearly equal to the annual demand in Europe. During the same period, mass mining of copper was developed to a very high degree. In mass mining, as indicated earlier, it is important to maintain production at high levels so that the fixed costs (overhead) can be spread over as many tons of production as possible. The copper industry thus became very vulnerable to decline in demand for the metal and developed a heavy stake in stabilizing prices and markets for copper. By 1926 the time seemed ripe for organization of a new cartel dominated by U.S. producers. U.S. mines then accounted for 53 percent of world copper production, and U.S. companies controlled 47 percent of world production outside the United States. American refineries accounted for 72 percent of world production of refined metal. Against this background the second cartel, Copper Exporters, Inc., was formed. It consisted of 18 American companies and 14 foreign companies, of which 4 were controlled by the United States. Together they controlled 95 percent of the copper production of the world. The justification given for the cartel was the threat of overproduction and destabilization of the copper industry.

During 1927 and early 1928 the new cartel operated reasonably well. The price of copper was actually allowed to decline to 13 cents per pound, lower than the 1913 price. Public reaction in the United States was good. There was, however, discontent in Europe, because the cartel eliminated the brokers who had previously handled orders in the European market. Later in 1928, the pricing policy of the cartel was changed. The price of copper was rapidly advanced, reaching 24 cents a pound in March of 1928. At this point, a buyer's strike began, and a chain of events was initiated that ended in disaster. Scrap copper poured into the market. The aluminum industry, just then coming into its own, cut the price of aluminum, thus stimulating extensive substitution of aluminum for copper. Copper from newly developed mines in Northern Rhodesia and Canada, not members of the cartel, came into the market in significant volume. At this point it became evident that the cartel had no effective machinery for controlling production and price. An attempt to hold the price to 18 cents a pound failed completely. The stock market crash in October 1929 and the ensuing depression finished the cartel. By 1932 the price of copper had fallen to 5.5 cents per pound. When a tariff of 4 cents a pound was levied on imports of the metal, the foreign members withdrew and the cartel disbanded.

The sorry history of the second copper cartel has had a lasting effect on American attitudes toward control schemes in minerals. Ever since its demise, the United States has generally been opposed to control schemes of any kind. In the Trade Expansion Act of 1962, it is true, the "orderly marketing section" gave the President power to negotiate with other countries to form international cartels in minerals and other commodities. Such cartels could set limits on annual exports of producing countries. However, only once in recent history has the United States participated in a cartel in minerals, the international tin cartel. It is worthwhile to review its history briefly. After World War I, the tin-mining industry of the world had to deal with severe fluctuations in demand and price, together with a chronic capacity for overproduction. Those problems led to the organization of several cartels during the period between the two world wars. The last of those, organized in 1931, was a cartel of governments of the tin-producing countries. It was still functioning when the Japanese invaded Southeast Asia. The same problems reappeared after World War II. Again there was overproduction and chronic distress in Bolivia, the Belgian Congo, and the tin-producing countries of Southeast Asia, all countries heavily dependent on income from the sale of tin. Protracted negotiations led to the signing of the first International Tin Agreement (ITA) in 1956. The agreement established an International Tin Council with one representative from each of 20 tin-producing and tin-consuming countries. The consumer group was assigned 1000 votes, prorated on the basis of annual consumption, and the producer group was also assigned 1000 votes, prorated on the basis of annual production. The council's function was defined as stabilizing the price of tin and assuring the consumer of adequate supplies of tin at reasonable prices. The agreement provided for a buffer stock of 25,000 tons of tin, a significant amount relative to an annual production of 200,000 to 250,000 tons of tin. It also provided for setting floor and ceiling prices. The manager of the buffer stock was authorized to sell tin when the price rose to the ceiling and to buy tin when the price sank to the floor. Floor and ceiling prices were to be adjusted periodically by the council. The Tin Council included all the major producers (Malaysia, Indonesia, Thailand, and Bolivia) except the Soviet Union and Mainland China and all the major consumers except the United States and West Germany. The agreement has been renewed periodically at roughly 5-year intervals.

The cartel established under the ITA has had a rather rocky career. The Soviet Union has never signed the agreement, and for a period that country was a source of disturbance to the tin market. Until 1975, the United States refused to join the cartel, and sales from the U.S. stockpile, which amounts to more than 200,000 tons, have at times been used to influence the market, much to the dissatisfaction of the cartel members. The United States signed the fifth agreement in 1975 but has not signed the sixth, which came into force in 1982. Bolivia, with high costs of tin production, has become increasingly dissatisfied with the prices set by the Tin Council and for a time refused to sign the sixth agreement. Brazil has become a significant producer but has not joined the cartel. The United Kingdom is a consumer member of the cartel, but not a producer member. Production from Cornwall has actually increased.

The future of the tin cartel is uncertain. Overcapacity for production is still a problem. World tin consumption has not changed very much since 1967, and prospects for future increase are poor. Large surpluses during the 1980–1982 recession led in 1982 to imposition of export and production controls authorized under the fifth agreement. The major tin producers are developing countries that still rely on income from tin mining, and the imposition of controls has caused economic distress in their tin-producing regions. Planned releases from the U.S. stockpile are expected to exacerbate the problem. The situation is further complicated by a proposal for an association of tin-producing countries initiated by Malaysia, Thailand, and Indonesia.

Most control schemes in minerals have been monopolies or cartels of producers, and this has been a persistent source of dissatisfaction among consuming nations. The tin cartel gives representation to both producers and consumers, and it was hoped at one time that it might serve as a model for other cartels and a solution to the perennial problem of instability in mineral markets. The results, however, have been rather discouraging. The cartel has not really resolved the problem of fluctuations in demand and price. It has not been able to create the conditions necessary for a stable, healthy tin-mining industry, and it has not eliminated periods of economic distress in the tin-producing districts of the world. It has not been able to provide for orderly, conservational development of the world's tin resources.

Perhaps the most extraordinary of present mineral control schemes is that in the diamond industry. It began as a monopoly but has evolved into a strange kind of cartel. It had its origin in the diamond mining that began with the discovery of diamond-bearing pipes in the Kimberley district, South Africa, in 1870. In the 10 years following discovery, individual miners swarmed into the area and opened a great number of small pits in the pipes. Each miner worked as fast as he could; it must have been a chaotic scene. The result was overproduction and cutthroat competition that soon brought the diamond industry to the brink of disaster. Between 1880 and 1914, however, holdings in the Kimberley field were consolidated under five companies, and a syndicate (actually a monopoly) was created to regulate production, control marketing, and set prices. By 1914 the syndicate controlled 95 percent of world production of diamonds.

Beginning in 1907, discoveries of alluvial diamonds (diamonds in river placer deposits) were made in the Belgian Congo, Southwest Africa, the Union of South Africa, Angola, the Gold Coast, and Sierra Leone. The alluvial deposits were far cheaper to work than pipe deposits, and production gradually shifted to the alluvial fields. By 1941, 94 percent of production came from such fields. This forced modification of the old syndicate, but production and marketing continued to be tightly controlled. The advantages of control appealed to the new producers, and each new field became incorporated in the control organization. At the producing level the organization became a cartel of governments and private interests. It is now a very complicated structure. At the marketing level it operates as a monopoly, through an arm known as the Central Selling Organization. As the African nations assumed independence, some chose to withdraw from the cartel, but the power of the control organization over production and marketing of natural diamonds was not seriously weakened. Zaire withdrew from the cartel in 1981, but found it expedient to return in 1983. New pipe deposits opened up in Botswana, Lesotho, and South Africa are under the control of the organization, and it has been announced that marketing of important new production in Australia will be handled by the selling organization. It does not, however, control marketing or production of synthetic diamonds, which now supply two thirds of world demand for industrial diamond products. Furthermore, it is faced with an oversupply of gem diamonds, now the main support of the cartel.

The long success of the control scheme in diamonds is due to several factors. It brought stability to an important industry and could be justified on that basis. It has provided a reliable and adequate supply of industrial diamonds at what were considered reasonable prices. Through the Diamond Research Corporation, it has developed many new and valuable applications of diamonds in industry and many improvements in the use of diamonds. Because of this, and its reasonable price policy, consumer reaction has generally been good. Gem diamonds are an investment item. The organization has had a stabilizing influence on the investment market and has provided rigid standards for various classes of gem diamonds that provide considerable protection to the consumer. The threat of overproduction of gem diamonds has almost constantly overhung the gem market. The investor in diamonds has no desire for a return to the chaotic conditions that characterized the early days of the diamond industry. Through its marketing arm the cartel has maintained rigid, even ruthless control over prices of gem diamonds even in periods of overproduction. It has been able to respond promptly to changes in diamond supply and demand. A final reason for success has been the involvement of governments in the control scheme; governments have a significant stake in maintaining a stable and financially sound industry. It is perhaps a commentary on the skill with which the control scheme has been managed that, in an industry that is comparable to the tin-mining industry in its potential for overproduction and erratic movements of demand and price, it has built a long record of continuing success. At present, however, it faces an uncertain future.

The monopolies and cartels discussed above are only a few of the many schemes that have operated in the past. They serve, however, to indicate the functions of such schemes and certain features essential to successful operation. First, there must be control over all or at least a very large proportion of total world production of a mineral commodity and over the marketing of the commodity. Second, the internal structure of the control organization must be such that it is capable of exercising its control and moving promptly to meet changing market conditions. Third, it must see to it that the world receives, within limits imposed by the nature of world reserves, adequate supplies of the commodity. Fourth, its price policy must be perceived as reasonable by the consumer. Unfavorable consumer reaction sets in motion events that can destroy a control scheme, sometimes very quickly. It stimulates a search for alternate sources and a search for substitutes and for technological adaptations that reduce demand for the commodity. It stimulates conservation, including recycling where that is possible. A control scheme that fails to meet one more of these requirements may survive in a period of rising demand but becomes extremely vulnerable whenever demand slackens.

PRESENT-DAY CARTELS IN MINERALS

In recent years, besides the tin cartel and the diamond monopoly-cartel, there have been cartels in petroleum, iron ore, copper, uranium, mercury, and bauxite. The most conspicuous has been the petroleum cartel, the Organization of Petroleum Exporting Countries (OPEC), currently consisting of Saudi Arabia, Iran, Iraq, Qatar, the United Arab Emirates, Ecuador, Kuwait, Libya, Algeria, Nigeria, Gabon, Indonesia, and Venezuela. OPEC was formed in 1960 as an organization for negotiating with the petroleum companies. During its life there have been progressive nationalization of companies and progressive movement of OPEC members into the refining and marketing of petroleum. The price of petroleum was advanced from $1.80 per barrel in 1970 to a peak of $35 in 1979. It declined to about $28 to $30 per barrel by 1985, but the effects of advance in price were felt throughout the world.

For a time the cartel was totally successful, operating in the context of rising world demand for petroleum and control of 48 to 55 percent of total world production of crude oil during the period 1973 to 1979. From its beginnings, however, there have been sources of weakness within the cartel. It is a loose association of oil-producing nations. It has no control over price or production save that given it by common consent of its members, the interests of which do not always coincide. It has no means of enforcing its decisions, and various members have from time to time disregarded its decisions on pricing and production. It has set prices that bear no relation to costs of production and are adjusted to whatever the traffic will bear. It or its members have used curtailment of production, or the threat of it, as in the Arab oil embargo, as a political weapon. These actions have had their inevitable results. Exploration in countries outside the cartel has developed alternate sources of production and diminished the importance of OPEC sources. Britain and Norway now produce 3.5 million bbl of crude oil per day, more than most of the OPEC nations do. Egypt, Oman, Cameroon, China, Zaire, Colombia, and Mexico, not members of the cartel, are now exporting oil. In 1977, the United States imported nearly half its petroleum supply. In 1983, imports fell to 33 percent, largely owing to conservation. In 1977, OPEC nations accounted for 66 percent of non-Communist production. In 1982 their share was 44 percent. The Soviet Union, the world's largest single producer and not a member of the cartel, has at times undercut the cartel prices. These developments have produced (1983–1985) a glut of production that has forced a reduction in price and curtailment of production. They have greatly reduced the income of OPEC countries from petroleum, from $270 billion in 1980 to $152 billion in 1984. OPEC no longer has real control over world markets and prices for petroleum. In setting prices it needs the cooperation of other oil-producing countries that are not members of OPEC; such cooperation has not been forthcoming.

There is currently, as a consequence, a tendency to overemphasize the troubles of the cartel and perhaps to feel that the cartel is no longer a threat to consumers of petroleum. What should not be overlooked, however, is that the cartel has not collapsed. Temporarily (early 1986) oil prices have plunged due to OPEC efforts to force reduced world production but the era of cheap energy is not likely to return. Some of the alternative sources of petroleum now being explored, such as the North Sea fields, the deeper onshore fields of the United States, and the offshore fields of North America, involve high costs of discovery and production. Decontrol of prices of natural gas in the United States, held for years at unrealistic levels, was necessary but has meant higher costs of energy from that source. Certain countries—Great Britain, Norway, Nigeria, and Mexico, for example—cannot afford a precipitous drop in oil prices. The cost of coal has been rising. The OPEC countries control a large share of the world's reserves of low-cost petroleum and natural gas. Time is on their side.

One of the results of the difficulties of OPEC and the decline in oil prices, and a result potentially damaging for the future, is that support for efforts in the United States to develop alternative sources of energy from oil shales and from synthetic fuels produced by liquefaction or gasification of coal has weakened. The financial assurances necessary to protect the enormous investments involved are not being provided by government, except in a few cases. In effect U.S. efforts to develop a long-term energy policy are being frustrated by an illusion of abundance and security of energy supplies.

In the copper industry, the instability that led to the two copper cartels has persisted. Periods of undersupply and high prices have alternated with periods of oversupply and depressed prices. The industry has been plagued by prolonged strikes in Peru, Zambia, the United States, and Chile, and by civil strife in Chile and Zaire. The situation has been aggravated by discovery and development of a number of major new deposits in various countries of the world. The threat of overproduction has almost constantly overhung the world copper market, and at present total world mine capacity is in excess of what the market can absorb even in times of high demand. There have indeed been periods of high prices, but in large part those have coincided with periods of strife, labor or otherwise, that have interrupted production in one or more countries.

In 1967 this chronic instability led Zambia, Zaire, Peru, and Chile to form the CIPEC (Conseil Intergovernmental de Pays Exportateurs de Cuivre), a new cartel in copper. Indonesia subsequently joined as a full member, and Australia, Mauretania, Yugoslavia, and Papua New Guinea joined as associate members. The purpose of the cartel was the usual one of stabilizing markets and prices. It has never been an effective organization, largely because the members, especially Chile, have not been willing to control production. Productive capacity for copper has increased far beyond the needs of the world, and copper prices have plunged to near-disastrous levels. Allowing for inflation, the price of copper is the same as during the Great Depression of the 1930s.

In 1974 there was organized an International Bauxite Association (IBA) consisting of Jamaica, Guinea, Australia, Haiti, Indonesia, Guyana, the Dominican Republic, Surinam, Sierra Leone, and Yugoslavia. These countries control roughly two thirds of Free World production. The association has acted chiefly as a consultative body, but since its inception prices of bauxite and alumina have been advanced by the member countries. The IBA, however, has not really attempted to control price or production of aluminum raw materials. Large increases in production capacity in various countries are likely to prevent any attempt to set prices at unreasonable levels.

An Association of Iron Ore Exporting Countries was formed in 1975. Like the bauxite association it is essentially a consultative body, consisting of up to nine member countries controlling about 25 percent of the world iron ore production and about 46 percent of the production that enters into international trade. It appears to have had little impact on world prices for iron ore.

As a final comment on control schemes it should be stated that they are not inherently either good or bad, despite the American distaste for such arrangements. Everything depends on the purposes for which a scheme is organized and the way in which it is managed. Some control schemes have been frankly predatory, but most control schemes of the 20th century have been responses to the instability of world mineral markets. There is clearly a need for stabilizing those markets. Instability defeats attempts at conservational development of the world's mineral resources, which is long overdue as an object of international concern. Instability has been a source of economic distress that time and again has plagued mineral-producing countries and regions. Cartels involve problems, but no alternative means of stabilization have yet been devised by man. It therefore seems unwise to condemn them categorically as instruments of international trade. In general there is no need to fear cartel arrangements. Badly managed or predatory cartels generate forces that lead to their destruction. The future may show, however, that the oil cartel is an exception to the general rule.

REFERENCES AND ADDITIONAL READING

Blair, J. M., 1976, *The Control of Oil.* Pantheon, New York, 441 pp.

Clarfield, K. W., Jackson, S., Keeffe, J., Noble, M. A., and Ryan, A. P., 1975, *Eight Mineral Cartels: The New Challenge to Industrialized Nations.* McGraw-Hill, New York, 177 pp.

Elliott, W. Y., May, E. S., Rowe, J. W. F., Skelton, A., and Wallace D. H., 1937, *International Control in the Non-ferrous Metals.* Macmillan, New York, 801 pp.

International Tin Study Group, 1950, *Tin: A Review of the World Tin Industry, 1949–50.* The Hague, Netherlands, 55 pp.

Leith, C. K., Furness, J. W., and Lewis, C., 1943, *World Minerals and World Peace.* Brookings Institution, Washington, D. C., pp. 107–134.

Van Rensburg, W. C. J., *Economics of the World's Mineral Industries.* McGraw-Hill, New York, pp. 210–226.

14 Minerals from the Sea

The great bulk of the world's mineral production has come from mineral deposits on the lands, and those deposits are still our main reliance for mineral supplies. In the past 35 years, however, there has been a growing attention to the mineral resources of the sea, and the manner of their development has become a problem of international concern. Resolution of the problem has become important both to the industrialized nations and to the less developed nations of the world.

Obtaining minerals from the sea is by no means new. Tin-bearing gravels have been mined for decades off the shores of Thailand and Indonesia. Mining offshore diamond deposits near the mouth of the Orange River in South Africa has been attempted. Sands containing iron oxides have been mined off the coast of Japan. Sand and gravel are mined off the coast of Great Britain and on a massive scale in Arctic waters for the construction of oil-well drilling platforms. However, the most important offshore mining operations of all are those that yield oil and gas from the continental shelves of the North Sea, Nigeria, Angola, the Pacific and Gulf Coasts of the United States, and Indonesia, Australia, and Mexico.

Most offshore mining operations have been in areas close to coastlines, areas over which bordering nations have asserted jurisdictional rights. Initially limited to 3 miles (the maximum range of ancient cannon), jurisdiction was later extended to 12 miles. Since World War II there has been a growing tendency to set the limits of jurisdiction still farther out. In 1945 President Truman proclaimed U.S. jurisdiction over the natural resources of the seabed of the continental shelf adjacent to the coastline of the United States. The rising importance of offshore petroleum resources was a major reason for this action. In 1946 Argentina proclaimed its sovereignty over the adjacent continental shelf, which extends up to 500 miles from the coast of Argentina and surrounds the Falkland Islands. Soon after, Costa Rica, Chile, Ecuador, Peru, and El Salvador, later joined by Colombia, passed laws proclaiming their exclusive rights to fish within a 200-mile offshore zone. The anchovy fisheries of Peru and Ecuador, for example, are of much economic importance to those countries. Nutrient-rich cold waters upwelling along their coasts feed a huge population of plankton, tiny plants and animals that support some of the densest fish populations in the world.

Though fishing rights were the major concern of most of the countries mentioned above, several developments after World War II brought the problem of offshore mining to the fore. Other nations joined the United States in recognizing the great importance of the continental shelves as sources of petroleum and natural gas. Deposits of metalliferous nodules were found to be spread over sizable portions of the deep ocean floors. In the Atlantis Deep of the Red Sea, a strange deposit of mud was found to contain large tonnages of zinc and copper, with recoverable silver, cobalt, and gold. Deposits of phosphate were found both on the continental shelves and on the deep sea floors. Oil and gas were discovered in one offshore area after another. Finally, successful mining on the continental shelves sparked interest in the possibilities of deep-sea mining.

Underwater mining is now technologically possible, within limits, but mining beneath the deep ocean waters raises some knotty questions of international law. Extending jurisdiction to the continental shelves was sanctioned by the Continental Shelf Convention of 1958, which was drafted by the International Law Commission, but the convention was not in force until 1964 (Auburn, 1977) and even then had not been ratified by all nations. The convention left unresolved various questions regarding the boundaries of the continental shelves of individual nations. Jurisdiction over the oil and gas resources of the North Sea, which is part of the continental shelf of Europe, had to be settled by negotiations involving the United Kingdom, Holland, Norway, West Germany, and Denmark. For the deep seas there was still no international law apart from the traditional law of the freedom of the seas. The jurisdictions asserted by the South American nations cited above, apart from Argentina, extended far beyond the limits of the narrow continental shelves of western Central America and western South America. There was no law to cover such cases as the Red Sea mud deposits, which are in the bottom of a deep that lies between the Sudan and Saudi Arabia; those two nations simply asserted joint sovereignty over the deposits.

The actions by various nations and the growing importance of ocean resources have fueled international controversy over rights to the use of the sea and exploitation of its resources. Not only minerals are involved; there are also questions of air and navigation rights. As indicated above, many nations are gravely concerned over the fisheries off their coasts. The advent of huge mechanized fishing fleets, Japanese and Russian in particular, has threatened depletion or even destruction of some of the world's best fishing grounds. The United States is particularly concerned about the fisheries off the Alaskan coasts and those of the Georges Bank, which lies east of Cape Cod, Massachusetts. The nearshore and offshore fisheries are dependent on breeding grounds that lie far outside the 12-mile limit and for a time were heavily fished by Russian and Japanese fleets. In 1976, the United States extended its jurisdiction over coastal waters of the Alaskan and Atlantic coasts to 200 miles offshore and assumed control over fishing within the 200-mile limit.

In the 1970s, discussion of rights to the seas developed into a confrontation between Free World industrialized nations and developing nations, specifically with reference to mineral resources. The former nations have the financial and technological resources necessary for ocean mining. They also see ocean mining as a means of making up for some of their deficiencies in mineral production. They would encourage development of deep-sea resources by private enterprise. The developing nations have two objectives with regard to ocean resources. All wish to share in the revenues of ocean mining. Some, such as Chile, Peru, and Zambia, wish to limit production from ocean deposits so as to protect markets for the products of their land-based mineral operations. The developing nations argue that the resources of the oceans are the common heritage of all nations. They want exploitation tightly controlled, and they want private and national enterprise excluded.

Although a number of kinds of mineral deposits are involved, much of the discussion of ocean mineral resources has centered on the deposits of manganese-rich nodules found on portions of the ocean floors, generally at depths ranging from 12,000 to 15,000 feet. The nodules were actually discovered by the Challenger Expedition of 1872–1876, but their potential importance as sources of metals was not recognized until the late 1950s, and the mining industry became interested in the nodules only in 1962. The nodules are black, rounded lumps, pebble-size to cobblestone-size, that are thickly strewn over some portions of the ocean floors (Figs. 14-1 and 14-2). They have been found on portions of the floors of all the ocean basins, but some areas are much richer in nodules than others, and the metal contents of the nodules vary from one area to another. Since 1962 a great deal of exploration, sampling, and testing of the nodules has been done. The most promising area is considered to be a belt that lies between the Clarion and Clipperton fracture zones and extends about 3000 miles eastward from a point south of Hawaii (Fig. 14-3). Nodules in the belt contain more than 1.8 percent nickel plus copper. McKelvey et al. (1979) conclude that about half the area contains nodules in concentrations (wet-weight units) greater than 5kg per square meter. Assuming that 20 percent of the nodules in this half of the area are recoverable, its potential recoverable resources total about 2.1 billion dry metric tons of nodules averaging about 25 percent manganese, 1.3 percent nickel, 1.0 percent copper, 0.22 percent cobalt, and 0.05 percent

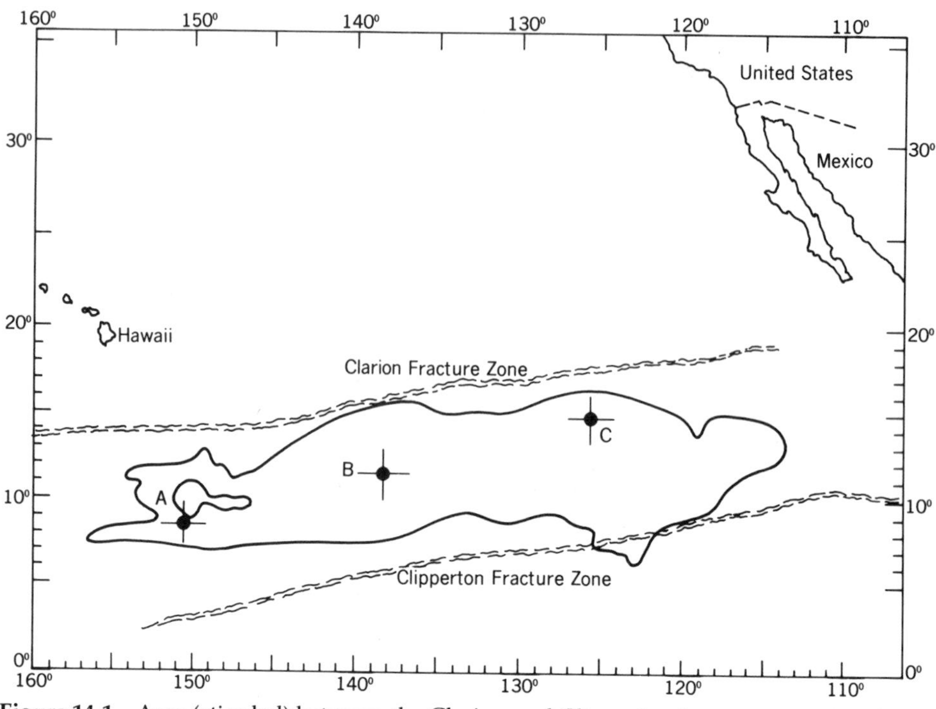

Figure 14-1 Area (stippled) between the Clarion and Clipperton fracture zones in which manganese nodules contain more than 1.8 percent nickel plus copper. The black circles are the approximate centers of areas A, B, and C covered by the Deep Ocean Mining Environmental Study (DOMES) of the National Oceanic and Atmospheric Administration and by the estimates of Sorem (1982). Based on maps by Horn et al. (1972) and McKelvey et al. (1979).

molybdenum. The tonnages of metals contained in the recoverable nodules would be, in millions of metric tons, manganese, 525; nickel, 27.3; copper, 21; cobalt, 4.62; and molybdenum, 1.05.

The tonnage of nodules present over a given portion of the sea floor must be calculated from photographs of the floor such as the one in Fig. 14-2, and grade must be estimated from analyses of nodules recovered from various points. Sorem (1982) studied more than 12,000 photographs of the sea floor in three areas within the Clarion-Clipperton belt (Fig. 14-1). The black circles A, B, and C are the approximate centers of areas about 110,000, 51,000, and 73,000 square kilometers in extent. Grades for the areas were calculated from analyses of 882 nodules from various points. The total tonnage of nodules in the three areas was calculated at 1.2 billion metric tons averaging 25.5 percent manganese, 1.3 percent nickel, 1.06 percent copper, and 0.25 percent cobalt. The indicated tonnages of the metals are respectively 306, 15.6, 12.7, and 3 million metric tons.

The figures given in the estimates may be compared with estimates of the reserve base for the metals in deposits on land. DeYoung et al. (1984) estimate the reserve base for manganese at about 5.1 billion metric tons.

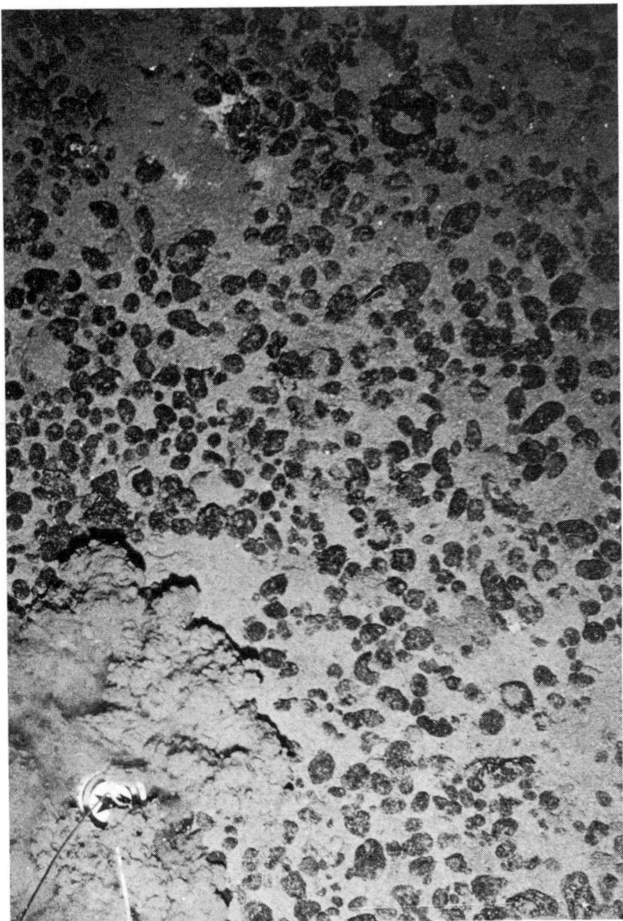

Figure 14-2 Manganese nodules on the floor of the Eastern Pacific Ocean, at about 15° N. latitude, 126° W. longitude. Depth about 4500 meters. The nodule concentration is about 14 kg per square meter. The compass head at the lower left is about 3.1 inches in diameter. Photograph furnished by R. K. Sorem. From Sorem and Fewkes (1979), by permission of Plenum Publishing Corp.

The U.S. Bureau of Mines estimates the reserve base for copper at about 510 million metric tons, for nickel at about 100 million metric tons, and for cobalt at about 8.3 million metric tons. In making comparisons, it should be understood that estimates for the nodules indicate only the order of magnitude of the resources present. Certainly, however, the nodules are a significant resource, and they could be of considerable importance in the future. For the United States they are possible sources of three highly strategic metals.

During 1962–1978, research and development were undertaken by five international firms or consortia, with investments of up to $100 million by each. At least two viable methods of recovering the nodules were

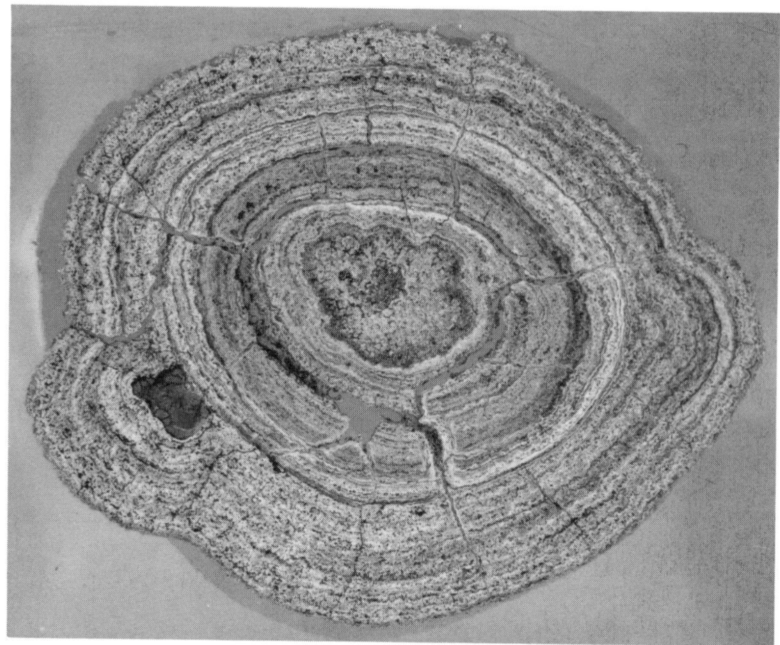

Figure 14-3 Polished section of a manganese nodule showing the concentric layering characteristic of deep-sea nodules. The nodule is about 1.3 inches long. Photograph furnished by R. K. Sorem. From Sorem and Fewkes (1979), by permission of Plenum Publishing Corp.

developed. One technique employs a continuous chain of buckets extending from a ship on the surface to the sea floor far below. The buckets would excavate the nodules and bring them to the surface. The other technique employs a suction dredge and airlift, so that the nodules would be sucked up through a pipe into a ship at surface. The device is really a monster vacuum cleaner that can be made to sweep over the ocean floor. The technical feasibility of deep-sea mining has thus been demonstrated, but the economics of nodule mining are unattractive at a time when the metals involved are a glut on world markets. Metals from the sea would have to compete with metals that can be produced from mines on land at considerably lower costs.

Deep-sea mining is still in the development stage. Establishing a deep-sea mining venture will require the further investment of hundreds of millions of dollars. A prospective mining area on the sea floor must be precisely characterized. This means determining the density of nodule distribution in more detail, the metal contents of the nodules, and the detailed topography of the sea floor in the area. A commercial mining system must be built, and necessary onshore facilities for receiving and processing the nodules and extracting the metals from them must be provided. The total effort might require 10 years; only then could return on

investment begin. Such enormous and costly effort cannot be undertaken without firm assurance of the right to mine a deep-sea deposit once it has been delineated and all the necessary investments have been made. It is the old question of security of tenure, without which no major mining venture can be undertaken. Unfortunately, security of tenure is the crux of the problem, for no such security is provided by traditional international law.

The controversy over rules of mineral development, fisheries, navigation, and other matters led in 1974 to the first of a long series of International Conferences on the Law of the Sea, under the auspices of the United Nations. The protracted negotiations during the conferences were a series of debates between the developing nations, plus the Soviet Union, and the industrialized nations of the Free World, led by the United States. The two groups agreed that the resources of the sea are the common heritage of the nations, but they disagreed sharply on the conditions under which development of resources should be permitted. The industrial nations wanted the oceans to be open to mining by their nationals. The developing nations wanted mining to be governed by an international authority. The developing nations prevailed in the drafting of the Law of the Sea Treaty that was put in final form during the conference in 1981. The treaty calls for establishment of an International Seabed Authority (ISA) under the United Nations. The ISA would have exclusive right to exploit deep-sea resources or to license their exploitation. All revenues from mining would go to the developing nations. The ISA would establish an international mining company called Enterprise. Treaty provisions would force industrialized nations to sell to Enterprise the technology necessary for deep-ocean mining. Private companies might be licensed by the ISA, but they would have to pay fees of up to a million dollars a year and would be subject to United Nations production controls and a taxation rate of up to 70 percent. Revenues would be distributed to underdeveloped nations at the discretion of the United Nations. Enterprise would be financed by low-cost loans from industrialized countries; the U.S. share would be 25 percent. The ISA would be a one-nation–one-vote international body, governed by an assembly and an executive council on which the Soviet Union and its satellites would have three seats whereas the United States would have to compete with its allies for representation. The U.N. General Assembly would be designated as the supreme authority over the ISA. Thus the developing countries, often hostile to the United States, would have control.

The terms were regarded as unacceptable by the Free World industrialized nations. In December 1982, the treaty was signed by 117 nations and rejected by 47, including the United States, Belgium, West Germany, England, Italy, and Japan. The treaty was to take effect when signed by legislatures of 60 nations but would be binding only on the ratifiers.

Meanwhile, in 1980 Congress passed the Deep Seabed Hard Minerals Act, under which the National Oceanic and Atmospheric Administration (NOAA) is authorized to license deep-sea mining operations by U.S. citizens or U.S.-controlled organizations. NOAA regulations governing seabed mining will include environmental protection rules. U.S. negotiations with England, France, Belgium, West Germany, Holland, and Japan led in 1984 to an agreement to respect each others' licensing actions. The United States will seek to develop mechanisms for deep seabed mining through domestic law rather than through the Law of the Sea Treaty.

It is indeed unfortunate that no real agreement on a revision of the international law of the sea could be reached. However, the treaty is not a reasonable document. It could not serve the interests of all the nations of the world, but only the self-interests of a large bloc of developing nations. One of the troublesome features of the treaty, insofar as it concerns mineral resources, is that it does not set firm rules for the future in terms of which orderly development of ocean resources might proceed. Instead, it would set up a body whose policies could be changed at any time at the whim of the United Nations General Assembly. No genuine security of tenure is provided.

The latest act on the part of the United States is the presidential proclamation of 1983 creating an Exclusive Economic Zone (EEZ) extending 200 nautical miles offshore from the United States (Fig. 14-4). All economic activity withn this zone will be controlled by the United States, with the provision that there will be no interference with air and sea navigation rights of other nations. Responsibility for management of mineral resources is to be shared by the National Oceanic and Atmospheric Administration and the Department of the Interior (see report of the National Advisory Committee on Oceans and Atmosphere, 1983).

The creation of the EEZ is a major act of policy for two reasons. One is that it opens to exploration and development of mineral resources areas totaling 3,900,000 square miles, more than the land area of the United States. In effect, it enlarges the area of the public domain to more than six times its present size. The second is that under the proclamation mining organizations can be guaranteed security of tenure, subject to environmental and other regulations laid down by government. The manganese nodules of the Clarion-Clipperton zone lie outside the EEZ. However, creation of the EEZ provides a stimulus to exploration for other mineral deposits that lie withn the zone. The United States has mounted a massive effort to survey the sea bottoms of its EEZ. On the Juan de Fuca Ridge, a major rift in the sea floor off the northern coast of California, deposits of zinc and copper sulfides are under investigation. Phosphate deposits are known on the Blake Plateau off the Carolina coast. Large seaward extensions of the phosphate deposits already being worked in the Aurora district, North Carolina (Fig. 14-5), have been recognized in

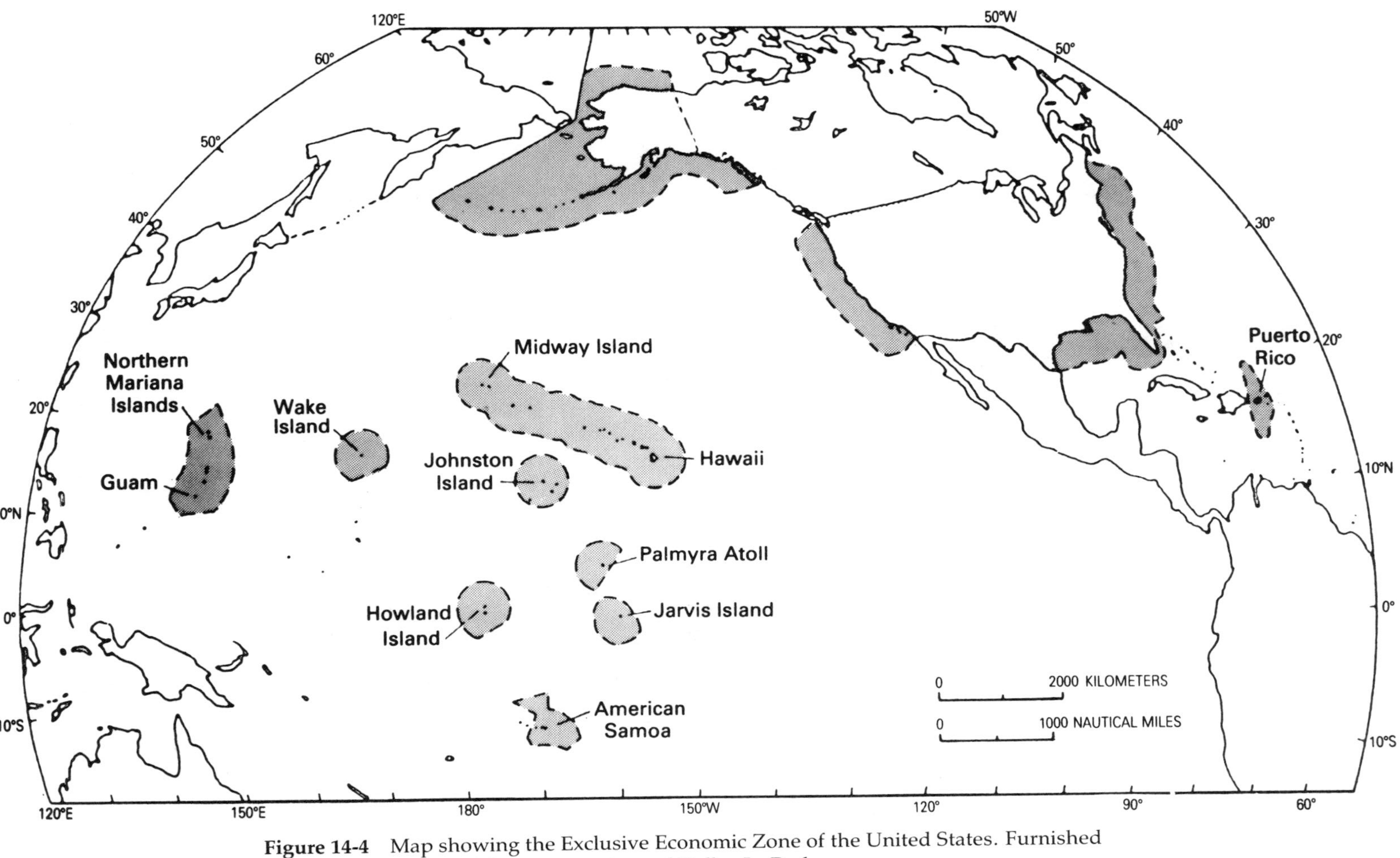

Figure 14-4 Map showing the Exclusive Economic Zone of the United States. Furnished by the U.S. Geological Survey, courtesy of Dallas L. Peck.

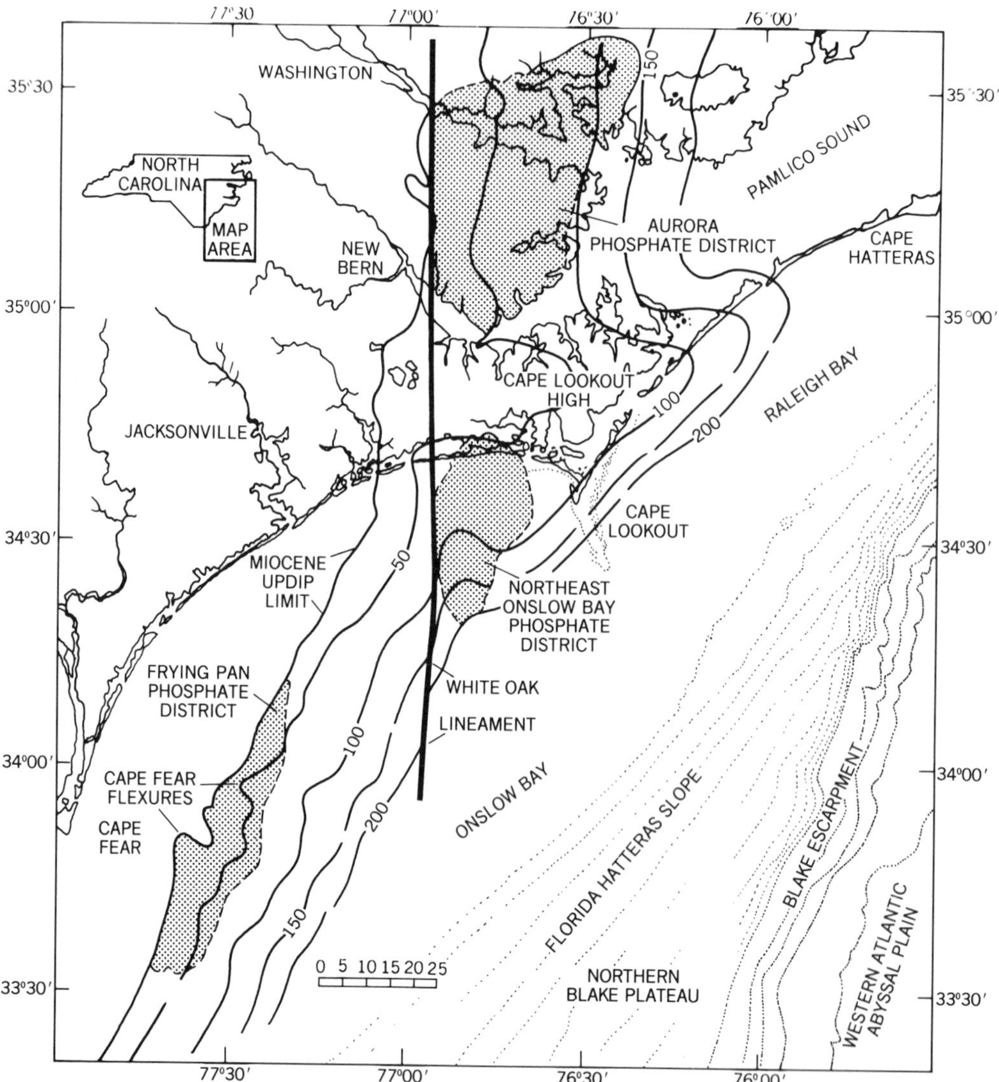

Figure 14-5 Map showing location of the onshore phosphate deposits of the Aurora district (Lee Creek) of North Carolina and the offshore deposits that occur in the same geologic formation. Simplified map based on Fig. 5 of Riggs et al. (*Economic Geology*, 1985), by permission.

shallow waters along the coast of North Carolina and are being studied (Riggs et al., 1985; Riggs and Manheim, 1985). Demonstrated resources in the deposits of the Aurora district have been estimated at 1.0+ billion metric tons (Cathcart et al., 1984) to 1.6+ billion metric tons (Zellers and Williams, 1978), and large additional resources are inferred. Preliminary estimates of phosphate (Riggs and Manheim, 1985) in the Frying Pan phosphate district (Fig. 14-5) suggest a potential yield of 3.7 billion metric tons of phosphate concentrate. Substantial lower-grade phosphate resources are present in the Northeast Onslow Bay phosphate district. In view of the mounting obstacles to further mining in the Florida phosphate district, the offshore deposits could be of great importance to American agriculture.

Cobalt-rich manganese crusts have been found on the tops and flanks of seamounts (submarine mountains) associated with the Hawaiian chain of islands. There are other possibilities—placer deposits of gold, platinum, chromite, titanium minerals, zircon, and rare-earth metals off certain parts of the Atlantic and Pacific coasts, and deposits of sand and gravel along the Atlantic coast.

Activity involving or aimed at ocean mining has been spreading around the world. EEZs have now been established by some 50 nations (Bernier, 1984), and within them a wide range of exploration and research programs has been undertaken. Offshore mining now accounts for half of Thailand's annual tin production. Production of tin offshore of four Indonesian islands continues, and both Indonesia and Thailand have undertaken additional exploration of offshore areas for tin. An international effort to appraise high-grade phosphate deposits east of New Zealand is in an advanced stage. Phosphate deposits off the coasts of Gabon, the Congo (Brazzaville), the Soviet Union, Peru, the Fiji Islands, and certain islands of French Polynesia are under investigation. Deposits of titanium minerals, gold, platinum, chromite, diamonds, and other minerals in various parts of the world are receiving attention. Sulfide deposits continue to be discovered. Exploration for petroleum off Indonesia continues, and petroleum is being sought in the China Sea. Development of means of recovering metals from the Red Sea muds appears to be well advanced. Few of the projects currently in progress are likely to result in commercial operations during the present century, but they have significant potential for the future.

No one knows what other deposits may be found in future decades on or beneath the ocean floors. Exploration for minerals on the lands of the United States has been actively in progress for nearly 150 years, yet we still do not know the full extent of the mineral resources present beneath our lands. At this early stage of offshore exploration it is scarcely likely that we know much about the mineral resources hidden by the sea, and this prompts a final word on negotiation of appropriate international law. So far as minerals are concerned, current negotiations are a poker game in which none of the players knows the number or value of the chips at the center of the table. This being so, the United States is well advised to proceed cautiously in negotiations that may have enormous influence on the future development and availability of minerals from the sea.

REFERENCES AND ADDITIONAL READING

Auburn, F. M., 1977, Legal aspects of nodule mining. In *Marine Manganese Deposits*, G. F. Glasby, ed., Elsevier, Amsterdam, pp. 439–460.

Bernier, L., 1984, Ocean mining activity shifting to exclusive economic zones. *Engineering and Mining Journal*, Vol. 185, No. 7, pp. 57–60.

Bilder, R. B., 1980, International law and natural resource policies. In *Resources and Development*, P. Dorner and M. A. El-Shafie, eds., University of Wisconsin Press, Madison, pp. 385–421.

Cameron, E. N., 1977, Our mineral problems—context of ocean mining. *Marine Mining*, Vol. 1, pp. 73–84.

Cathcart, J. B., Sheldon, R. P., and Gulbrandsen, R. A., 1984, Phosphate-rock resources of the United States. U.S. Geological Survey Circular 888, 48 pp.

DeYoung, Jr., J. H., Sutphin, D. M., and Cannon, W. F., 1984, International strategic minerals inventory summary report—Manganese. U.S. Geological Survey Circular 930-A.

Granville, A., 1975, The recovery of deep-sea minerals: Problems and prospects. *Minerals Science and Engineering*, Vol. 7, No. 3, pp. 170–188.

Horn, D. R., Horn, B. M., and Delach, M. N., 1972, *Ferromanganese Deposits of the North Pacific*. Office of the International Decade for Ocean Exploration, National Science Foundation, Washington, D.C., 77 pp.

Hubred, G., 1975, Deep-sea manganese nodules, a review of the literature. *Minerals Science Engineering*, Vol. 7, No. 1, pp. 71–85.

McKelvey, V. E., Wright, N. A., and Rowland, R. W., 1979. Manganese nodule resources in the northeastern equatorial Pacific. In *Marine Geology and Oceanography of the Pacific Manganese Nodule Province*, J. J. Bischoff and D. Z. Piper, eds., Plenum, New York, pp. 747–762.

Mero, J. L., 1965, *The Mineral Resources of the Sea*. Elsevier, Amsterdam, 342 pp.

Moore, J. R., 1984, Alternative sources of strategic minerals from the seabed. In *American Strategic Minerals*, G. J. Mangone, ed., Crane Russak, New York, pp. 85–108.

National Advisory Committee on Oceans and Atmosphere, 1983. *Marine Minerals: An Alternative Mineral Supply*. U.S. Government Printing Office, Washington, D.C., 33 pp.

Oxman, B. H., Caron, D. D., and Buderi, C. L. O., eds., 1983, *Law of the Sea, U.S. Policy Dilemma*. Institute for Contemporary Studies, San Francisco, 184 pp.

Post, A. M., 1983, *Deepsea Mining and the Law of the Sea*. Martinus Nijhoff, The Hague, 358 pp.

Riggs, S. R., and Manheim, F. T., 1985, Mineral resources of the continental margin. In *Volume of Decade of North American Geology*, Geological Society of America, in press.

Riggs, S. R., Snyder, S. W. P., Hine, A. C., Snyder, S. W., Ellington, W. D., and Mallette, P. M., 1985, Geologic framework of phosphate resources in Onslow Bay, North Carolina continental shelf. *Economic Geology*, vol. 80, pp. 716–738.

Shanks, W. C. III, 1983, Economic and exploration significance of Red Sea metalliferous brine deposits. In *Cameron Volume on Unconventional Mineral Deposits*, W. C. Shanks III, ed., American Institute of Mining, Metallurgical and Petroleum Engineers, New York, pp. 157–171.

Sorem, R. K., 1982, Polymetallic resource estimates for east equatorial Pacific manganese nodule deposits. Collected abstracts, Sixth Symposium, International Association on the Genesis of Ore Deposits, pp. 310–311.

Sorem, R. K., and Fewkes, R. H., 1979, *Manganese Nodules*. Plenum, New York, 723 pp.

U.S. News and World Report, 1982, The struggle for the bottom of the sea. March 15, 1982, pp. 69–70.

Zellers, M. E., and Williams, J. M., 1978, Evaluation of the phosphate deposits of Florida using minerals availability system. U.S. Bureau of Mines, Open-File Report No. 112-78, 106 pp.

15 Some Thoughts on Our Mineral Future

THE FUTURE OF MINERAL EXPLORATION IN THE UNITED STATES

In Chapter 7 it was shown that if present levels of U.S. mineral production are to be maintained through the year 2005, additions to the U.S. reserve base for a number of mineral commodities will have to be made. Additions through advancement in the technology of mineral extraction and processing can certainly be expected, but additions must come in substantial part through discovery and development of mineral deposits yet unknown. Given the growing difficulty and rising costs of finding new deposits, a political, social, and economic framework unusually favorable to mineral exploration and development will have to be provided. Otherwise U.S. reserves will steadily decrease, and rates of production will inevitably decline.

It should be clear from the succeeding chapters that the existence of such a framework is far from assured. Developments of the last two decades are not encouraging:

1. The Mining Law of 1872 is increasingly ill suited to the requirements of modern mineral exploration.
2. Land policy has moved steadily toward restricting access to the public domain for purposes of exploration.
3. Access to those lands that are still legally open to exploration is made difficult by the proliferation of administrative rules and regulations.
4. Requirements for environmental protection have added to costs of exploration and development and have introduced a new element of uncertainty and risk.

5. Certain parts of American mining industry are increasingly vulnerable to competition from abroad and are financially less able to undertake exploration than in the past.
6. Rising levels of taxation add to the problems of all industry, the mining industry included.
7. Some segments of American mining industry must compete in the international marketplace with government-supported operations abroad. Government encouragement of U.S. mining industry, in contrast, is very limited.

The net result of these developments is that during the past two decades the domestic social, political, and economic framework has become less favorable to domestic mineral exploration than in the past. It is therefore difficult to be optimistic about the future of mineral exploration in the United States, and thus about prospects for maintaining present levels of production during the next 20 years. Increasing decay of domestic smelting, refining, and mineral-processing industries is to be expected. U.S. production of metals peaked in 1970 and has been declining irregularly since. In today's environment a reversal of that trend does not seem likely.

In considerable part the developments reviewed above have their origins in change in social attitudes toward the natural resources of the United States (Cameron, 1981; Ostrom, 1983). The change began with the conservation movement of the early decades of the present century. Interest in conservation lapsed during World War II, and in the late 1940s and the 1950s the nation was chiefly concerned with replenishing resources that were seriously depleted by the demands of war and with providing energy and mineral raw materials required for a great expansion of American industry. For a time neither industry nor the general public was much concerned with conservation of mineral resources, and there was little concern over the impact of industrial activity on the environment. This attitude changed rapidly during the 1960s, and an intense concern with conservation and all aspects of the environment arose.

The result of changing social attitudes is a new set of priorities for the use of natural resources. Development of mineral resources has a much lower priority than in the past and often fares poorly in competition with use of lands for wildlife refuges, wilderness areas, wild and scenic rivers, and other scenic and historical attractions. Perhaps the new set of priorities has found its ultimate expression in the massive withdrawals of lands in Alaska, the last great wilderness frontier but also the last great frontier of mineral exploration in America.

To those who are concerned over the deteriorating mineral position of the United States, the new priorities are difficult to reconcile with the continuing large-scale use of minerals in the national economy, with our growing dependence on mineral supplies from abroad, with our growing deficits in international trade, with the loss of some of our mineral-based industries, and with the related weakening of American power and influence in international affairs. Yet the change in priorities is real. Its impact has been felt throughout American mineral industry; it will be reflected in the future availability of mineral supplies from domestic sources.

In the early part of this century, development of mineral resources was considered essential to the economic welfare of the United States. In the 1980s the development of mineral resources seems to be viewed by the general public (with active encouragement of environmental groups) primarily as a source of quick profits to large mining companies, as a source of disturbance of the environment, and as an activity in conflict with more attractive pursuits. The fact that mining industry is basic to the American economy (and the world economy) is not widely perceived. The lessons of the mineral shortages of two world wars are long since forgotten. The energy "shocks" of the 1970s have not even led to design of a national energy policy.

This is not a healthy situation. Agriculture, energy, and mineral raw materials are still the pillars of the economic structure of a nation. If any one of these pillars is weakened, the repercussions will ultimately be felt throughout the structure. In recent years there has been much discussion of a transition of the United States from an industrial economy to a service economy and of the elaborate structure of service "industries" that has been developed. These "industries," however, are not industries at all, in the true sense of organizations that produce the raw materials and manufactured goods on which service *organizations* and all society depend. We will do well to remember that a skyscraper cannot stand securely upon a weak foundation.

On the whole, the results of environmental concerns have been extremely beneficial, and protection of the environment clearly deserves the continuing support of the American people. There are, however, some aspects of the environmental movement that are not healthy and not in the long-term interest of the nation. Perhaps the principal one is the tendency, actively fostered by some environmentalists, to identify industries in general and the mining industry in particular as enemies of society. Mining *has* caused damage to the environment, and in certain areas the damage has been severe. However, since 1970 the industry, in response to public pressure and environmental legislation, has taken effective measures to reduce the environmental consequences of mining and to restore lands after mining to forms that make them useful for other purposes. The abuses that stirred so much anger in the 1960s and early 1970s are largely a thing of the past, and to continue to berate industry for those abuses does not serve the interests of a nation that is heavily dependent on discovery and development of its mineral resources. The mining industry performs a basic function for society, and it deserves the encouragement of the American public. Certainly its activities must be monitored to ensure that they do not prevent achievement of other objectives of society. There should be, however, a careful distinction between monitoring and actions that are merely the placing of obstacles in the way of development of mineral resources.

THE MINING INDUSTRY IN A CONSUMER SOCIETY

During the present century the United States has changed from a producer-oriented economy to one that is strongly consumer-oriented. Producers are now a political minority; they can fare badly in the resolution of issues that affect consumer groups. This has a strong bearing on the health of the domestic mining industry. As we have seen, markets for many minerals have a long history of instability. If mining industry is to have adequate return on investment, losses or lack of profits during hard times must be offset by high profits when times are good. In a consumer-oriented society this is difficult to do. Increases in prices during periods of high demand are fiercely resisted by the consumer, and actions are taken to reduce "excess" or "windfall" profits. If such actions were offset by actions to create profits or at least to reduce losses during times of depression in mineral markets, a reasonable balance might be achieved. No such offsetting actions are taken. The consumer accepts low prices as his due; he has no concern with the consequences of his actions. In the early days of mining in the United States, high profits from successful ventures could be kept by the producer and used to offset losses due to the many failures that are a part of all mineral exploration and development. Heavy taxation of profits has diminished that capability. This is a factor in the attempts of major mining organizations to diversify into other industries in which the risks are less. If the trend continues it can only lead to further decline of American mining industry, to the detriment of those very consumers who have created a climate unfavorable to mining ventures. Producers of minerals in other countries will not be slow to take advantage of shortages of mineral supplies from U.S. sources.

Part of the current American attitude toward mining is a carry-over from the 19th century, when there were spectacular successes in some districts of the West. Mining became identified as a quick source of easy profits. Those days are long since gone, although there was a brief revival during the uranium mining boom of the late 1940s and 1950s. Mining today is a highly competitive industry, in which profit margins are low. It is capital-intensive, yet the profit margins and the long lead times between discovery and first production make it difficult to attract capital funds in competition with other industries in which returns on investment are higher and can be realized in much shorter periods of time.

There is concrete evidence of the declining attractiveness of American metal mining in particular. The 1970s found a number of major American metal-mining companies in weakened financial condition. The major petroleum companies perceived an opportunity to secure control of mining companies on what seemed very favorable terms. A number of oil companies moved into mining industry, including such giants as Exxon, Chevron, Sohio, Atlantic Richfield, and Pennzoil. Most of them have been severely disappointed in the results, and one oil company after another has liquidated its mining interests. Meanwhile, the financial condition of remaining independent mining companies has deteriorated. Deterioration began in the 1970s and accelerated during the recession of the early 1980s. As Table 15-1 indicates, earnings have declined, total debt and debt as a percentage of capitalization have risen sharply, and return on equity, which for many years has been lower for mining than for all other industries, has plunged. The metal-mining industries have not shared in the recovery of 1982–1985. The capital funds necessary for replacement of depreciating plant, for exploration, and for technological advancement can no longer be generated out of cash flow but must be obtained by borrowing. This is the significance of the rise in debt. The latest to feel the effects are the major domestic aluminum producers, which are scarcely represented in Table 15-1. In the face of competition from alumina plants, smelters, and refineries newly established in bauxite-producing countries during the 1970s and early 1980s, the position of the United States as the principal Free World producer of aluminum is being eroded. The United States has become an importer of aluminum.

TABLE 15-1 Composite Data for 10 Leading Independent Metal-Mining Companies: Amax, Asarco, Cleveland-Cliffs, Cominco, Falconbridge, Hanna, Inco, Newmont, Noranda, and Phelps Dodge

	1979	1980	1981	1982	1983
			Millions of Dollars		
Net earnings	$1770	$1872	$539	$(963)	$(778)
Total debt	5086	5814	6773	7876	7744
			Percent		
Debt as percent of capitalization	34%	34%	36%	42%	43%
Return on equity	18	16	5	(9)	(8)

Source: Atchison, S. D., et al. (1984). The table was compiled from company financial reports.

THE STATUS OF MINERAL CONSERVATION

Despite the growing problems of the mineral industry, in a number of aspects of mineral conservation the twentieth century has brought remarkable progress—the discovery and development of new mineral deposits, improvement in technology of mineral extraction, diversification in mineral use, and more efficient use of mineral raw materials. Without that progress, the rise in material standards of living that makes the century unique in history could not have taken place, and deterioration of American metal mining industry would have begun at a much earlier time. Despite the phenomenal enlargement of the scale of mineral use both in the United States and in the rest of the world, technology has created mineral reserves in the world as a whole faster than they have been consumed, and this is a most noteworthy achievement in mineral conservation. Yet conservation in the fullest sense requires not just ability to create new reserves of minerals but advances in the art of managing mineral resources. Here the record is not so bright.

In its management of the petroleum industry in the 1930s, 1940s, and 1950s, government showed what could be done toward conservation of petroleum and natural gas in the United States and how the interests of both industry and the general public could be served in the process. Stabilization of the industry through adjustment of supply to demand was the key to the success of those efforts. In sharp contrast is the treatment of the coal industry. Given the importance of the coal resources of the United States, failure to achieve stabilization of the coal industry is very difficult to comprehend or to condone. Why is there no interstate coal compact through which state agencies could collaborate with one another and with the Federal Government in stabilizing the coal industry and encouraging conservational development? We have public-service commissions that are charged with ensuring orderly generation of heat and electricity from coal and other fuel at reasonable costs. Why do we have no means of ensuring orderly development of the largest and most valuable of all fossil energy resources?

For some years, protection of the environment from damage due to coal mining has been high on the list of national priorities. Management of coal resources to ensure their orderly development should be just as high on the priority list. The United States faces a progressive decline in domestic production of oil and natural gas. The development of oil shale resources is virtually at a standstill. Energy from nuclear fission is being rejected as an alternative source. The potential of solar and geothermal energy is limited. Energy from nuclear fusion is far in the future, and there is no assurance that it will become an economically viable source. Future production of energy is likely to make even heavier demands on our coal resources than in the past. It is essential that coal resources be developed with full attention to conservation. This could be done by a stable, healthy, profitable coal-mining industry. It is not likely to be done if the instability of the past 60 years is allowed to continue.

As we have seen, the coal industry is by no means the only mining industry in which mineral conservation has been hampered by instability of demand and price and by strife between labor and private management. The metal-mining industries have been particularly affected, because they are vulnerable to the vagaries of international markets. The problem is complicated by depletion of higher-grade ores in the United States and by rising costs of discovery, development, and extraction in a period when prices of metals are depressed. In such a situation, as profits dwindle, research and development find only limited support. The exploration needed to provide the metals sources of the future is not done. It may be that the restructuring of domestic mining industry now in progress will alleviate these problems. If not, the health of domestic metal-mining industry will depend on government action and support. Government, in effect, will have to assume a role in mineral resource management, to assure that the economy of the nation rests on an adequate domestic mineral base. We cannot afford to have development of some of the nation's most important mineral resources fall prey to the vicissitudes of the international marketplace. This will mean some difficult decisions with regard to the kinds and amounts of support to be given and their economic consequences. If a decision is ever made to stabilize mining industries, or even just key mining industries, the support structure must be designed so that incentives for continued research and development are maintained. Otherwise stabilization will encourage complacency and perhaps stagnation. In the long run that could prevent the orderly and efficient utilization of the nation's resources that we have defined as the objective of mineral conservation.

On the international scene, stabilization requires concerted international action. Our review of international control schemes indicates the problems and pitfalls involved. Thus so far so little progress has been made that one must conclude that international stabilization of mining industry will not be one of the accomplishments of the twentieth century.

During most of the period since 1934 the United States has been committed to the promotion of free international trade. The resultant intense competition in world markets has had certain positive consequences. It has been a powerful stimulus to research and development of new technology of mineral extraction and use. For the short run it has offered the consumer the lowest possible prices for mineral raw materials. Yet it has its own price. Given the cyclical nature of the world economy, competition in international markets guarantees the recurring instability that has steadily hampered the orderly development of the world's resources. Underlying the pursuit of lowest possible prices there is a premise that deserves some rather sober reflection, the more so because it is so seldom recognized. The premise is that there is no limit to the availability of mineral supplies at reasonable costs and that for this reason we can afford to ignore the loss of resources of subeconomic material that inevitably results from intense competition, particularly in times of depressed prices.

Let us look at just one concrete example of loss of resources. In an earlier chapter, the Steelpoort chromite seam of South Africa was described as one of the world's great mineral deposits. The seam ranges from 38 to 46 inches thick. A few feet above it is the Leader Seam, equal in quality but only 14 to 16 inches thick. A conservative estimate of chromite in the Leader Seam would be 100 million tons, equivalent to a 10-year supply of chromite at the 1980 world production rate. As the Steelpoort Seam is mined, the rocks above it, including the Leader Seam, slowly settle into the workings. To come back later to mine it would be extremely difficult and costly. Conservation would seem to require that the Leader Seam be mined at the same time as the Steelpoort Seam. Attempts have been made to do this, but the added costs of mining and disposing of the waste rock between the two seams have made these efforts uneconomic in what is a highly competitive industry. In effect, the Leader Seam is a lost resource of chromite. Many analogous examples could be cited from the world's mining operations. In our relentless competitive pursuit of the lowest possible prices for today's consumers, we are assuring higher prices and smaller mineral supplies for the consumers of the future.

CONSERVATION: THE ULTIMATE PROBLEM

We have given a good deal of attention to the problem of mineral conservation, but that is only one part of the broader problem of conservation of natural resources, a problem that has been the source of bitter conflict in the United States during the past 20 years. Concern with this problem mounts as population growth brings increasing pressure on our natural resources. In a very real sense we are trying to have our cake and eat it too. On the one hand we value the almost infinite variety with which this country is endowed—its mountains, plains, and prairies; its lush forest regions; its deserts; and its wildlife of remarkable diversity. We see areas where those resources have been destroyed or severely marred. We feel that the progress of such degradation must somehow be arrested. On the other hand, for millennia the human race of which we are a part has been striving to better the conditions of its existence. It has made great progress in the past 5000 years. From an existence that was a simple struggle for survival in a frequently hostile environment, life for some segments of the world population has become life in which comfortable living for long time spans is taken for granted. What was once a hard-earned privilege is now perceived as a right, to such a degree that the struggles of the past are almost forgotten. We forget that success has come because man has learned to use the resources—mineral, forest, wildlife, soil, water, and air—that nature has provided.

The human race has learned the means of utilizing natural resources to modify and control the nature of its environment and the conditions under which it lives. However, it has not learned to control itself. In particular it has not learned to control its capacity for procreation, and this looms as a major problem confronting the human race. Already population growth in many developing nations is negating efforts toward improvement of social and economic conditions. That has not yet happened in the United States, where there is still much open land and where population increase is at a relatively low rate. But surely the problems we are having with environmental damage are related to the growth of our population from 90 million in 1900 to more than 230 million in 1984.

If population growth continues, the struggle to preserve the environment is doomed to failure. It is not realistic to expect that in the United States, or elsewhere, man will abandon his efforts to maintain and improve the conditions of his existence. Despite nostalgic longings, he will not go back to hunting and gathering, or to a primitive agricultural or pastoral existence. He will not stop using large amounts of forest products, mineral raw materials, and mineral sources of energy. Too many human beings would starve or freeze to death. On the other hand, if he will not control his numbers, the environmental damage that concerns him now will be followed by environmental devastation.

Nearly 200 years ago, Malthus predicted a gloomy future for the human race because, he said, population would always expand to the limits of the means of subsistence. The prophecy has thus far been negated in the United States and certain other countries by increases in the efficiency of agriculture and industry, and it has become fashionable to heap scorn on the Malthusian conclusions. Yet there are many countries in Africa, Southeast Asia, and South and Central America where the prophecy has been fulfilled or is very close to fulfillment. Ethiopia is only one tragic example. Sickness, malnutrition, and even starvation are the results. In some of these countries devastation of the environment is already under way. Current aid programs treat the symptoms but have not yet successfully addressed the cause of the malaise. Cultural and religious barriers have thus far proved insuperable. There is something tragic in cultures that call for charity and mercy to the poor and at the same time permit or even encourage the explosion of population that is the root of poverty. One wonders how, in considering human life, sanctity can be divorced from quality.

Malthus was concerned with the pressure of population on the means of subsistence. Now we must be concerned with pressure of population on all the resources of the earth. There is just so much room on earth, no more. Man has acquired the power to occupy as much of the earth as he may choose. He has exercised that power freely and has only recently begun to visualize the end result and to take measures intended to forestall it. Unfortunately our present efforts, though impressive in some immediate results, are pathetically deficient for the future. They are putting the cork in a bottle in which a gas-producing reaction is taking place, without doing anything to stop the reaction that will inevitably blow the cork.

The conclusions presented here have been forced upon the author during nearly 50 years of geological work in Canada, the United States, Mexico, western Europe, Cuba, Madagascar, various countries of Africa, and certain hinterlands of eastern and northeastern Brazil. Three things have become apparent. First, protection of the environment is possible only in a vigorous, progressive, and economically healthy society. Second, such a society can long exist only through wise and efficient use of all its resources. Third, even the most advanced societies now existing on the earth are threatened by the growth of population. Simple preservation is not the answer to protection of the environment, because it flies in the face of human desires and human needs. The answer can only lie in judicious use combined with relief of population pressure. How to achieve this is the ultimate problem of conservation of the resources of the earth.

AT THE CROSSROADS: MINERAL POLICY AND THE NATIONAL SECURITY

As noted earlier, there is a direct correlation between the rise of civilization during the past 8000 years and the increasing diversity and scale of use of minerals for energy and mineral raw materials. In an industrial era, the power, the influence, and the security of a nation depend on assured supplies of food, energy, and mineral raw materials and on possession of an industrial structure capable of converting mineral raw materials into essential manufactured goods. Despite this, in the United States there is no real commitment to maintaining a strong domestic mineral base. This lack is reflected in failure to develop a long-term coherent mineral policy, although the National Materials and Mineral Policy Act of 1980, discussed in Chapter 12, establishes the legal basis for development of such a policy. What mineral policy we have is on little more than a year-to-year basis, and in foreign affairs it is frequently subordinated to other policy considerations. The use of mineral sanctions against Rhodesia, as described in Chapter 12, is a conspicuous example.

Quite different is the attitude of the Soviet Union toward the development of its mineral resources. Almost since the October Revolution there has been a national commitment to the development of a strong mineral base through exploration and development of Soviet mineral resources. Progress was slow prior to World War II, and Russia remained dependent on imports for a number of important mineral commodities, particularly the nonferrous metals. However, after the war an intensive program of mineral exploration was undertaken in both European and Asiatic Russia. By Western standards the effort has apparently not been cost-efficient, but it has been so successful that, as shown in Fig. 7-6, the Soviet Union is nearly self-sufficient in minerals and is an important exporter of oil and natural gas as well as certain nonfuel minerals. At the same time, the Soviets have developed a vertically integrated, broad industrial structure that includes all elements from mining and mineral processing to production of finished goods. The Soviet Union is now able to pursue her adventuristic policy abroad unhampered by concern over interruptions of supply of either mineral raw materials or finished goods.

The Soviet view of the importance of mineral industry is thus in marked contrast to that of the United States. In the United States the importance of an economic activity is evaluated mainly in terms of its percentage contribution to the gross national product. Under such a rating scheme mineral industry ranks low, because if the value of raw minerals and fuels is used in calculating the percentage, as is commonly done, mineral industry contributes only a few percent of the total GNP. This gives a distorted view. Only a small part of the total cost of an automobile is the cost of raw materials. The rest represents costs of manufacturing parts, assembly, finishing, and marketing, and profits at various stages of the process. Yet without the necessary mineral raw materials the automobile would not exist, and without mineral fuels it would never move from the factory to the showroom, much less from the showroom to the road. The value of minerals to society must be measured in terms of the value of all economic activity that is made possible by mineral use, including, for the United States, those industrial activities that are essential to the security of the nation.

A key issue in all attempts to develop a mineral policy is the role of imports in U.S. mineral supply. During the entire period of industrial development in the United States, the mineral supply problem has been conceived as that of meeting whatever demands for minerals are created by an expanding and increasingly versatile technology. If a new product is developed, mineral raw materials for it must be found. If domestic mines cannot furnish the minerals at competitive prices, they must be sought from sources abroad. Under this concept of the problem of mineral supply, demand for minerals is controlled by technology that develops without reference to availability of minerals from domestic resources. Pursuit of this concept has led automatically to inadequacy of domestic mineral supplies. One has only to match the current demands of technology against U.S. reserves of minerals to see that self-sufficiency is impossible so long as technology controls demand for minerals.

In meeting the demands of technology, international market forces have been accepted as the prime determinant of the ratio of supply from domestic mines to supply through imports. In effect, a high priority has been set on the acquisition of mineral supplies at lowest possible costs, regardless of their source. The maintenance of the freest possible international trade has therefore been a cardinal objective of U.S. foreign policy for many decades. Thus far the United States has accepted the consequences of this in the decline of important segments of domestic mining and processing industries and the transfer of mineral-based industries to other countries. With this erosion of the mineral and industrial basis of the security of the nation, there has been a progressive weakening of U.S. power and influence in international affairs.

There has been much discussion in recent years of the need for a strong defense capability, so that the United States can negotiate with the Soviet Union from a position of strength. That position is discussed mainly in terms of numbers of guns, missiles, tanks, airplanes, ships, and other weapons available to the armed forces. There is too little discussion of the broad foundation of military power—secure domestic supplies of mineral and other raw materials and possession of an industrial structure capable of converting raw materials into military equipment. Stockpiles of minerals would help in an emergency, but they are at best a poor substitute for supplies of minerals from domestic sources and for the mineral-based industries that develop and flourish out of domestic mining activity. Stockpiles do nothing to check the transfer of mineral-based industries to other countries.

In many quarters deterioration of the mineral position and industrial strength of the United States is accepted as inevitable, so that the United States must stake its future on the availability of mineral materials and many essential manufactures from the remainder of an unstable world. One wonders why. There is no natural law requiring that demand be controlled by an unrestrained technology. Why should we not tailor our technology, so far as possible, to the availability of minerals from domestic sources? Must we rely on market forces and allow price to be an overriding policy consideration? The energy problem is an example of what happens when we do this. We have huge energy resources in the United States, but it has been easier and in the short run cheaper to import petroleum and natural gas than to learn to use our far larger resources of coal and oil shale, not to mention fissionable materials, economically and acceptably. Can we really afford the easier course? Imports of oil are a palliative, not a cure. They are a postponement of the energy problem, at the price of insecurity and increased difficulties with our international balance of trade. They have retarded research and development of our own large energy resources and are setting the stage for future energy crises. The same is true for certain of our deficient metals.

The flight of mineral-based industries is one of the more serious problems facing the United States. It is unlikely to be checked so long as American industry is heavily dependent on use of minerals that are not supplied, or are inadequately supplied, from domestic mineral deposits. Dependence cannot be relieved in terms of the present U.S. approach to its mineral problems. However, there is an alternative approach, an approach that would combine active encouragement and support of domestic mining industry with adaptation of industrial technology, so far as possible, to the use of domestic mineral resources.

U.S. policies affecting mineral industry should be designed to stimulate discovery and development of new domestic mineral deposits, to encourage improvement in the technologies of mineral extraction and use, including recycling, and to eliminate uses of strategic minerals that are unnecessary or even frivolous. Changes in the policies and procedures of land management are required. Tax and other incentives must be provided. Effective environmental regulation is essential, but it should be more efficient, less costly, and less time-consuming. Stabilization of domestic mining industry should be a major policy objective; this means modifying trade policy with respect to minerals and devising means of adjusting domestic supply to domestic demand.

Technology should be adapted toward employment of new industrial materials that will make maximum use of nonmetallic minerals abundantly available from domestic deposits. The aim should be to reduce the use of metals and the relatively few nonmetallic mineral commodities in which the United States is deficient. In this way basic industries that are based on mineral imports would be changed, so far as possible, into industries that are based solidly on domestic mineral resources. This is hardly an idle dream; through materials science substantial progress toward this end is already being made or is within reach. Some of the progress has been described in an earlier chapter—the use of ceramic materials and composites that will perform functions usually assigned to metals, the use of fiber optics to replace copper in communication systems, the partial replacement of cobalt and chromium by less strategic metals. These are examples of the kinds of things that need to be done. It is unlikely that use of the strategic metals that are the core of U.S. dependence on mineral imports could be eliminated entirely. However, by adaptation it could be reduced to more manageable proportions for metals for which domestic reserves are completely or almost completely lacking. As to the rest, provided discovery and development of new domestic deposits are vigorously pursued, domestic demand might be adjusted to the productive capacity of domestic mines. There is much written today about replacing heavy industry by industries based on high technology, by which the electronic and related industries are generally understood. But materials science is also high technology, technology that could enable the United States to maintain the kind of broad industrial base that is essential to a strong economy and to national security.

If the alternative approach considered here were adopted, it is possible that the costs of some consumer goods would rise. But what is the economic cost of transfer of mining and mineral-based industries to other countries in terms of reduced employment, loss of the contributions of those industries to the economy, decreased tax revenues, and diminished breadth and strength of the nation's industrial base? And what is the cost in diminished national security?

The alternative approach suggested here could not be implemented overnight; many years of directed effort would be required. The effort would have to be supported by a long-term, coherent mineral and materials policy. It would be no easy undertaking, but it offers an alternative to progressive deterioration of the mineral position and industrial strength of the United States. Complete self-sufficiency for the United States is probably unattainable, but a much nearer approach to it seems possible if we discard the working principles that have led to declining self-sufficiency in the past. There might or might not be a decreased versatility of American technology and industry, but certainly there would be a reduced fragility and a reduced vulnerability to the hazards of a divided world. Certainly we have never given the alternative a trial. We have not addressed wholeheartedly the problem of development and use of our national mineral resources. Thus far it has been easier and cheaper to go abroad for deficient minerals. It may be true that this approach has served our interests in the past, but perhaps it is time to appraise its potential future costs and to consider whether there is not, indeed, a viable alternative that gives more assurance of future national security and economic strength. In adopting it, the United States would be assuming leadership along the path that the world must ultimately take as global reserves of the less abundant minerals are progressively depleted.

The alternatives discussed above are the reasons for the title of this book. At a crossroads one must choose which path to follow. We stand now at a crossroads along the path of our economic and industrial development. We can choose to continue as before, following the road of progressive deterioration in our mineral position and increasing dependence on sources of dubious reliability abroad, or we can take a new road that offers promise of an improved position. The economic and technologic resources accumulated in this nation since the turn of the century are without parallel in the history of the world. Surely we can use them to alleviate our mineral problems.

REFERENCES

Atchison, S. D., Crawford, M., Houston, P., Norman J. R., and Ryser, J., 1984, Death of mining. *Business Week,* December 17, 1984, pp. 64–80.

Cameron, E. N., 1981, Changes in the political and social framework of United States mineral resource development, 1905–1980. In *Economic Geology, 75th Anniversary Volume,* B. J. Skinner, ed., pp. 955–964.

Ostrom, M. E., 1983, Two centuries of mineral policy in Wisconsin. In *International Minerals,* A. F. Agnew, ed., Westview Press, Boulder, Colo., pp. 21–36.

Author Index

Alexandersson, G., 106, 142
Allsman, P. T., 202
Alter, H., 202
Arthur Andersen & Co., 62
Arthur D. Little, Inc., 233, 236
Atchison, S. D., 290, 300
Atwood, G., 236
Auburn, F. M., 274, 284
Aus, R. M., 249, 260
Averitt, P., 45, 79

Bailly, P., 206, 208, 210, 219
Barney, G. O., 160
Barton, P. B., 232, 236
Bates, R. L., 106
Bernier, L., 283, 284
Bilder, R. B., 284
Blair, J. M., 272
Borden, G. S., 238, 245
Bosson, R., 19, 20
Brobst, D. A., 14, 20, 106, 142, 183
Brock, S. M., 236
Brooks, D. B., 202
Bundtzen, T. K., 219
Burton, I., 202

Cameron, D. E., 150, 151
Cameron, E. N., 16, 17, 20, 79, 163, 183, 284, 287, 300
Carpenter, R. A., 236
Cathcart, J. B., 282, 284
Clarfield, K. W., 272
Committee on Foreign Relations, U.S. Senate, Subcommittee on African Affairs, 258, 260
Committee on Interior and Insular Affairs, U.S. House of Representatives, Subcommittee on Mines and Mining, 258, 260
Congressional Research Service, 260
Conrad, R. F., 245
Cuff, D. J., 79
Curlin, J. W., 260

Dayton, S., 216, 218, 219
Dempsey, S., 204, 212, 219
Department of Agriculture, 219
Department of Commerce, 36
Department of Energy, 69, 70, 201
Department of the Interior, 220, 224, 227, 228, 231, 236
DeYoung, J. H., Jr., 117, 143, 184, 233, 236, 242, 243, 245, 276, 284
Dulaney, T., 232, 236
Duncan, D. C., 66, 67, 79

Eckes, A. E., Jr., 260
Elliott, W. Y., 264, 272
Ellis, P. L., 203
Ely, N., 211, 220, 243, 245
Emigh, G. D., 95, 107
Energy Information Administration, 39, 43, 45, 46, 52, 53, 54, 55, 56, 63, 79, 237
Energy Research and Development Administration, 14, 16, 69
Engineering and Mining Journal, 71, 216, 220, 239, 245, 250, 260

Fantel, R. J., 145, 160
Fischman, L. L., 160
Flawn, P. T., 20, 203, 220, 235, 237

General Accounting Office, Energy and Minerals Division, 245
General Services Administration, 261
Goeller, H. E., 78, 79, 203
Gordon, E., 14, 16, 20
Gordon, R. L., 261
Govett, G. J. S., 160
Graham, A. P., 239, 245

Granville, A., 284
Griffitts, W. R., 16, 20

Hawley, C. C., 217
Hewett, D. F., 172, 184
Holland, T. H., 252, 253, 261
Hooker, J. R., Jr., 253, 261
Horn, D. R., 276, 284
Hubbert, M. K., 56, 57, 58, 59, 79, 147
Hubred, G., 284
Huddle, F. P., 261

International Tin Study Group, 272
Interstate Oil Compact Commission, 195, 203

Johnson, W., 228, 237

Kellogg, H. H., 79, 203
Kilgore, 116, 143
Krauskopf, K. B., 2, 20
Kulcinski, G. L., 75, 79

Lacy, W. O., 37
Landsberg, H. H., 203
LeFond, S., 107
Leith, C. K., 203, 262, 272
Lemons, J. F., Jr., 119, 143, 160
Leontief, W., 151, 156, 160
Liedtke, J. H., 250
Lovejoy, W. F., 203
Lutjen, G. P., 237

McGraw-Hill, Inc., 37
McKelvey, V. E., 14, 17, 20, 184, 201, 203, 275, 276, 284
Mackenzie, B. W., 243, 245
McRory, R. E., 64, 79
Malenbaum, W., 151, 156, 160
Maley, T. S., 220
Marsh, S. P., 214, 220
Mason, B., 97, 107
Meadows, D. H., 160
Mero, J. L., 284
Merrill, C. W., 37
Mining Engineering, 189, 203
Molotch, M. L., 237
Moore, J. R., 284
Morgan, J. D., Jr., 166, 168, 169, 184, 255, 261

National Academy of Sciences, 20, 59, 70, 79, 147, 160, 161, 184, 188, 203
National Advisory Committee on Oceans and Atmosphere, 280, 284
National Bureau of Standards, 259, 261
National Commission on Materials Policy, 248, 250, 261
National Geographic Society, 79
National Materials Advisory Board, 261
National Resources Board, 186, 203
Norton, J. J., 16, 20

Office of Technology Assessment, 215, 220
Ohle, E. L., 37, 230, 231, 237
Oil and Gas Journal, 237
Ostrom, M. E., 287, 300
Oxman, B. H., 284

Pehrson, E. W., 250, 261
Peterson, G. R., 131, 143
Phizackerley, P. H., 79
Poland, J. F., 237
Post, A. M., 284
President's Materials Policy Commission, 184, 248, 261
Pretorius, D. A., 138, 140, 143, 147, 161

Reagan, R., 258, 261
Rickard, T. A., 37, 184
Ridge, J. D., 197, 203
Ridker, R. G., 151, 156, 161
Riggs, S. R., 282, 284
Risser, H. S., 237
Roberts, W. A., 245
Rose, J. G., 190, 203
Ross, J. R., 143
Rosta, J., 143
Rowe, R. B., 23, 37
Ruedesili, L. C., 79

Schmidt, R. A., 45, 79
Shanks, W. C., III, 284
Skinner, B. J., 143, 160, 161
Smith, G. H., 203
Snow, G. G., 37
Sorem, R. K., 276, 277, 278, 284, 285
Spencer, V. E., 163, 184
Squires, A. M., 79
Steinhart, E. C., 79
Strauss, S. D., 233, 237

Tennessee Valley Authority, 201
Tilton, J. E., 161
Time, 92
Toombs, R. B., 20
Trainer, F. E., 184

U.S. Bureau of Mines, 13, 15, 20, 39, 82, 93, 95, 96, 98, 99, 102, 107, 113, 118, 120, 123, 125, 129, 131, 132, 133, 134, 135, 139, 143,

146, 151, 152, 153, 154, 155, 156, 159, 161, 163, 164, 165, 166, 167, 168, 169, 170, 173, 175, 178, 179, 181, 182, 184, 192, 203, 229, 233, 237
U.S. Geological Survey, 13, 15, 20, 59, 62, 201
U.S. News and World Report, 285
University of Wisconsin, 75

Van Rensburg, W. C. J., 245, 261, 272

Wall Street Journal, 62, 233, 237
Weinberg, A. M., 203
Wells, H., 232, 237

Zellers, M. E., 282, 285
Zwartendyk, K. J., 16, 20

Subject Index

Abrasives, 109–111
  corundum, 91
  diamond, 90–91
  emery, 89
  garnet, 91
  quartz sands, 89
  synthetic, 92
Abyssinia, 252
Acquired lands, 205
Adaptation, *see* Technology, adaptation of
Africa, phosphate, 94
Ajo, 233
Alabama, iron ore, 114
Alaska:
  allocation of lands, 217
  chromite, 119
  fisheries, 275
  gold, 137
  lead deposits, 132
  mineral potential, 216–218
Alaska Coal Lands Leasing Act, 187, 211
Alaskan National Interests Land Act, 217
Alaska Native Claims Settlement Act, 205, 217
Alaska Native Regional Corporations, 205
  allocation of lands to, 217
Alaska Statehood Act, 205, 217
Albania, chromite production and reserve base, 120
Alberta tar sands, sulfur in, 102. *See also* Heavy oils
Algeria, 67
  phosphate, 94
  production and reserve base, 95
Almadén mercury deposits, 135
Alsace, potash, 97
Aluminum, 78, 129–131, 155
  alternative sources, 131
  energy requirements, 130, 131, 189
  extraction from bauxite, 130
  recycling, 131, 192
  world production, 131, 146, 154, 155
  see also specific countries
American Gas Association, 51, 63
American Petroleum Institute, 50, 52, 63
American Revolution, 8
American Universities Field Staff, 253
Ammonia, 93
Anatolia, 86
Andes Copper Corporation, 18
Anglo-Iranian Oil Company, 252
Angola, diamonds, 90, 268
  offshore oil, 273
Anthracite, 41, 42
Antimony, 78, 137
  world production, 146, 154, 155, 159
  world reserve base, 146, 159
  see also United States
Appalachian Regional Development Act, 227
Appalachians, chromite, 119
Appalachian states, coal, 41
Arab oil embargo, 58, 59, 252
Araxà, phosphate, 95
Argentina, jurisdiction over continental shelves, 273
Arizona, 10
  manganese deposits, 116
  mining law, 204
Arkansas, manganese deposits, 116
Arsenic, 137
ASARCO, Tacoma smelter, 233
Asbestos, 104
  world production, 146, 153, 155, 159
  world reserve base, 146, 159
Atacama desert, 93, 100
Athabaska, *see* Heavy oils
Atlantis Deep, 274
Atomic Energy Commission, 69
Aurora district, 280
Australia, 173, 271
  diamonds, 90, 268

Australia (*Continued*)
  iron ore, 112, 114
  kaolin, 87
  offshore oil, 273
  phosphate, 94
  production and reserve base:
    bauxite, 132
    coal, 43, 46
    copper, 129
    iron ore, 113
    lead, 133
    manganese ore, 117
    nickel, 123
    silver, 139
    tin, 135
    zinc, 134
  uranium, 70
    production, 71
Availability and price, 58, 148

Ball clays, 87
Baltimore Canyon, 59
Baring Bros., 263
Barite, 88
  world production, 146, 153, 155, 159
  world reserve base, 146, 159
  see also United States
Barium, 78
Basic oxygen process, 110
Bauxite, 4, 129–130
  world production and reserve base, 131, 146
Beaufort Sea, 59
Bedford limestone, 83
Belgian Congo, 265
Belgium, 279, 280
Bentonite, 78, 87–88
Beryllium, 78, 137
  resources, 16
Bessemer process, 110
Bingham, 124
Bismuth, 137
Bituminous coal, 41–42
Bituminous Coal Act, 198
Bituminous Coal Conservation Act, 198
Blake Plateau, 280
Bolivia, 266, 267
  production and reserves:
    tin, 135
    silver, 139
Bonneville Power Administration, 130
Bootleg oil, 195
Boron, 99, 100, 155, 156
  world production, 146, 153, 155
  world reserve base, 146
  see also United States
Botswana:
  diamonds, 268
  nickel production and reserve base, 123
Boycotts, *see* Mineral sanctions
Brazil:
  diamonds, 90
  ferrochrome, 173
  iron ore, 112
  manganese ore, 116
  nickel laterites, 98
  potash, 98
  production, reserve base:
    bauxite, 133
    chromite, 120
    iron ore, 112
    manganese ore, 117
    silicon, 118
    tin, 135
Breeder reaction, 69
Brick, 81, 86
British Columbia, molybdenum, 125
Bromine, 78, 99, 100
Buffer stock, tin, 267
Building Stone Act, 210
Bunker Hill and Sullivan Co., 200
Bureau of Land Management, 205, 207, 217, 218
Bureau of Land Management Conservation System, 217
Burgan oil field, 50
Burma, tin production and reserve base, 135
Bushveld Complex:
  chromite, 120
  vanadium, 123, 124
Butte:
  history of, 171
  labor strife, 200

Cabinet Council on National Resources and Environment, 259
Cadmium, 78
  world production, 154
California, 85, 104
  asbestos, 104
  gold, 140
  mercury, 136
  mining law, 204
  talc, 103
Cameroon, oil, 270
Canada, 62
  aluminum production, 131, 132
  asbestos, 104
  copper, 266
  heavy oil, 64–65
  lead deposits, 132
  peat, 40

phosphate, 94
platinum production, 142
production, reserve base:
copper, 129
iron ore, 113
lead, 133
molybdenum, 125
nickel, 123
potash, 98
silver, 139
zinc, 134
silicon production, 118
sulfur, 102
uranium, 70
production, 71
Candu reactor, 69
Carajas, 114
Cartels in minerals, 248, 262–263, 265–272
basic requirements, 269
definition, 263
evaluation, 272
examples:
bauxite, 271
copper, 264–266, 271
diamonds, 266–268
iron ore, 272
petroleum, 269–271
tin, 266–267
Carter, President, 47
Caspian Sea, 99
Cement materials, 83–85
gypsum, 84
lime, 84
natural cement, 84
portland cement, 84
world production, 146, 152, 155, 158
world reserve base, 146, 158
see also United States
Central Africa, uranium, 71
Central America, sulfur, 101
Centralia, coal fires, 227
Central Selling Organization, 268
Central states, coal, 41
Ceramic materials, 88–90
feldspars, 88–89
world production, 146, 153, 155, 158
nepheline, 88
quartz, 89
see also Clays; United States
Cesium, 78
Challenger Expedition, 275
Chase Manhattan Bank, 219
Chevron Resources, 232
Chile, 18, 167, 265, 271
fishing rights, 273
lithium reserves, 137
nitrate, 93
production, reserve base:
copper, 129
molybdenum, 125
silver, 139
China:
aluminum production, 131
coal production, 46
reserves, 43
petroleum, 270
reserves, 53
production, reserve base:
iron ore, 113
manganese ore, 117
phosphate, 95
tin, 135
tin, 267
tungsten, 125
Chromite, 119–121
deposits, 119–120
world production, 120, 146, 155, 158
world reserve base, 120, 146, 158
see also specific countries
Chromium, 118–120
world production, 154
Clarion-Clipperton zone, 275, 276, 280
Clays, 81, 85–88
ball clays, 87
bentonites, 87
fireclays, 87
kaolin, 86–87
minerals, 86
world production, 146, 152, 155, 158
world reserve base, 146, 158
see also United States
Climax mine, 124
Coal, 12–14, 38–48
composition, 41–42
deposits, 40–41
environmental damage from, 222–223
formation of, 39–40
gasification of, 47
heating value, 41
liquefaction of, 48
recovery percentage, 43, 45
resources and reserves, 42–45
as source of other fuels, 47–48
world production, 46
world reserves, 47
see also specific countries
Coal Act, 210
Coal mine fires, 227
Coast Ranges, chromite, 119
Cobalt, 78, 126
in deep-sea nodules, 275–276
on seamounts, 281

Cobalt (*Continued*)
 world production, 146, 154, 155, 158
 world reserve base, 146, 158
 see also specific countries
Coeur d'Alene district, 139
Cold Lake, *see* Heavy oils
Colombia:
 fishing rights, 273
 oil, 270
Colorado Plateau, vanadium, 123
Common Varieties Act, 211
Comptoir Escompte, 263
Comstock Lode, 138
Congo (Brazzaville), 283
Congressional Research Service, 258
Connally Act, 195
Conseil Intergovernmental de Pays Exportateurs de Cuivre, 271
Conservation, *see* Mineral conservation
Conservation movement, 186–187
Constructional materials, 82–87. *See also* United States
Continental Shelf Convention, 274
Continental shelves, jurisdiction over, 273–274
Copper, 127–128
 in deep-sea nodules, 275–276
 deposits of, 10, 12, 127–128
 exploration for, 128
 production, world, 129, 146, 154, 155, 159
 reserve base, world, 129, 146, 159
 see also specific countries
Copper Export Association, 264–265
Copper Exporters, Inc., 265–266
Cornwall, tin, 267
Costa Rica, fishing rights, 273
Crandon deposit, 234
Creighton mine, 31
Crust of the earth, 2
 composition of, 3
Cuba, nickel production and reserve base, 123
Czechoslovakia:
 coal production, 46
 kaolin, 87

Dead Sea:
 bromine, 100
 potash, 98
Death Valley, 100
Deep Ocean Mining Environmental Study, 276
Deep Seabed Hard Minerals Act, 280
Defense Production Act, 251, 255
Department of Agriculture, 218
Department of Commerce, 36
Department of Defense, 205
Department of Energy, 36, 48, 69, 70, 201
Department of the Interior, 227
Depletion allowance, 241–242
Depletion of mineral deposits, definition, 47
Deposits, *see* Mineral deposits
Destin Dome, 59, 61
Deuterium, 71
Development of mineral deposits, time lag, 19–20
Diamond Research Corporation, 268
Diamonds:
 cartel in, 266–268
 see also Abrasives
Diatomite, 104–105
 world production, 153, 155, 158
 world reserve base, 158
 see also United States
Discovery of mineral deposits, prospects for, 200–202
Dismal Swamp, peat, 40
Dominican Republic, nickel laterites, 121
Ducktown, 222
Duluth gabbro:
 exploration of, 234
 nickel deposits in, 122

Eastern Europe, coal reserves, 43
East Germany, coal production, 46
East Texas oil field, 50
Ecuador, fishing rights, 273, 274
Egypt, oil, 270
Eisenhower, President, 250
Elements, abundance in crust, 2–3
Elk Point, 50
El Salvador, fishing rights, 273
Embargoes, *see* Mineral sanctions
Embargo on oil, *see* Arab oil embargo
Endangered Species Act, 219
Energy, *see* Coal; Geothermal energy; Natural gas; Nuclear energy; Petroleum; Solar energy
 outlook, 76–77
 systems, 77–78
Energy costs, nonfuel minerals, 189
 impact on conservation, 190
Energy Information Administration, 43, 45, 46, 52, 53, 54, 55, 56
Energy Research and Development Administration, 69
 estimates of uranium resources, 14, 16
Energy systems, 77–78
 minerals and elements used in, 78
England, *see* United Kingdom
Environmental damage, 294–295
Environmental impact of mining, 221–231
 areas affected, 227–229
Environmental protection, 288–289
Environmental Protection Agency, 45, 232

Environmental regulation of mining, 231–232
  and conservation, 190, 233–234, 235–236
  costs of, 232–233
Ethiopia, invasion of, 252
Europe, coal deposits, 43
European Economic Community, 249
  reliance on imports, 167, 168
Exclusive Economic Zone, 280, 281

Faber process, 93
Falkland Islands, 273
Federal Emergency Management Agency, 256
Federal Land Management Policy Act, 210, 211, 214
Feldspar, 88–89
  world production, 146, 153, 155, 158
  world reserve base, 146, 158
  see also United States
Ferroalloy metals, 115–126. *See also specific metals*
Ferrochromium, 119, 254
  sanctions on, 254
  in Southern Rhodesia, 254
  tariff, 249
Ferrosilicon, 117
Fertilizer minerals, 92–98
  use in agriculture, 92–93
  see also Gypsum; Lime; Nitrogen and Nitrates; Phosphate; Potash
Fiji Islands, phosphate, 283
Finland, chromite production and reserve base, 120
Fireclays, 87
Fishing rights, 273, 274, 275
Fish and Wildlife Service, 217
Fixed carbon, 41
Flask, of mercury, 3
Florida, 85
Florida phosphate deposits, 94, 96, 223, 283
  fluorine in, 103
Fluorine, from phosphate rock, 191
Fluorine and fluorspar, 102–103
  world production, 146, 153, 155, 158
  world reserve base, 146, 158
  see also Mexico; South Africa; United States
Fossil fuels, *see* Coal; Natural gas; Oil shale; Petroleum
France, 279, 280
  iron ore production and reserve base, 113
  potash production and reserve base, 98
  silicon production, 118
  Super Phenix reactor, 69
  uranium production, 71
Frying Pan phosphate district, 282
Fuller's earth, 87

Gabon, 54
  iron ore, 112
  manganese ore, production and reserve base, 116, 117
  phosphate, 282
  potash, 98
Gallium, 78
Garnet, 78
Gas, *see* Natural gas
Gates of the Arctic National Park, 218
General Agreements on Tariff and Trade, 247, 248, 251
General Leasing Act, *see* Mineral Leasing Act
General Services Administration, 256
Geological mapping, 22
Geophysical prospecting, 22
Georges Bank, 59, 275
Georgia:
  kaolin deposits, 86, 131
  manganese deposits, 116
  marble, 83
Geothermal energy, 73–76
Germanium, 78
Germany, synthetic petroleum, 47
Getty Oil Company, 64
Geysers area, California, 74
Ghana, manganese ore, production and reserve base, 117
Giant oil fields, 50
Glacial lakes, 40
Glass sands, 89
Gogebic Range, 114
Gold, 78, 137–141
  deposits, 140–141
  properties, 1, 138
  recent discoveries, 202
  reserves, 140
Gold Coast, diamonds, 268
Goldfield, 139
Governors' Conference, 186
Grade of ore, 7
Graphite, 78, 104
Gravel, *see* Sand, gravel, and stone
Great Britain, *see* United Kingdom
Great Plains gasification project, 47
Great Salt Lake, potash, 98
Greece:
  bauxite production and reserve base, 133
  nickel laterites, 121
Green River formation, 100
Growth rates:
  U.S. nonfuel mineral consumption, 156
  world nonfuel mineral consumption, 155
Guano, 95
Guatemala, nickel laterites, 121
Gulf of Alaska, 59
Gulf Coast, sulfur, 101

Gulf of Mexico, petroleum, 59, 60
Guyana, 271
Gypsum, 84, 96, 99
  world production, 146, 152, 155, 158
  world reserve base, 146, 158
  see also United States

Hafnium, 78
Haiti, 271
Halite, *see* Salt
Healy, coal, 216
Heavy oils, 64–65
  in Alberta and Saskatchewan, 64
  in United States, 65
  in Venezuela, 65
Henderson mine, 124
Hitler, A., 252
Holland, 280
Holland, T. H., sanctions proposed by, 252–253
Hong Kong, 174
Hubbert, M. K., 56–57, 147
  estimates of U.S. recoverable reserves of petroleum, 58
  forecasts of U.S. petroleum production, 57
Hull-Rust-Mahoning mine, 114
Hungary:
  bauxite production and reserve base, 133
  manganese ores, production and reserve base, 117
Hunker Creek, 141
Hydrogen, 78

Idaho, phosphate, 94
Illinois Geological Survey, 199
Import quotas, 249, 250
India, 112
  coal production, 46
  ferrochrome, 173
  production and reserve base:
    bauxite, 133
    chromite, 120
    iron ore, 113
    manganese ore, 117
  silver, 140
Indium, 78
Indonesia, 53, 269, 271
  nickel production and reserve base, 123
  petroleum, 54
  tin, 267
    production and reserves, 135
Industrial Revolution, 43, 114
Institute for Economic Analysis, 151
International Bauxite Association, 271
International Law Commission, 274
International Monetary Fund, 250
International Nickel Co., 234
International Seabed Authority, 279
International Tin Agreement, 266, 267
International Tin Council, 266
International Trade Commission, U.S., 250
International Trade Conferences, 247
International Trade Organization, 248
Interstate Oil Compact Commission, 195
Iran, 54, 269
  nickel laterites, 121
  sulfur, 101
Iraq, 54, 269
Ireland, peat, 39
Iridium, *see* Platinum-group metals
Iron, 78
  cast, 110
  pig, 110
  wrought, 110
Iron formations, 111–112
Iron ore, 112–115
  world production, 113, 146, 155, 158
  world reserve base, 113, 146, 158
  see also specific countries
Iron and steel, 109–115
  industry, U.S., 111–112
  technology, 110–111
Israel:
  phosphate deposits, 94
  phosphate production, 95
  potash production and reserve base, 98
Italy, 72, 279
  geothermal energy, 75
  potash production and reserve base, 98
  sanctions against, 252
  silicon production, 118
  sulfur, 101

Jamaica, 167, 171
  bauxite production and reserve base, 133
Japan, 72, 174, 252, 254, 279, 280
  geothermal energy, 75
  offshore mining, 273
  pyrophyllite, 104
  reliance on imports, 167, 168
  silicon production, 118
  silver production and reserve base, 139
  steel production, 111
  sulfur, 101
Johnson, President, 256
Jordan:
  phosphate production and reserve base, 94–95
  potash production and reserve base, 98
Juan de Fuca Ridge, 280

Kansas, severance tax, 240
Kaolin, 85–86
Kaolin minerals, 86

Karaboghas Gulf, 99
Kennecott Copper Co., 231
Kennedy, President, 248
Kentucky:
  ball clays, 87
  severance tax, 240
Kidd Creek, 128, 139
Kimberley district, 90, 91, 267
Kiruna-Gällivare iron ore deposit, 114
Kola Peninsula, phosphate, 94
Kuwait, 50, 54, 269

Labrador, iron ore, 112
Ladysmith copper deposit, 231, 234
Lake Superior district, 112, 223
Land use, United States, 229
Laterite deposits of nickel, 121, 122
  of aluminum, 130
  of iron, 122
Law of the Sea, Conferences on, 279
  Treaty, 279
Lead, 78, 109, 132–133
  recycling of, 132, 192
  world production, 146, 154, 155, 159
  world reserve base, 133, 146, 159
  see also specific countries
League of Nations, 252
Lee Creek, 94
Lesotho, diamonds, 268
Liberia, iron ore production and reserve base, 113
Libya, oil reserves, 53, 54
Lignite, 41, 42, 44, 222
Lik deposit, 216
Lime, 81, 84
  world production, 146, 152, 155, 158
  world reserve base, 146, 158
  see also United States
Lithium, 16, 100, 136–137
  in Atacama Desert, 100
  in brines, 100
  in nuclear fusion, 71
  resources of, 16
  at Silver Peak, 100
  see also United States
Little Rock, 130
Lode claims, *see* Mining claims
Long ton, 3
Louisiana:
  severance taxes, 240
  sulfur, 101
Louisville, 84

McKittrick oil field, 64
Magma, 4
Magnesium, 136
  world production, 146, 154, 155, 158
  world reserve base, 155, 158
  see also United States
Magnesium compounds, 99, 136
Malaysia, tin, 267
  production and reserve base, 135
Malthus, T. R., 294
Manchuokuo, invasion of, 252
Manchuria, iron ore, 112
Manganese, 78, 116–117
  in deep-sea nodules, 275–276
  ore:
    world production, 117, 146, 158
    world reserve base, 117, 146, 158
  tariffs on, 247
  world production, 154, 155
  see also specific countries
Manganese nodules, 33, 278–279
  importance of, 277
  metals in, 275–277
Marine mining, 273, 277–278, 279, 283
  jurisdiction over, 272–275
Marquette Range, 114
Mauretania, 271
Mercury, 78, 135–136
  world production, 135, 146, 154, 159
  world reserve base, 146, 159
Mesabi Range, 113, 114, 243
Metal mining industry, United States, condition of, 290
Metals:
  history of use, 108–109
  use in relation to price, 148
  value, 109
  see also specific metals
Metric ton, 3
Mexico, 62, 174, 270
  fluorspar, 103
  manganese ores, 116
  petroleum, 53, 270, 273
  production and reserve base:
    lead, 133
    manganese ore, 117
    molybdenum, 125
    silver, 139
    zinc, 134
  sulfur, 101
Mica, 78
Michigan, 85
Middle East:
  natural gas reserves, 53
  petroleum reserves, 53
Milwaukee, 84, 173
Mineral abrasives, *see* Abrasives
Mineral-bearing land, value of, 231
Mineral conservation:
  in coal industry, 197–199
  current status in United States, 291–292

Mineral conservation (*Continued*)
  definition, 186
  impact of economic instability on, 194–199
  international, 292–293
  in nonfuel mineral industries, 199–200
  in petroleum industry, 195–197
Mineral conservation practices:
  adaptation of design, 193–194
  efficient mineral use, 190–191
  improved extractive technology, 188–190
  recycling, 191–193
  substitution, 193–194
Mineral consumption, world, 145
  per capita, 149–150, 151
  see also specific countries and commodities
Mineral deposits:
  definition, 3
  kinds, 3–6
  sizes, 5
  shapes, 5, 6
  valuation, 19–20
  value, 18–19
Mineral development in United States:
  environment of, 286–287
  social attitudes toward, 287–289
Mineral exploration, 21–24
  geochemical prospecting, 22
  geological mapping, 22
  geophysical prospecting, 22, 23
  history of, in United States, 21
  importance of, in mineral conservation, 187–188
  need for, 218–219
Mineral industry, nature, 36
  and national security, 37
  role in U.S. economy, 36, 296
  transfer of, 172–174
  in United States, history of, 162–166
Mineral Leasing Act, 187, 211
Mineral materials processing, 34–35
  concentration, 34
  smelting, 35
Mineral policy, 258, 259, 295–299
  in Soviet Union, 295–296
Mineral production, world, 146, 152–159
  per capita, 150, 151
  see also specific countries and commodities
Mineral reserves, 9–12
  classes of, 11–12
  estimation of, 9–12
  importance of estimates, 8–9
  incompleteness of estimates, 12
  see also specific countries and commodities
Mineral-resource potential:
  of Alaska, 216–218
  assessment of, 214–215
Mineral resources, 13–17
  classes of, 13–15
  problems of estimation, 14, 16–17
  value, 18–20
Mineral rights, 204, 205
Minerals:
  availability of, 8–12, 35, 144–160, 179–183
  definition, 1, 2
  rates of production, 147
  sources of, 2–8
  see also Nonfuel minerals
Mineral sanctions, 252–254
  boycotts, 252
  embargoes, 252
  against Italy, 252
  under League of Nations, 252
  against North Korea, 253
  against Southern Rhodesia, 253–254
  of T. H. Holland, 252
  under United Nations, 253–254
Mineral stockpiles, 254–258
  selection of materials for, 256–257
  of United States, 257
  U.S. expenditures on, 255
  use for influencing markets, 256
Mineral supply, control by technology, 298
Mines, 24–33
  dredging operations, 33
  open pit, 24–27
    advantages of, 29
  strip, 27
    advantages of, 29–31
  underground, 34–35
    hazards of, 29–31
  wells, 32–33
Mine Safety Act, 219
Minimills, steel, 112
Mining:
  hazards of, 31
  rates of, 35–36, 147
  recovery percentages in, 33
  see also Environmental regulation of mining; Marine mining
Mining claims, 205–207
Mining industry:
  in consumer society, 289–290
  current status of, 290
  nature of, 36
  role in U.S. economy, 37, 288
Mining law:
  administrative, 212
  on private lands, 204–205
  on public domain, 205–211
Mining Law of 1872, 205–212
  apex provision, 207
  defects of, 207–212

Minnesota:
peat in, 40
taxes on iron ore, 243
Mississippi, ball clays, 87
Missouri, 85
Molybdenum, 13, 78, 124–125
world production, 125, 146, 154, 155, 159
world reserve base, 125, 146, 159
see also specific countries
Monopoly:
in copper, 263–264
definition of, 263
in sulfur, 263
Montana:
manganese deposits, 116
phosphate, 94
severance tax on coal, 240
talc, 103
Morocco, production and reserve base:
lead, 133
phosphate, 95
Most-favored-nation clause, 248
Mother Lode, 138, 140
Mukluk prospect, 59
Multinational corporations, 174
Multiple Use Act, 211
MX missile sites, 202

Namibia:
diamonds, 90, 269
uranium production, 71
National Academy of Sciences, 188
estimate of petroleum, 59
National Advisory Committee on Oceans and Atmosphere, 280
National Bituminous Coal Commission, 198
National Commission on Materials Policy, 248, 250
National Conservation Commission, 186
National Environmental Protection Act, 219
National Forests, exploration in, 213
National Materials and Mineral Policy Act, 258, 260, 295
National Oceanic and Atmospheric Administration, 276, 280
National Parks, 212, 217
exclusion from mineral entry, 213
National Park Service, 205, 207, 217
National Recovery Act, 195
National Wilderness Preservation System, 213
Native Regional Corporations, 205
allocation of lands to, 217
Natural gas, 48–49, 273, 274
for manufacture of ammonia, 93
proved reserves of, definition, 63–64
sulfur in, 102
world production, 63
world reserves, 53
see also United States
Nauru, phosphate, 95
Nepheline, 88
Nevada, 4
manganese deposits, 116
mercury, 136
New Brunswick, potash, 98
New Caledonia:
nickel deposits, 121, 122
nickel production and reserve base, 123
Newfoundland, iron ore, 114
New Mexico, mining law, 204
New York, talc, 103
New Zealand:
geothermal energy, 75
phosphate, 283
sulfur, 101
Nickel, 78, 121–123
in deep-sea nodules, 275–276
deposits of, 121, 122
world production, 123, 146, 154, 155, 159
world reserve base, 123, 146, 159
see also specific countries
Nigeria:
oil, 54, 269, 273
tin, production and reserve base, 135
Nine-Power Treaty, 252
Niobium, 78
deposits, discovery of, 202
Nitrogen and nitrates, 78, 93–94
deposits, Chile, 93
world consumption, 93
see also Chile; United States
Noatak National Preserve, 218
Non-ferrous metals, 126–137. *See also specific metals*
Non-fuel mineral industries, conservation in, 199–200
Nonfuel Mineral Policy Review, 258
Nonfuel minerals:
availability of:
from U.S. resources, 175–183
from world resources, 145–160
consumption of:
United States, 163, 165, 170, 175–177
world, per capita, 149–150, 151
production of, world, 146, 152–159
growth rates of, 155
reserve base/consumption indices for, United States, 178, 179
reserve base/production indices for, world, 146, 158–159
see also specific commodities

Nonmetallic minerals:
  availability of, future, 106
  groups of, 81
  use of, ancient, 80–81
  see also specific countries and commodities
North Dakota, 47
Northern Rhodesia, copper, 266
North Korea, sanctions against, 253
North Sea:
  jurisdiction over, 274
  offshore oil, 273
Norway:
  aluminum production, 131, 132
  silicon production, 118
Nuclear energy, 68–73
  advantages, 72
  breeder reaction, 69
  from nuclear fission, 68–71
  from thermonuclear fusion, 71–73

Occupational Safety and Health Act, 219
Ocean Island, phosphate, 95
Ocean mining, *see* Marine mining
Offshore mining, *see* Marine mining
Oficina-Temblador area, 64
Oil, *see* Petroleum
Oil fields, *see* Petroleum
Oil shales, 65–67
  in Brazil, 67
  in Latvia, 67
  in United States, 65–66
Oil wells:
  costs of drilling, 62
  depth of, 61
Okefenokee Swamp, peat in, 40
Oman, oil, 270
Onslow Bay phosphate district, 283
Open pit mining, 24–28, 29
  of coal, 45
Orange River, offshore diamonds, 273
Orderly marketing provision, 248
Ore:
  definition, 7
  factors determining, 7–8
  grade, 7
  minimum grades of, 7
Ore body, 7
Oregon, chromite, 119
Organization of Petroleum Exporting Countries (OPEC), 54, 55, 269, 270, 271
  importance as source of petroleum, 54–55
  members, 54
Osmium, 78. *See also* Platinum-group metals
Outer Continental Shelves Act, 211

Pachuca-Real del Monte, 139
Palabora, phosphate, 95
Palladium, *see* Platinum-group metals
Papua-New Guinea, 271
Paradox Valley, potash, 33, 97
Peace River, *see* Heavy oils
Peat, 40–41
  change to coal, 41
Pedis possessio, 210
Pennsylvania, 85
Per capita consumption, *see* Nonfuel minerals
Perlite, *see* United States
Peru:
  fishing rights, 271
  phosphate, 283
  production and reserve base:
    copper, 129
    lead, 133
    molybdenum, 125
    silver, 139
    zinc, 134
Petroleum, 48–62
  alternate sources of, 64–68
  cost of discovery of, 61–62
  exploration for, 59
  exploration targets, 70
  formation and accumulation of, 49–50
  future availability of, 67–72
  imports by United States:
    amounts, 66
    cost, 54
    sources, 55
  industry, 195–197
  mining of, 64
  offshore, 273, 274
  proved reserves, definition of, 50–52
  recovery from reservoirs, 51–52, 196
  reserves, by regions, 53
  supply forecasts, 56–60
  world production, 55
  world reserves, 53
  see also specific countries
Phelps Dodge, 233
Philippines:
  chromite, production and reserve base, 120
  geothermal energy, 73
  nickel, production and reserve base, 123
Phosphate, 94–96
  deposits of, 4, 94–95, 273, 280–282
  world production, 94, 146, 152, 155, 158
  world reserve base, 95, 146, 158
  see also specific countries
Phosphate mining, environmental aspects of, 223–224
Pickett Act, 213
Pittsburg coal seam, 42
Placer claims, *see* Mining claims
Placer deposits:
  diamonds, 268

gold, 140
Platinum, 78. *See also* Platinum-group metals
Platinum-group metals, 142
world consumption, 142
world reserves, 142
Plutonium, 69
Point Arguello, *see* Santa Maria Basin
Poland:
coal production, 46
production and reserve base:
copper, 129
potash, 139
silver, 139
sulfur, 101
Polynesia, phosphate, 283
Population growth, 92, 294–295
Porphyry deposits:
of copper, 127
of molybdenum, 124
Potash, 96–98
cartel in, 96
deposits of, 96
world production, 98, 146, 153, 155, 158
world reserve base, 98, 146, 158
see also specific countries
Potosí, 139
Potrerillos, 18
Precious metals, 137–142. *See also* Gold; Platinum-group metals; Silver
President's Materials Policy Commission, 248
Price, effect on availability of minerals, 58, 148
Processing of mineral materials, 34–35
Prospecting, *see* Mineral exploration
Protective tariffs, *see* Tariffs
Prudhoe Bay, 50, 53, 57, 66, 216
Public domain, 186
jurisdiction over, 205
mineral potential of, 214–215
mining law on, 205–211
Public Lands, access to, 215
Public Lands Leasing Act, *see* Mineral Leasing Act
Pyrophyllite, *see* Talc and pyrophyllite

Qatar, 54
Quartz, 78, 89, 90
Quartz Hill, 124
Quebec, iron ore, 112
Quotas:
on lead and zinc imports, 250
on oil imports, 250
on steel imports, 249

Rare earths, 78
Reagan, President, Materials Policy Report by, 258
Reciprocal Trade Agreements Act, 247–248
escape clause, 249
most-favored-nation clause, 248
Reclamation of mined lands, 223–226, 227, 228, 230
Recovery percentages in mining, 29, 43–45, 57
Recycling of mineral materials, 142, 191–193
Red Dog deposit, 216
Red Sea muds, 274, 283
Reliance on imports:
by European Economic Community, 168
by Japan, 168
by Soviet Union, 169
by United States, 165–167, 258, 298
Republic of South Africa, *see* South Africa
Reserve base:
definition of, 17
vs. production rates, 148–149
world, 146, 158–159
see also specific countries and commodities
Reserve base/consumption index, 178
Reserve base/primary consumption index, 178. *See also* United States
Reserve base/production index, 145, 147
Reserve base/production indices for various commodities, world, 146, 156–157, 158–159
Reserves, *see* Mineral reserves
Reserves/maximum rate of production, 147
Resources, *see* Mineral resources
unidentified, uncertainties of, 15–16
Rhenium, 78
Rhodium, 78. *See also* Platinum-group metals
Riddle, Oregon, nickel deposits, 122
Rocky Mountains phosphate:
fluorine in, 103
vanadium in, 123
Romania, 47
Roofing granules, 83
Roosevelt, President Theodore, 186
Roosevelt Hot Springs, 75
Rosendale, 84
Rothschilds, 263
Rubidium, 78
Ruthenium, 78. *See also* Platinum-group metals

Saline brines, 99, 100
Saline minerals, 99–100
Salt, 96–98
world production, reserve base, 146, 152, 155, 158
see also United States
Salt domes, sulfur in, 101
Salt lakes, deposits in, 99–100
Salton Sea basin, 75

Sanctions, *see* Mineral sanctions
Sand, gravel, and stone, 82–83. *See also* United States
Santa Maria Basin, 50, 59
Saskatchewan, potash, 98
Saudi Arabia, 54, 62, 274
  Ghawar oil field, 50
  Petroleum reserves, 53
Scandium, 78
Searles Lake, 34, 97
Seawater, composition of, 97
Sécrétan corner, 264
Sekulu, phosphate, 95
Selective mining, 239, 240
Selenium, 78
Self-sufficiency in minerals, *see* Reliance on imports
Senegal, phosphate production and reserve base, 95
Sherman Antitrust Act, 198, 264, 265
Sicily, sulfur deposits, 101
Sierra Leone, 90, 268, 271
Sierra Nevada, 119
Silicomanganese, 116
Silicon, 78, 117–118
  world production and reserve base, 118, 154, 155, 158
  see also specific countries
Silver, 78, 138, 139–140
  world production, 139
  world reserve base, 139
  see also specific countries
Silver Peak, lithium, 100
Smoot-Hawley Act, 247
Soda ash (sodium carbonate), 99, 100
  world production and reserve base, 146, 153, 155, 158
  see also United States
Sodium, 78
Sodium sulfate, 99
  world production and reserve base, 146, 153, 155, 158
Solar energy, 73
South Africa, 254
  coal production, 46
  coal reserves, 43
  diamond industry, 268
  diamonds, 90. *See also* Cartels in minerals
  ferrochrome, 120, 173
  fluorspar, 103
  gold deposits, 140
  iron ore, 112
  minerals, importance of, 258
  platinum reserves, 142
  production and reserve base:
    chromite, 120
    iron ore, 113
    lead, 133
    manganese ore, 117
    nickel, 123
    phosphate, 95
  silicon production, 118
  synthetic petroleum, 47
  uranium production, 71
  vanadium, reserve base, 123
South America, phosphate, 94
South Carolina, kaolin deposits, 86, 131
Southern Rhodesia, sanctions against, 253–254
South Korea, steel, 111
Soviet Union, 72, 247, 254, 270
  aluminum production, 131
  asbestos, 104
  breeder reactor, 69
  diamonds, 90
  ferrochromium, 120
  iron ore, 112
  kaolin, 87
  mineral policy, 295–296
  phosphate, 94, 283
  platinum-group metals, 142
  production and reserve base:
    coal, 43, 46
    chromite, 120
    copper, 129
    manganese ore, 116
    phosphate, 95
    potash, 98
    silver, 139
    tin, 135
  reliance on imports, 167, 169
  silicon production, 118
  steel production, 111
  sulfur, 101
  tin, 267
  vanadium reserves, 123
Spain:
  kaolin, 87
  potash, 98
    production and reserve base, 98
  silicon production, 118
Spiegeleisen, 116
Spor Mountain, beryllium, 137
Stainless steel, 119
Stassfurt, potash, 97
Steel, *see* Iron and steel
Steel industry, protective measures for, 249
Steelpoort chromite seam, 120, 293
Stillwater Complex:
  chromite in, 119
  platinum in, 142
Stockpiles, 254–259, 297
Stone, *see* Sand, gravel, and stone
Stratabound deposits of copper, 127

Strategic and Critical Materials Stockpiling Acts, 255, 256
Strategic and Critical Materials Stockpiling Revision Act, 256
Strategic mineral problem:
  impact of industrial change on, 259
  impact of materials science on, 259
Strategic minerals:
  definition of, 255
  in Korean War, 255–256
  in U.S. stockpile, 257
  in World War I, 254
  in World War II, 255
Strategic Petroleum Reserve, 54, 256
Strip mining, environmental damage from, 27–28, 223
Strontium, 78
Subbituminous coal, 222
Subcommittee on Mines and Mining, report of, 258
Submerged Lands Leasing Act, 211
Sub-Saharan Africa, population growth, 92
Substitution, role in conservation, 193–194
Sudan, 274
Sudbury, 122, 222
Sulfur, 100–102
  in coal, 42, 45, 102, 222–223
  deposits of, 101–102
  from natural gas, 102
  resources, 102
  from sulfides, 101–102
  world production and reserve base, 146, 153, 155, 159
  see also specific countries
Super Phenix reactor, 69
Surface mining, environmental consequences of, 223
Surface Mining and Reclamation Act, 219
Surface Resources Act, 211
Surinam, 271
  bauxite production and reserve base, 133
Swansea, 171
Sweden, iron ore production and reserve base, 113
Synthetic Fuels Administration, 47
Synthetic petroleum, 47–48

Taconite, 115
Taconite Amendment, 243
Taiwan, 174
Talc and pyrophyllite, 103–104
  world production and reserve base, 146, 153, 155, 159
  see also United States
Tanganyika, diamonds, 90
Tantalum, 78
Tariff policy of United States, 247–251
Tariffs on imports, 246–251
  effects of, 247
  on manganese, 247
Taxes on mining:
  in Canada, 243–244
  effect on mineral availability, 242–245
  income, 239, 240, 241–242
  on petroleum industry, 244
  property, 238, 239
  sales, 239, 240
  severance, 239, 240
  Social Security, 242
  on taconite, 244
  sales, 239, 240
  in Wisconsin, 244
Technology, adaptation of, 193–194, 297
Tennessee:
  ball clays, 87
  phosphate, 4, 94
  sandstone, 83
Tennessee Valley Authority, 130
Texas, 85
  public lands, 204
  sulfur, 102
  talc, 103
Texas Railroad Commission, 195, 196
Thailand, 120
  potash, 98
  tin, 267, 273, 283
    production and reserve base, 135
Thallium, 78
Thermonuclear fusion, *see* Nuclear energy, from thermonuclear energy, 71–73
Thickness of ore, influence on mining, 31
Thompson district, nickel deposits, 122
Thorium, 68, 78
Three-Mile Island, 70
Tin, 78, 135
  U.S. stockpile of, 267
  world production, 135, 146, 154, 155, 159
  world reserve base, 135, 146, 159
  see also specific countries
Titanium and titanium minerals, 104
  world production and reserve base, 146, 153, 155, 159
  see also United States
Togo, phosphate production, reserve base, 95
Tokamak reactor, 72
Ton, 3
Trade Act of 1974, 249
Trade barriers, world, 251
Trade deficit, 251
Trade Expansion Act, 248
Trade policy, consequences of, 296–297
Trail Ridge, 33
Transfer of mineral industry, 170–174, 298

Trigger prices on steel, 249
Tritium, 71
Trona, *see* Sodium carbonate
Troy ounce, 137
Truman, President, 273
Tungsten, 3, 78, 125–126
  world production, 146, 154, 155, 159
  world reserve base, 146, 159
  see also specific countries
Tunisia, phosphate, 94
  production and reserve base of, 95
Turkey:
  chromite production and reserve base, 120
  nickel laterites, 121

Underground coal gasification, 48
Union Carbide Corporation, 253
United Arab Emirates, 54
United Kingdom, 48
  coal production, 46
  kaolin, 87
  offshore mining, 273
  potash, production and reserve base, 98
  tin, 267
    production and reserve base, 135
United Nations, 253, 279
United Nations Security Council, 253
United States:
  chromium supply, 172
  coal exports, 46
  coal fields, 44
  coal industry, 197–199
  cobalt resources, 126
  consumption of nonfuel minerals, forecasts of, 156
  copper deposits, need for additional discoveries of, 188
  copper industry, condition of, 128
  diamonds, 90
  energy, sources of, 39
  energy resources, 297. *See also specific fuels*
  fluorine and fluorspar, 102–103
  gold deposits, 139, 140, 141
  importance as producer and consumer of minerals, 168
  kaolin deposits, 86
  land use, 229
  and Law of the Sea, 279, 280
  manganese:
    resources of, 116
    supply of, 172
  mineral industry:
    history of, 162–174
    nature of, 171–172
  mineral policy, objectives of 298–299
  mineral position, changes in, 164–174
  mineral production, consumption, reserve base:
    aluminum, 136, 177
    antimony, 177, 179, 182
    asbestos, 104, 175, 178, 181
    barite, 176, 178, 181
    bauxite, 130, 133
    boron, 99, 175, 178, 181
    bromine, 99
    cement, 82, 176, 178, 181
    chromite, 119
    chromium, 177, 182
    clays, 82, 85, 176, 178, 181
    coal, 43–44, 45, 46
    cobalt, 177, 179, 182
    constructional materials, 82
    copper, 177, 179, 182
    diatomite, 175, 178, 181
    feldspar, 175, 178, 181
    fluorine and fluorspar, 175, 178, 181
    gold, 141
    gypsum, 82, 176, 178, 181
    iron ore, 113, 177, 179, 182
    lead, 133, 177, 179, 182
    lime, 82, 176, 178, 181
    lithium, 137, 177, 179, 182
    magnesium, 136, 177, 179, 182
    magnesium compounds, 99
    manganese, 177, 179, 182
    mercury, 177, 179, 182
    molybdenum, 125, 177, 179, 182
    natural gas, 63
    nickel, 123, 177, 179, 182
    nitrogen and nitrates, 93
    perlite, 175, 178, 181
    petroleum, 52
    phosphate, 95, 176, 178, 181
    potash, 98, 176, 178, 181
    salt, 99, 176, 178, 181
    sand, gravel, and crushed stone, 99, 164, 178, 181
    silicon, 118, 177, 179, 182
    silver, 139
    soda ash (sodium carbonate), 99, 176, 178, 181
    sodium sulfate, 99
    steel, 111–112
    sulfur, 176, 178, 181
    talc and pyrophyllite, 175, 178, 181
    tin, 135, 177, 179
    titanium, 177
    tungsten, 125, 177, 179, 182
    uranium, 69, 71
    vanadium, 177, 179, 182
    zinc, 134, 177, 179, 182

mineral production, value of, 36, 296
nickel deposits, 122
offshore oil, 59, 60, 61, 273
petroleum, imports of, 270
phosphate, rates of production of, 147
platinum-group metals, deposits of, 142
production of nonfuel minerals, 163, 164, 165, 170, 181–182
reserve base, nonfuel minerals, 179, 181, 182
reserve base/primary consumption indices, various minerals and metals, 178–179
reserve base/production indices, nonfuel minerals, 181–182
silver supply, 139–140
sources of energy for, 39
steel industry, 111–112, 249
tariff policy, 247–251
trade deficit, 54
zinc mining industry, 172
U.S. Bureau of the Census, 239
U.S. Bureau of Mines, 17, 36, 165, 167, 199, 213, 214, 229, 277
resources classification, 13–15
U.S. Congress, 247, 248
U.S. Forest Service, 205, 206, 217, 218
U.S. Geological Survey, 44, 201, 213, 214
estimates of recoverable petroleum, 59
resource classification, 13–15
U.S. International Trade Commission, 249, 250
U.S. Public Health Service, 232
U.S. Tariff Commission, 249, 250
University of Tennessee Experiment Station, 223
Ural Mountains, nickel laterites, 121
Uranium, 78
energy in, 68
isotopes of, 68
minerals of, 68
world resources, 70
see also United States
Utah, 4
phosphate, 94
Utah Power and Light Company, 75

Vanadium, 123–124
in phosphate rock, 123
in uranium deposits, 123
world production, 146, 154, 155, 158
world reserve base, 146, 158
Venezuela, 173
heavy oil sands, 64
iron ore, 112
production and reserve base, 113
petroleum reserves, 53
Vermont:
asbestos, 104
marble, 83
talc, 103
Vernal, phosphate mining, 232
Victoria, gold deposits, 140
Volcanogenic deposits, 127

Wabasca area, *see* Heavy oils
War Industries Board, 263
Washington state, chromite, 119
Water gas, 47
Watt, J., 280
Webb-Pomerene Act, 264
West Africa, diamonds, 90
Western Europe, coal reserves, 43
Western states, impact of mining on, 230
West Germany, 72, 267, 279, 280
aluminum production, 131
coal production, 46
potash production and reserve base, 98
White Pine, 127
Wilderness Act, 213–214, 215
Wilderness areas:
acreage of, 214
mineral potential of, 214–215
Wildlife Refuges, 205, 213, 217
Windfall Profits Tax, 244
Wisconsin, taxes on mining, 244
Wisconsin Department of Natural Resources, 234
Withdrawals of land from mineral entry, 212–214
Witwatersrand, 137, 140
World:
growth rates of mineral production, 149–150, 154–155
nonfuel mineral production, 146, 152–154
projected, 158–159
per capita mineral consumption, 150–151
reserve base, 146
reserve base/production indices, 146, 158–159
see also specific commodities
Wyoming, 115
bentonites, 88
phosphate, 94
severance tax, 240

Yellowstone National Park, 212
Yugoslavia, 271
bauxite production and reserve base, 133
coal production, 46
lead production and reserve base, 133

Zaire, 167
  copper production and reserve base, 129
  diamonds, 90, 268
  lithium resources, 270
  petroleum, 270
  tin production and reserve base, 135
Zambia, 167, 271, 275
  cobalt, 126
  copper, 127
    production and reserve base, 129
Zeolites, 104
Zimbabwe:
  chromite, 104, 120
    production and reserve base, 120
  ferrochrome, 120, 173
Zinc, 78, 133–134
  recycling of, 133, 192
  world production, 134, 146, 154, 155, 159
  world reserve base, 134, 146, 159
Zinc industry, in United States, 172
Zirconium, 78